*INTERNATIONAL SERIES OF MONOGRAPHS IN*
**PURE AND APPLIED MATHEMATICS**



---

VOLUME 49

# DIFFERENTIAL GEOMETRY ON COMPLEX AND ALMOST COMPLEX SPACES

# Differential Geometry on Complex and Almost Complex Spaces

KENTARO YANO

*Professor of Mathematics*
*Tokyo Institute of Technology*

A Pergamon Press Book

THE MACMILLAN COMPANY
NEW YORK
1965

THE MACMILLAN COMPANY
60 Fifth Avenue
New York 11, N.Y.

This book is distributed by
THE MACMILLAN COMPANY
pursuant to a special arrangement with
PERGAMON PRESS LIMITED
Oxford, England



Library of Congress Catalog Card Number 63-10097

# CONTENTS

# PREFACE

It was J. A. Schouten and D. van Dantzig [1], [2]† who in 1930 first tried to transfer the results in differential geometry of spaces with Riemannian metric and with affine connexion to the case of spaces with complex structure.

In their papers there appeared a Hermite space with so-called symmetric unitary connexion. The space with the same connexion was also found independently by E. Kähler [1] in 1933 and such a space is now called a Kähler space. S. Bergmann [1] applied the theory of such a space to the theory of functions of two complex variables by means of his kernel functions.

On the other hand, in his beautiful book on harmonic integrals published in 1941 W. V. D. Hodge [1] applied his theory of harmonic integrals to derive certain results on algebraic varieties. But as is pointed out by Hodge himself and by A. Weil, many of the results obtained by Hodge depend only on local properties of the metric used and some of them depend only on the fact that an algebraic variety is capable of carrying a Kähler metric and not on the actual metric selected, and consequently the results of Hodge on algebraic varieties could be extended to any Kähler manifold. That this is actually the case was shown by B. Eckmann and H. Guggenheimer ([1], [2], [3], [4]) in 1949.

The more differential geometric aspects of Kähler spaces were studied by S. Bochner, H. Guggenheimer, A. Lichnerowicz and others.

A. Weil [1], [2] pointed out in 1947 that there exists in a complex space a tensor field $F$ of type (1,1) whose square

† See the Bibliography at the end of the book.

is minus unity. C. Ehresmann [1], [2] defined in 1947 an almost complex space as an even-dimensional differentiable manifold which carries a tensor field $F$ whose square is minus unity. A complex space is of course an almost complex space but the converse is not necessarily true.

The condition for an almost complex structure to be induced by a complex structure was studied and given by C. Ehresmann and P. Libermann [1], B. Eckmann and A. Frölicher [1], E. Calabi and D. C. Spencer [1] in the case where the almost complex structure is of class $C^\omega$ and by A. Newlander and L. Nirenberg [1] in 1957 in the case where the almost complex space is merely differentiable.

In this problem a tensor introduced by A. Nijenhuis [1] played a very important rôle.

The differential geometry of complex and almost complex spaces is studied extensively by M. Apte, M. Ako, S. Bochner, W. M. Boothby, E. Calabi, S. S. Chern, N. Coburn, R. Couty, D. van Dantzig, B. Eckmann, C. Ehresmann, A. Frölicher, T. Fukami, P. R. Garabedian, S. I. Goldberg, H. Guggenheimer, J. Hano, N. S. Hawley, W. V. D. Hodge, S. Ishihara, S. Kotō, S. Kobayashi, G. Legrand, A. Lichnerowicz, P. Libermann, Y. Matsushima, H. Mizusawa, I. Mogi, A. Nijenhuis, K. Nomizu, M. Obata, T. Ōhtsuki, E. M. Patterson, S. Sasaki, S. Sawaki, J. A. Schouten, D. C. Spencer, S. Tachibana, Y. Tashiro, H. Wakakuwa, H. C. Wang, W. J. Westlake, the present author and others.

In the opinion of the present author the complex and almost complex spaces are vast and very fruitful fields for differential geometry. The principal purpose of the present book is to introduce the readers into this very interesting branch of differential geometry and encourage them to exploit this new domain.

To reduce the pre-requisites from differential geometry to the minimum and to fix our notation, the author tries in the first Chapter to give a survey of present-day differential geometry.

Since Green's theorem plays a very important rôle in our discussions of complex and almost complex spaces, the author gives in Chapter II not only the proof of Green's theorem but also various applications of this theorem to problems appearing in Riemannian geometry in the large. Some of them play rather important parts in the following discussions.

The author has the hope that the first part of the present book might serve as an introduction to the modern differential geometry.

Chapter III is devoted to the definition of a complex space and to the discussions of vectors, tensors, metric and affine connexions in it.

In Chapter IV we define a Kähler space and discuss curvatures of the space, harmonic tensors in it, contravariant and covariant analytic vectors and tensors, infinitesimal transformations, a theorem of Matsushima and its generalization by Lichnerowicz.

Chapter V is devoted to the definition of almost complex space and to the discussions of the so-called integrability condition of the almost complex structure and Chapter VI to the discussions of various affine connexions which can be introduced in an almost complex space.

In a Hermite space there are a Riemannian metric $g$ and an exterior differential form $F$ of the order 2, defined in terms of the complex structure. One of the characteristic properties of a Kähler space is the fact that the form $F$ is closed and this implies the vanishing of covariant derivative of $F$. An almost Kähler space is defined as an almost complex space with a Hermite metric the associated form $F$ of the almost complex stucture being closed. This does not imply the vanishing of the covariant derivative of $F$. But we can prove that $F$ is harmonic. Chapter VII is devoted to the discussions of such almost Kähler spaces.

S. Tachibana studied [4] in 1959 an almost Hermite space whose associated differential form $F$ is a Killing form instead of a harmoinc form. We discuss such spaces in Chapter VIII.

Chapter IX is devoted to the discussions of the most general Hermite spaces.

A. G. Walker [1] started in 1955 the study of the so-called almost product spaces.

In such a space there exists a mixed tensor field $F$ whose square is unity instead of being minus unity as in the case of an almost complex space.

Chapter X is devoted to the study of locally product spaces and Chapter XI to that of almost product spaces. The author tries to make as clear as possible the analogy between the almost complex space and the almost product space.

In the last Chapter we state the theory of holomorphically projective transformations mainly developed by S. Ishihara, S. Tachibana and Y. Tashiro.

The first draft of this book was based on the series of lectures which the author gave at the University of Hong Kong. The author expresses his deepest gratitude to the authority of the University of Hong Kong and especially to Professor Y. C. Wong who invited the author as Visiting Professor to the eminent University and gave him a pleasant time in Hong Kong which allowed him to concentrate his effort on writing this book. The author also appreciates very much the collaboration of the member staff of the University of Hong Kong and of his Japanese colleagues M. Ako, S. Ishihara, S. Kotō, Y. Mutō, T. Nagano, M. Okumura, T. Takahashi and S. Sawaki for the preparation of the manuscript of the sent book.

It is a very pleasant duty for the author to express here his deep gratitude to Professor E. T. Davies of the University of Southampton and Professor I. N. Sneddon of the University of Glasgow who suggested the publication of this book at Pergamon Press, who took all possible care in production.

KENTARO YANO

CHAPTER I

# REVIEW OF FUNDAMENTAL CONCEPTS AND FORMULAE IN RIEMANNIAN GEOMETRY†

## § 1. Riemannian spaces

We consider an $n$-dimensional Riemannian space M of class $C^\infty$ with the fundamental positive definite metric

$$ds^2 = g_{ji}(\xi) d\xi^j d\xi^i, \tag{1.1}$$

covered by any system of coordinate neighbourhoods $(\xi^h)$, where the indices $h, i, j, \ldots$ run over the range $1, 2, \ldots, n$. In formula (1.1) we understand that the repeated indices $j$ and $i$ stand for the summation over the range of these indices.

If, from any system of coordinate neighbourhoods covering the space, we can choose a finite number of coordinate neighbourhoods which cover the whole space, we say that the space is *compact*.

If we can cover the whole space by a system of coordinate neighbourhoods in such a way that the Jacobian

$$\Delta = \left| \frac{\partial f^{h'}}{\partial \xi^h} \right| \tag{1.2}$$

† See, for example, E. Cartan [1]; S. S. Chern [2], [3]; L. P. Eisenhart [1]; J. A. Schouten [3]; T. J. Willmore [4]; K. Yano [11]; K. Yano and S. Bochner [1].

of the coordinate transformation

$$\xi^{h'} = f^{h'}(\xi) \tag{1.3}$$

occurring in the intersection of two overlapping coordinate neighbourhoods $(\xi^h)$ and $(\xi^{h'})$ is always positive, then we say that the space is *orientable*.

Since the quadratic differential form $g_{ji}(\xi)d\xi^j d\xi^i$ is supposed to be positive definite, the determinant

$$\mathfrak{g} = |g_{ji}| \tag{1.4}$$

formed by the covariant components $g_{ji}$ of the fundamental tensor is always positive. The contravariant components $g^{ih}$ of the fundamental tensor satisfy

$$g_{ji}g^{ih} = A_j^h \tag{1.5}$$

where $A_j^h$ denotes the so-called unit tensor whose components are given by

$$A_j^h = \delta_j^h = \begin{cases} 1 & \text{when} \quad h = j, \\ 0 & \text{when} \quad h \neq j. \end{cases} \tag{1.6}$$

We use the covariant components $g_{ji}$ and contravariant components $g^{ih}$ of the fundamental tensor to lower and to raise the indices of components of a tensor, for example

$$T_{jih} = T_{ji}{}^a g_{ah}$$

and

$$T_{ji}{}^h = T_{jia} g^{ah}.$$

Consider two symbols $\varepsilon_{i_1 i_2 \ldots i_n}$ and $\varepsilon^{i_1 i_2 \ldots i_n}$ defined by

$$\varepsilon_{i_1 i_2 \ldots i_n} = \varepsilon^{i_1 i_2 \ldots i_n} = \begin{cases} +1 & \text{when } (i_1, i_2, \ldots, i_n) \text{ is an even permutation of } (1, 2, \ldots, n), \\ -1 & \text{when } (i_1, i_2, \ldots, i_n) \text{ is an odd permutation of } (1, 2, \ldots, n), \\ 0 & \text{otherwise.} \end{cases} \tag{1.7}$$

The symbols $\varepsilon_{i_1 i_2 \ldots i_n}$ and $\varepsilon^{i_1 i_2 \ldots i_n}$ define a relative tensor of weight $-1$ and $+1$ respectively.

Since $\mathfrak{g}$ given by (1.4) is a relative scalar of weight $+2$ in an orientable space, the quantities defined by

$$e_{i_1 i_2 \ldots i_n} = \sqrt{\mathfrak{g}}\, \varepsilon_{i_1 i_2 \ldots i_n} \tag{1.8}$$

and

$$e^{i_1 i_2 \ldots i_n} = \frac{1}{\sqrt{\mathfrak{g}}}\, \varepsilon^{i_1 i_2 \ldots i_n} \tag{1.9}$$

are both skew-symmetric tensors, and we have

$$e^{h_1 h_2 \ldots h_n} = e_{i_1 i_2 \ldots i_n} g^{i_1 h_1} g^{i_2 h_2} \ldots g^{i_n h_n}.$$

We define a skew-symmetric tensor $e^{h_1 \ldots h_p}_{i_1 \ldots i_p}$ by

$$e^{h_1 \ldots h_p}_{i_1 \ldots i_p} = \frac{1}{(n-p)!}\, e_{i_1 i_2 \ldots i_p i_{p+1} \ldots i_n} e^{h_1 h_2 \ldots h_p i_{p+1} \ldots i_n}. \tag{1.10}$$

This tensor has the following property:

$$e^{h_1 \ldots h_p}_{i_1 \ldots i_p} = \begin{cases} +1 & \text{when } (h_1 \ldots h_p) \text{ is an even permutation of } (i_1 \ldots i_p), \\ -1 & \text{when } (h_1 \ldots h_p) \text{ is an odd permutation of } (i_1 \ldots i_p), \\ 0 & \text{otherwise.} \end{cases} \tag{1.11}$$

Take a skew-symmetric tensor:

$$w: \qquad w_{i_1 i_2 \ldots i_p}.$$

The skew-symmetric tensor defined by

$$*w: \qquad \frac{1}{p!}\, e_{i_1 i_2 \ldots i_p i_{p+1} \ldots i_n} w^{i_1 i_2 \ldots i_p} \tag{1.12}$$

is called the dual of $w$. For the dual $**w$ of the dual $*w$, we have

$$**w: \quad (-1)^{p(n-p)} w_{i_1 i_2 \ldots i_p}. \tag{1.13}$$

In the case where the space is compact and orientable, we can consider the integral of a scalar function

$$\int_M f(\xi)\sqrt{\mathfrak{g}}\, d\xi^1 \wedge d\xi^2 \wedge \ldots \wedge d\xi^n \tag{1.14}$$

taken over the whole space M, where the so-called volume element

$$d\sigma = \sqrt{\mathfrak{g}}\, d\xi^1 \wedge d\xi^2 \wedge \ldots \wedge d\xi^n \tag{1.15}$$

has a definite sign.

We now construct the Christoffel symbols

$$\left\{ {h \atop j\ i} \right\} = \frac{1}{2} g^{ha}(\partial_j g_{ia} + \partial_i g_{ja} - \partial_a g_{ji}), \tag{1.16}$$

where $\partial_i$ stands for the partial differentiation with respect to the coordinate $\xi^i$. Then the covariant derivatives of a scalar $f(\xi)$, of a contravariant vector $v^h$, of a covariant vector $w_i$ and of a general tensor $T_{ji}{}^h$, for example, are given by

$$\nabla_i f = \partial_i f, \tag{1.17}$$

$$\nabla_j v^h = \partial_j v^h + \left\{ {h \atop j\ i} \right\} v^i, \tag{1.18}$$

$$\nabla_j w_i = \partial_j w_i - \left\{ {h \atop j\ i} \right\} w_h \tag{1.19}$$

and

$$\nabla_k T_{ji}{}^h = \partial_k T_{ji}{}^h + \left\{ {h \atop k\ a} \right\} T_{ji}{}^a - \left\{ {a \atop k\ j} \right\} T_{ai}{}^h - \left\{ {a \atop k\ i} \right\} T_{ja}{}^h \tag{1.20}$$

respectively.

The covariant derivative of a relative scalar $\mathfrak{f}$ of weight $p$ is given by

$$\nabla_i \mathfrak{f} = \partial_i \mathfrak{f} - p \left\{ {a \atop i\ a} \right\} \mathfrak{f}. \tag{1.21}$$

It is a well-known fact that the covariant derivatives of $g_{ji}$, $g^{ih}$, $A_i^h$, $\sqrt{\mathfrak{g}}$, $e_{i_1 i_2 \ldots i_n}$ and $e^{i_1 i_2 \cdots i_n}$ all vanish:

$$\nabla_k g_{ji} = \partial_k g_{ji} - \left\{ {a \atop k\ j} \right\} g_{ai} - \left\{ {a \atop k\ i} \right\} g_{ja} = 0, \tag{1.22}$$

$$\nabla_j g^{ih} = \partial_j g^{ih} + \left\{ {i \atop j\ a} \right\} g^{ah} + \left\{ {h \atop j\ a} \right\} g^{ia} = 0, \tag{1.23}$$

$$\nabla_j A_i^h = \partial_j A_i^h + \left\{ {h \atop j\ a} \right\} A_i^a - \left\{ {a \atop j\ i} \right\} A_a^h = 0, \tag{1.24}$$

$$\nabla_i \sqrt{\mathfrak{g}} = \partial_i \sqrt{\mathfrak{g}} - \left\{ {a \atop i\ a} \right\} \sqrt{\mathfrak{g}} = 0, \tag{1.25}$$

$$\nabla_j e_{i_1 i_2 \ldots i_n} = \partial_j e_{i_1 i_2 \ldots i_n} - \left\{ {a \atop j\ i_1} \right\} e_{a i_2 \ldots i_n} - \ldots$$

$$\ldots - \left\{ {a \atop j\ i_n} \right\} e_{i_1 i_2 \ldots i_{n-1} a} = 0 \tag{1.26}$$

$$\nabla_j e^{i_1 i_2 \cdots i_n} = \partial_j e^{i_1 i_2 \cdots i_n} + \left\{ {i_1 \atop j\ a} \right\} e^{a i_2 \cdots i_n} + \ldots$$

$$\ldots + \left\{ {i_n \atop j\ a} \right\} e^{i_1 \cdots i_{n-1} a} = 0. \tag{1.27}$$

Take a covariant vector $w_i$, then we have

$$\nabla_j w_i = \partial_j w_i - \left\{ {a \atop j\ i} \right\} w_a,$$

from which

$$\nabla_j w_i - \nabla_i w_j = \partial_j w_i - \partial_i w_j, \tag{1.28}$$

which is independent of Christoffel symbols. This is called the *rotation* of the covariant vector field $w_i$ and is denoted by

$$\text{Rot } w: \quad \nabla_j w_i - \nabla_i w_j.$$

A covariant vector $w_i$ may be regarded as the coefficients of an invariant differential form

$$w = w_i d\xi^i.$$

The exterior differential of $w$ is defined as the exterior differential form

$$dw = \frac{1}{2}(\partial_j w_i - \partial_i w_j) d\xi^j \wedge d\xi^i = \frac{1}{2}(\nabla_j w_i - \nabla_i w_j) d\xi^j \wedge d\xi^i .$$

Thus we sometimes denote the rotation by

$$dw: \qquad \nabla_j w_i - \nabla_i w_j . \tag{1.29}$$

For a general skew-symmetric tensor $w_{i_1 i_2 \ldots i_p}$, we have

$$\nabla_j w_{i_1 i_2 \ldots i_p} = \partial_j w_{i_1 i_2 \ldots i_p} - \begin{Bmatrix} a \\ j \; i_1 \end{Bmatrix} w_{a i_2 \ldots i_p} - \begin{Bmatrix} a \\ j \; i_2 \end{Bmatrix} w_{i_1 a \ldots i_p} - \ldots - \begin{Bmatrix} a \\ j \; i_p \end{Bmatrix} w_{i_1 i_2 \ldots a} ,$$

from which

$$\begin{aligned} &\nabla_j w_{i_1 i_2 \ldots i_p} - \nabla_{i_1} w_{j i_2 \ldots i_p} - \ldots - \nabla_{i_p} w_{i_1 i_2 \ldots j} \\ &= \partial_j w_{i_1 i_2 \ldots i_p} - \partial_{i_1} w_{j i_2 \ldots i_p} - \ldots - \partial_{i_p} w_{i_1 i_2 \ldots j} , \end{aligned} \tag{1.30}$$

which is independent of the Christoffel symbols. This is also called the rotation of the skew-symmetric tensor $w_{i_1 i_2 \ldots i_p}$ and is also denoted by

$$\text{Rot } w: \quad \nabla_j w_{i_1 i_2 \ldots i_p} - \nabla_{i_1} w_{j i_2 \ldots i_p} - \ldots - \nabla_{i_p} w_{i_1 i_2 \ldots j} .$$

A skew-symmetric tensor $w_{i_1 i_2 \ldots i_p}$ may be regarded as the coefficients of an invariant exterior differential form

$$w = \frac{1}{p!} w_{i_1 i_2 \ldots i_p} d\xi^{i_1} \wedge d\xi^{i_2} \wedge \ldots \wedge d\xi^{i_p} .$$

The *exterior differential* of $w$ is defined as the exterior differential form

$$\begin{aligned} dw &= \frac{1}{(p+1)!} (\partial_j w_{i_1 i_2 \ldots i_p} - \partial_{i_1} w_{j i_2 \ldots i_p} - \ldots - \partial_{i_p} w_{i_1 i_2 \ldots j}) \\ &\qquad d\xi^j \wedge d\xi^{i_1} \wedge \ldots \wedge d\xi^{i_p} \\ &= \frac{1}{(p+1)!} (\nabla_j w_{i_1 i_2 \ldots i_p} - \nabla_{i_1} w_{j i_2 \ldots i_p} - \ldots - \nabla_{i_p} w_{i_1 i_2 \ldots j}) \\ &\qquad d\xi^j \wedge d\xi^{i_1} \wedge \ldots \wedge d\xi^{i_p} . \end{aligned}$$

Thus we sometimes denote the rotation by

$$dw: \quad \nabla_j w_{i_1 i_2 \dots i_p} - \nabla_{i_1} w_{j i_2 \dots i_p} - \dots - \nabla_{i_p} w_{i_1 i_2 \dots j}. \tag{1.31}$$

Take next a contravariant vector $v^h$, then we have

$$\nabla_j v^h = \partial_j v^h + \left\{ {h \atop j\ i} \right\} v^i, \tag{1.32}$$

from which

$$\nabla_a v^a = \frac{1}{\sqrt{\mathfrak{g}}} \partial_a (\sqrt{\mathfrak{g}}\, v^a)$$

which depends only on $\sqrt{\mathfrak{g}}$. This is called the *divergence* of the contravariant vector $v^h$ and denoted by

$$\text{Div } v: \quad \nabla_i v^i. \tag{1.33}$$

If we start from a covariant vector $w_i$ and then form a contravariant vector $g^{ji} w_i$, we have

$$\text{Div } w: \qquad g^{ji} \nabla_j w_i.$$

We sometimes call the Div $w$ the *codifferential* of the differential form

$$w = w_i d\xi^i$$

and denote it by

$$\delta w: \qquad g^{ji} \nabla_j w_i. \tag{1.34}$$

For a general skew-symmetric tensor $w_{i_1 i_2 \dots i_p}$, we define its divergence as

$$\text{Div } w: \qquad g^{ji} \nabla_j w_{i i_2 \dots i_p}.$$

We sometimes call the Div $w$ the codifferential of the exterior differential form

$$w = \frac{1}{p!} w_{i_1 i_2 \dots i_p} d\xi^{i_1} \wedge d\xi^{i_2} \wedge \dots \wedge d\xi^{i_p},$$

and denote it by

$$\delta w: \qquad g^{ji} \nabla_j w_{i i_2 \dots i_p}. \tag{1.35}$$

The $\delta w$ is equal to $(-1)^{n(p-1)} * d * w$.

It is easily verified that the operators $d$ and $\delta$ have the properties

$$ddw=0, \tag{1.36}$$

$$\delta\delta w=0. \tag{1.37}$$

Starting from a scalar function $f(\xi)$, we first form its covariant derivative $\nabla_i f$ and then its divergence. Then we obtain

$$\Delta f=g^{ji}\nabla_j\nabla_i f. \tag{1.38}$$

The differential operator $\Delta=g^{ji}\nabla_j\nabla_i$ or $\nabla^i\nabla_i$ is sometimes called the *Laplacian.*

A skew-symmetric tensor $w$: $w_{i_1 i_2 \dots i_p}$ which satisfies

$$\text{Rot } w=0 \quad \text{or} \quad dw=0, \tag{1.39}$$

$$\text{Div } w=0 \quad \text{or} \quad \delta w=0 \tag{1.40}$$

is called a *harmonic tensor.*

If the covariant derivative of a skew-symmetric tensor vanishes, then it is a harmonic tensor. Thus the tensor $e_{i_1 i_2 \dots i_n}$ is a harmonic tensor.

## § 2. Curvature tensors

When we can choose, in a Riemannian space M, a system of coordinate neighbourhoods in such a way that the components $g_{ji}$ of the fundamental metric tensor are always constant, then we say that the Riemannian space M is *locally Euclidean.*

A necessary and sufficient condition for a Riemannian space to be locally Euclidean is

$$K_{kji}{}^h=0, \tag{2.1}$$

where

$$K_{kji}{}^h=\partial_k\left\{{h \atop j\ i}\right\}-\partial_j\left\{{h \atop k\ i}\right\}+\left\{{h \atop k\ a}\right\}\left\{{a \atop j\ i}\right\}-\left\{{h \atop j\ a}\right\}\left\{{a \atop k\ i}\right\} \tag{2.2}$$

is the *Riemann–Christoffel curvature tensor.*

The curvature tensor appears also in the *Ricci identities:*

$$\nabla_k\nabla_j v^h - \nabla_j\nabla_k v^h = K_{kji}{}^h v^i, \tag{2.3}$$

$$\nabla_k\nabla_j w_i - \nabla_j\nabla_k w_k = -K_{kji}{}^h w_h, \tag{2.4}$$

$$\nabla_l\nabla_k T_{ji}{}^h - \nabla_k\nabla_l T_{ji}{}^h = K_{lka}{}^h T_{ji}{}^a - K_{lkj}{}^a T_{ai}{}^h - K_{lki}{}^a T_{ja}{}^h. \tag{2.5}$$

Now consider in a Riemannian space M a curve

$$\xi^h = \xi^h(t) \qquad t_1 \leqq t \leqq t_2$$

and a vector field $v^h(t)$ defined along the curve. If the vector field $v^h(t)$ satisfies

$$\frac{\delta}{dt}v^h = \frac{dv^h}{dt} + \left\{ {h \atop j\ i} \right\} \frac{d\xi^j}{dt} v^i = 0\dagger \tag{2.6}$$

along the curve, then we say that the vector $v^h(t)$ is *parallel* along the curve.

Since the equations of geodesics in M are given by

$$\frac{d^2\xi^h}{ds^2} + \left\{ {h \atop j\ i} \right\} \frac{d\xi^j}{ds}\frac{d\xi^i}{ds} = 0, \tag{2.7}$$

$s$ being the arc length, we can see that the tangent $\frac{d\xi^h}{ds}$ to a geodesic is always parallel along the geodesic.

To transport a vector $v_1^h$ given at one end point $\xi^h(t_1)$ of the curve parallelly along the curve to the other end point $\xi^h(t_2)$ of the curve, we integrate the simultaneous ordinary differential equations

$$\frac{dv^h}{dt} + \left\{ {h \atop j\ i} \right\} \frac{d\xi^j}{dt} v^i = 0$$

with the initial condition $v^h = v_1^h$ for $t = t_1$. Let the unique solution be $v^h = v^h(t)$, then $v^h(t)$ is the vector field parallel

† The symbol $\delta$ means the absolute differential

$$\delta v^h = dv^h + \left\{ {h \atop j\ i} \right\} d\xi^j v^i.$$

along the curve and coincides with $v_1^h$ at one end point $\xi^h(t_1)$ of the curve. Thus the vector to be found is $v^h(t_2)$ at the other end point $\xi^h(t_2)$ of the curve.

By a parallel displacement of vectors along a curve, the length of a vector $v^h$

$$\sqrt{g_{ji}v^jv^i} \tag{2.8}$$

and the angle $\vartheta$ between two vectors $v^h$ and $w^h$ given by

$$\cos\vartheta = \frac{g_{ji}v^jw^i}{(\sqrt{g_{ji}v^jv^i})(\sqrt{g_{ji}w^jw^i})} \tag{2.9}$$

are both unchanged.

We notice that the parallel displacement of a vector from a point $P_1$ to another point $P_2$ depends in general upon the curve joining $P_1$ and $P_2$.

Suppose that the parallel displacement of vectors from one point to another does not depend on the curve joining these points. Fix a point $P_0$ and take an arbitrary vector $v_0^h$ at $P_0$, then transport $v_0^h$ parallelly to another point. Since the parallel displacement does not depend on the curve, we get a vector field $v^h(\xi)$ which is a vector at the point $(\xi^h)$ and is obtained from $v_0^h$ by a parallel displacement of $v_0^h$ from $P_0$ to $(\xi^h)$ along any curve joining them. For this vector field we have

$$\frac{\delta}{dt}v^h = \frac{d\xi^j}{dt}\nabla_j v^h = 0.$$

But, $\frac{d\xi^j}{dt}$ being arbitrary, we have

$$\nabla_j v^h = 0,$$

from which

$$0 = \nabla_k\nabla_j v^h - \nabla_j\nabla_k v^h = K_{kji}{}^h v^i.$$

Since this is true for any contravariant vector $v^h$, we have

$$K_{kji}{}^h = 0.$$

Thus if the parallel displacement of a vector along a curve in a Riemannian space does not depend on the curve, the space is locally Euclidean.

The curvature tensor satisfies the following algebraic identities:

$$K_{kji}{}^{h} = -K_{jki}{}^{h}, \tag{2.10}$$

$$K_{kji}{}^{h} + K_{jik}{}^{h} + K_{ikj}{}^{h} = 0. \tag{2.11}$$

Introducing covariant components

$$K_{kjih} = K_{kji}{}^{a} g_{ah} \tag{2.12}$$

of the curvature tensor, we have

$$K_{kjih} = -K_{jkih}, \tag{2.13}$$

$$K_{kjih} + K_{jikh} + K_{ikjh} = 0. \tag{2.14}$$

Moreover these covariant components satisfy

$$K_{kjih} = -K_{kjhi}, \tag{2.15}$$

$$K_{kjih} = K_{ihkj}. \tag{2.16}$$

The tensor

$$K_{ji} = K_{aji}{}^{a} = g^{ba} K_{bjia} = g^{ba} K_{jbai} \tag{2.17}$$

is called the *Ricci tensor*. It is a symmetric tensor:

$$K_{ji} = K_{ij}. \tag{2.18}$$

The scalar

$$K = g^{ji} K_{ji} \tag{2.19}$$

is called the *curvature scalar*.

The covariant derivative of the curvature tensor satisfies

$$\nabla_l K_{kji}{}^{h} + \nabla_k K_{jli}{}^{h} + \nabla_j K_{lki}{}^{h} = 0 \tag{2.20}$$

which is called the *Bianchi identity*. From this we find by contraction

$$\nabla_a K_{kji}{}^{a} = \nabla_k K_{ji} - \nabla_j K_{ki} \tag{2.21}$$

and from this by contracting with $g^{ji}$

$$\nabla_a K^{\ a}_{h} = \frac{1}{2} \nabla_h K, \tag{2.22}$$

which is a useful identity in the theory of relativity.

Now, for a 2-dimensional Riemannian space, the *Gaussian curvature* $k$ is given by

$$k = -\frac{K_{1212}}{\mathfrak{g}}. \tag{2.23}$$

Consider two linearly independent vectors $v^h$ and $w^h$ at a point $(\xi^h)$ and the 2-dimensional plane $\pi$ determined by them. Then all the geodesics passing through the point $(\xi^h)$ and tangent to $\pi$ form a 2-dimensional subspace of M. The Gaussian curvature of this 2-dimensional subspace at the point $(\xi^h)$ is given by

$$k = -\frac{K_{kjih} v^k w^j v^i w^h}{(g_{ki} g_{jh} - g_{ji} g_{kh}) v^k w^j v^i w^h} \tag{2.24}$$

and is called the *sectional curvature* of the space at $(\xi^h)$ with respect to the section $\pi$ determined by two linearly independent vectors $v^h$ and $w^h$.

Suppose that the sectional curvature at a point $(\xi^h)$ does not depend on the section. Then the curvature tensor must have the form

$$K_{kjih} = k(g_{kh} g_{ji} - g_{jh} g_{ki}) \tag{2.25}$$

at the point $(\xi^h)$. If this is true for any point of the space, the curvature tensor must be of the form (2.25) or

$$K_{kji}{}^h = k(A_k^h g_{ji} - A_j^h g_{ki}) \tag{2.26}$$

at every point of the space. Here $k$ might depend on the point. But, substituting (2.26) into the Bianchi identity, we can see that $k$=constant for $n > 2$.

A space whose curvature tensor is of the form (2.26) is called a *space of constant curvature*. The relation between the curvature scalar $K$ and the Gaussian curvature $k$ is given by

$$K = n(n-1)k. \tag{2.27}$$

Take $n$ mutually orthogonal unit vectors $v_a^h$ at a point $(\xi^h)$, then we have

$$g_{ji} v_b^j v_a^i = \delta_{ba} = \begin{cases} 1, & b = a, \\ 0, & b \neq a, \end{cases} \tag{2.28}$$

from which

$$g^{ji} = \sum_{a=1}^{n} v_a^j v_a^i. \tag{2.29}$$

The sectional curvature $k_{ba}$ with respect to the section determined by $v_b^h$ and $v_a^h$ is given by

$$k_{ba} = -K_{kjih} v_b^k v_a^j v_b^i v_a^h$$

and hence

$$\sum_{b=1}^{n} k_{ba} = K_{ji} v_a^j v_a^i, \tag{2.30}$$

from which

$$\sum_{b=1}^{n} \sum_{a=1}^{n} k_{ba} = K \tag{2.31}$$

by virtue of (2.29).

Equation (2.30) shows that if we fix a unit vector $v^h$ at a point $(\xi^h)$, the sum of $n - 1$ sectional curvatures with respect to $n - 1$ sections determined by $v^h$ and $n - 1$ unit vectors which are orthogonal to $v^h$ and mutually orthogonal is given by

$$K_{ji} v^j v^i \tag{2.32}$$

and is independent of the choice of $n - 1$ unit vectors orthogonal to $v^h$ and mutually orthogonal. The scalar $K_{ji} v^j v^i$ is called the *Ricci curvature* with respect to the direction $v^h$.

The equation (2.31) shows that the sum of the Ricci curvatures with respect to $n$ mutually orthogonal unit vectors is given by $K$ and is independent of the choice of these vectors.

The Ricci curvature with respect to a direction $v^h$ is given by

$$c = \frac{K_{ji}v^j v^i}{g_{ji}v^j v^i}. \tag{2.33}$$

If the Ricci curvature does not depend upon the direction $v^h$, we must have

$$K_{ji} = cg_{ji}. \tag{2.34}$$

If this is true at every point of the space, the Ricci tensor must have the form (2.34) at every point of the space. The factor $c$ might depend on the point, but substituting (2.34) into (2.22), we find that $c$ is a constant.

A space whose Ricci tensor has the form (2.34) is called an *Einstein space*. The relation between the curvature scalar $K$ and $c$ is given by

$$K = nc. \tag{2.35}$$

## § 3. Lie derivatives†

Consider a geometric object field $\Omega(\xi)$ given in an $n$-dimensional manifold M, whose law of transformation under a coordinate transformation

$$\xi^{h'} = \xi^{h'}(\xi) \tag{3.1}$$

is given by

$$\Omega'(\xi') = F(\xi, \ \xi', \ \Omega(\xi)), \tag{3.2}$$

with an inverse transformation

$$\Omega(\xi) = F(\xi', \ \xi, \ \Omega'(\xi')). \tag{3.3}$$

Consider then a point transformation

$${}'\xi^h = f^h(\xi) \tag{3.4}$$

† For details, see J. A. Schouten [3], K. Yano [1], [11].

which carries a point $\xi^h$ into $'\xi^h$. The geometric object field $\Omega(\xi)$ will have the value $\Omega('\xi)$ at the point $('\xi)$.

If we effect a coordinate transformation

$$\xi^{h'} = '\xi^h = f^h(\xi), \tag{3.5}$$

then the point which had $\xi^h$ as coordinates in the old coordinate system will have $\xi^{h'}$ as coordinates, in the new coordinate system.

The inverse image of $\Omega('\xi)$ by (3.4) is given by

$$'\Omega(\xi) = F(\xi', \ \xi, \ \Omega('\xi)). \tag{3.6}$$

The difference

$$'\Omega(\xi) - \Omega(\xi) \tag{3.7}$$

is called the *Lie difference* of $\Omega(\xi)$ with respect to the point transformation (3.4).

Suppose that (3.4) is an infinitesimal transformation

$$'\xi^h = \xi^h + v^h(\xi)\Delta t. \tag{3.8}$$

In this case we call

$$\underset{v}{\mathfrak{L}}\, \Omega = \lim_{\Delta t \to 0} \frac{1}{\Delta t} ['\Omega(\xi) - \Omega(\xi)] \tag{3.9}$$

the *Lie derivative* of $\Omega(\xi)$ with respect to (3.8).

The Lie derivatives of a scalar $f(\xi)$, a contravariant vector $u^h$, a covariant vector $w_i$, a general tensor $T_{\ i}^{\ h}$, for example, and the Christoffel symbols $\left\{ {h \atop j\ i} \right\}$ with respect to (3.8) are respectively given by

$$\underset{v}{\mathfrak{L}}\, f = v^i \nabla_i f, \tag{3.10}$$

$$\underset{v}{\mathfrak{L}}\, u^h = v^a \nabla_a u^h - u^a \nabla_a v^h, \tag{3.11}$$

$$\underset{v}{\mathfrak{L}}\, w_i = v^a \nabla_a w_i + w_a \nabla_i v^a, \tag{3.12}$$

$$\underset{v}{\mathfrak{L}}\, T_{ji}^{\ \ h} = v^a \nabla_a T_{ji}^{\ \ h} - T_{ji}^{\ \ a} \nabla_a v^h + T_{ai}^{\ \ h} \nabla_j v^a + T^{\ \ \ h} \nabla_i v^a \tag{3.13}$$

and

$$\underset{v}{\mathfrak{L}}\left\{ {h \atop j\ i} \right\} = \nabla_j \nabla_i v^h + K_{kji}{}^h v^k. \tag{3.14}$$

These formulae are written in terms of covariant derivatives, but they are all independent of the metric.

Commuting the differential operators $\nabla_i$ and $\underset{v}{\mathfrak{L}}$, we find

$$\underset{v}{\mathfrak{L}}\,(\nabla_j u^h) - \nabla_j\left(\underset{v}{\mathfrak{L}}\, u^h\right) = \left(\underset{v}{\mathfrak{L}}\left\{ {h \atop j\ i} \right\}\right) u^i, \tag{3.15}$$

$$\underset{v}{\mathfrak{L}}\,(\nabla_j w_i) - \nabla_j\left(\underset{v}{\mathfrak{L}}\, w_i\right) = -\left(\underset{v}{\mathfrak{L}}\left\{ {h \atop j\ i} \right\}\right) w_h, \tag{3.16}$$

$$\begin{aligned} &\underset{v}{\mathfrak{L}}\,(\nabla_k T_{ji}{}^h) - \nabla_k\left(\underset{v}{\mathfrak{L}}\, T_{ji}{}^h\right) \\ &= \left(\underset{v}{\mathfrak{L}}\left\{ {h \atop k\ a} \right\}\right) T_{ji}{}^a - \left(\underset{v}{\mathfrak{L}}\left\{ {a \atop k\ j} \right\}\right) T_{ai}{}^h - \left(\underset{v}{\mathfrak{L}}\left\{ {a \atop k\ i} \right\}\right) T_{ja}{}^h. \end{aligned} \tag{3.17}$$

On the other hand, by a straightforward calculation, we can prove

$$\underset{v}{\mathfrak{L}}\left\{ {h \atop j\ i} \right\} = \frac{1}{2} g^{ha}\left[\nabla_j\left(\underset{v}{\mathfrak{L}} g_{ia}\right) + \nabla_i\left(\underset{v}{\mathfrak{L}} g_{ja}\right) - \nabla_a\left(\underset{v}{\mathfrak{L}}\, g_{ji}\right)\right], \tag{3.18}$$

$$\nabla_k\left(\underset{v}{\mathfrak{L}}\left\{ {h \atop j\ i} \right\}\right) - \nabla_j\left(\underset{v}{\mathfrak{L}}\left\{ {h \atop k\ i} \right\}\right) = \underset{v}{\mathfrak{L}}\, K_{kji}{}^h. \tag{3.19}$$

Now, if the transformation (3.4) or (3.8) does not change the distance between two nearby points in a Riemannian space M, it is called a *motion*.† The condition for (3.4) to be a motion is given by

$$g_{ji}('\xi) d'\xi^j d'\xi^i = g_{ji}(\xi) d\xi^j d\xi^i$$

or

$$'g_{ji}(\xi) d\xi^j d\xi^i = g_{ji}(\xi) d\xi^j d\xi^i$$

or

$$'g_{ji}(\xi) = g_{ji}(\xi). \tag{3.20}$$

---

† See S. B. Myers and N. E. Steenrod [1]

In the case where (3.4) reduces to (3.8), this gives

$$\underset{v}{\mathfrak{L}}\, g_{ji} = v^a \nabla_a g_{ji} + g_{ai} \nabla_j v^a + g_{ja} \nabla_i v^a = \nabla_j v_i + \nabla_i v_j = 0. \tag{3.21}$$

The equations (3.21) are called *Killing equations.* (W. Killing [1]).

Conversely if a vector satisfies Killing equations, it generates a local one-parameter group of motions.

A vector satisfying Killing equations is called a *Killing vector*.

Since we have $\underset{v}{\mathfrak{L}}\, g_{ji} = 0$ for a Killing vector, we have also

$$\underset{v}{\mathfrak{L}} \left\{ {h \atop j\ i} \right\} = 0, \tag{3.22}$$

$$\underset{v}{\mathfrak{L}}\, K_{kji}{}^h = 0, \tag{3.23}$$

$$\underset{v}{\mathfrak{L}}\, \nabla_l K_{kji}{}^h = 0, \tag{3.24}$$

. . . . . . . . . . . . . . .

Equations (3.23), (3.24), . . . are integrability conditions of Killing equations.

If the transformation (3.4) or (3.8) does not change the parallelism of the space, that is, the Christoffel symbols, it is called an *affine motion*†.

The condition for (3.4) to be an affine motion is given by

$$'\left\{ {h \atop j\ i} \right\} = \left\{ {h \atop j\ i} \right\}. \tag{3.25}$$

In the case where (3.4) reduces to (3.8), this gives

$$\underset{v}{\mathfrak{L}} \left\{ {h \atop j\ i} \right\} = \nabla_j \nabla_i v^h + K_{kji}{}^h v^k = 0. \tag{3.26}$$

Conversely, if a vector $v^h$ satisfies (3.26), it generates a local one-parameter group of affine motions.

---

† See K. Nomizu [1], [4], [6].

A vector $v^h(\xi)$ satisfying (3.26) is called an *affine Killing vector.*

From (3.26), we get

$$\underset{v}{\mathfrak{L}} K_{kj\,i}{}^{h}=0, \qquad \underset{v}{\mathfrak{L}} \nabla_l K_{kji}{}^{h}=0, \;\ldots\ldots \tag{3.27}$$

which are integrability conditions of (3.26).

If the transformation (3.4) or (3.8) does not change the geodesics of the space, it is called a *projective motion.*

The condition for (3.4) to be a projective motion is given by

$$'\left\{{h \atop j\;i}\right\}=\left\{{h \atop j\;i}\right\}+A_j^h p_i+A_i^h p_j \tag{3.28}$$

where $p_i$ is a certain gradient vector.

In case where (3.4) reduces to (3.8), this gives

$$\underset{v}{\mathfrak{L}}\left\{{h \atop j\;i}\right\}=\nabla_j\nabla_i v^h+K_{kji}{}^{h}v^k=A_j^h p_i+A_i^h p_j. \tag{3.29}$$

Conversely if a vector $v^h$ satisfies (3.29), it generates a local one-parameter group of projective motions.

A vector $v^h$ satisfying (3.29) is called a *projective Killing vector.*

From (3.29), we get

$$\underset{v}{\mathfrak{L}} K_{kji}{}^{h}=-A_k^h\nabla_j p_i+A_j^h\nabla_k p_i \tag{3.30}$$

from which

$$\underset{v}{\mathfrak{L}}\left(-\frac{1}{n-1}K_{ji}\right)=\nabla_j p_i. \tag{3.31}$$

Substituting (3.31) into (3.30), we find

$$\underset{v}{\mathfrak{L}} P_{kji}{}^{h}=0, \tag{3.32}$$

where

$$P_{kji}{}^{h}=K_{kji}{}^{h}-\frac{1}{n-1}\,(A_k^h K_{ji}-A_j^h K_{ki}) \tag{3.33}$$

is the so-called *projective curvature tensor.*

Equation (3.32) gives the integrability conditions of (3.29).

If the transformation (3.4) or (3.8) does not change the angle of two directions, it is called a *conformal motion*.

The condition for (3.4) to be a conformal motion is given by

$$'g_{ji}(\xi) = \varrho^2(\xi) g_{ji}(\xi), \tag{3.34}$$

where $\varrho^2(\xi)$ is a positive scalar function.

In the case where (3.4) reduces to (3.8), this gives

$$\underset{v}{\mathfrak{L}} g_{ji} = \nabla_j v_i + \nabla_i v_j = 2\Phi g_{ji}, \tag{3.35}$$

where $\Phi$ is a certain scalar function.

Conversely if a vector $v^h$ satisfies (3.35), it generates a local one-parameter group of conformal motions.

A vector $v^h(\xi)$ satisfying (3.35) is called a *conformal Killing vector*.

From (3.35), we get

$$\underset{v}{\mathfrak{L}} \left\{ {h \atop j\ i} \right\} = A_j^h \Phi_i + A_i^h \Phi_j - \Phi^h g_{ji}, \tag{3.36}$$

where

$$\Phi_i = \nabla_i \Phi. \tag{3.37}$$

From (3.36), we find

$$\underset{v}{\mathfrak{L}} K_{kji}{}^h = -A_k^h \nabla_j \Phi_i + A_j^h \nabla_k \Phi_i - (\nabla_k \Phi^h) g_{ji} + (\nabla_j \Phi^h) g_{ki}, \tag{3.38}$$

from which

$$\underset{v}{\mathfrak{L}} C_{ji} = \nabla_j \Phi_i, \tag{3.39}$$

where

$$C_{ji} = -\frac{K_{ji}}{n-2} + \frac{K g_{ji}}{2(n-1)(n-2)}. \tag{3.40}$$

Substituting (3.39) into (3.38), we find

$$\underset{v}{\mathfrak{L}} C_{kji}{}^h = 0, \tag{3.41}$$

where $C_{kji}{}^h$ is the *Weyl conformal curvature tensor* given by

$$C_{kji}{}^h = K_{kji}{}^h + A_k^h C_{ji} - A_j^h C_{ki} + C_k{}^h g_{ji} - C_j{}^h g_{ki}. \qquad (3.42)$$

Equation (3.41) expresses the integrability conditions of (3.35).

A conformal motion for which $\varrho$ is constant is called a *homothetic motion.*† Equation (3.18) shows that a homothetic motion is an affine motion.

For any geometric object $\Omega$ appearing above, we can prove††

$$\underset{u}{\mathfrak{L}}\,\underset{v}{\mathfrak{L}}\,\Omega - \underset{v}{\mathfrak{L}}\,\underset{u}{\mathfrak{L}}\,\Omega = \underset{[u,\,v]}{\mathfrak{L}}\,\Omega, \qquad (3.43)$$

where

$$[u,\ v]^h = u^i \partial_i v^h - v^i \partial_i u^h = \underset{u}{\mathfrak{L}}\, v^h. \qquad (3.44)$$

Thus the set of all Killing, affine Killing ,projective Killing, conformal Killing or homothetic Killing vectors constitutes a Lie algebra.

---

† H. Hiramatu [1]; M. S. Knebelman and K . Yano [1]; E. B. Shanks [1]; K. Yano [1], [2], [11].

†† K. Yano [1], [11].

CHAPTER II

# THEOREM OF GREEN AND ITS APPLICATIONS

## § 1. Theorem of Green

Consider a contravariant vector field $v^h(\xi)$ defined on the whole Riemannian space M. Its divergence is given by

$$\nabla_a v^a = \frac{1}{\sqrt{g}}\, \partial_a(\sqrt{g}\, v^a). \tag{1.1}$$

The famous Theorem of Green is then stated as follows:

THEOREM 1.1. *In a compact orientable Riemannian space* M, *we have*

$$\int_M (\nabla_a v^a) d\sigma = 0 \tag{1.2}$$

*for an arbitrary vector field* $v^h(\xi)$, *where* $d\sigma$ *is the volume element*

$$d\sigma = \sqrt{g}\, d\xi^1 \wedge d\xi^2 \wedge \ldots \wedge d\xi^n. \tag{1.3}$$

We shall here reproduce a proof given by S. Bochner [1]†.

First of all, if a bounded set D is contained in a coordinate neighbourhood, then we have

$$\int_D (\nabla_a v^a) d\sigma = \int_D \partial_a(\sqrt{g}\, v^a) d\xi^1 \wedge \ldots \wedge d\xi^n.$$

Suppose now that R is a "rectangle" $a^h \leqq \xi^h \leqq b^h$ and that the vector $v^h(\xi)$ vanishes on the boundary of R. Then

† See also T. Y. Thomas [1]; K. Yano and S. Bochner [1].

we have

$$\int_{a_1}^{b_1} \partial_1(\sqrt{g}\ v^1)d\xi^1 = \ldots = \int_{a_n}^{b_n} \partial_n(\sqrt{g}\ v^n)d\xi^n = 0$$

and consequently

$$\int_{R} (\nabla_a v^a) d\sigma = 0. \tag{1.4}$$

But since the integral of $\nabla_a v^a$ is zero over any open set on which $v^h$ vanishes, the above equation shows that (1.2) is true if $v^h$ vanishes outside some "rectangle" R.

Since the space M is compact, we can cover it by a finite number of coordinate neighbourhoods

$$U_1, \quad U_2, \quad \ldots, \quad U_N$$

whose closures are contained in "rectangles"

$$R_1, \quad R_2, \quad \ldots, \quad R_N$$

respectively and moreover such that corresponding to each $\alpha$, $\alpha = 1, 2, \ldots, N$, there exists a coordinate neighbourhood $W_\alpha$ containing $R_\alpha$. Then we can easily find a coordinate neighbourhood $V_\alpha$ between $U_\alpha$ and $R_\alpha$ and a non-negative scalar function $g_\alpha$ of class $C^1$ in $R_\alpha$ such that

$$g_\alpha \geqq 1 \quad \text{in} \quad U_\alpha \quad \text{and} \quad g_\alpha = 0 \quad \text{outside} \quad V_\alpha.$$

Completing the function $g_\alpha$ by values zero outside $R_\alpha$, we have, throughout M,

$$g_1 + g_2 + \ldots + g_N \geqq 1.$$

Thus, if we put

$$f_\alpha = \frac{g_\alpha}{g_1 + g_2 + \ldots + g_N},$$

then functions $f_\alpha$ are of class $C^1$ and vanish outside the "rectangle" $R_\alpha$, their sum being

$$f_1 + f_2 + \ldots + f_N = 1.$$

Now, if we put

$$v_{\alpha}^{h} = f_{\alpha} v^{h},$$

then $v_{\alpha}^{h}$ vanishes outside the "rectangle" $R_{\alpha}$. Thus we have

$$\int_{M} \nabla_{a}(v_{\alpha}^{a}) d\sigma = 0. \tag{1.5}$$

But, on the other hand, we have

$$v^{h} = v_{1}^{h} + v_{2}^{h} + \ldots + v_{N}^{h}$$

and consequently

$$\nabla_{a} v^{a} = \nabla_{a}(v_{1}^{a}) + \nabla_{a}(v_{2}^{a}) + \ldots + \nabla_{a}(v_{N}^{a}).$$

Thus integrating this over the whole space and taking account of (1.5), we find

$$\int (\nabla_{a} v^{a}) d\sigma = 0.$$

Take a scalar function $f(\xi)$ defined over the whole space M, then the Laplacian

$$\Delta f = g^{ji} \nabla_{j} \nabla_{i} f$$

is the divergence of the vector $g^{ji}\nabla_{i} f$. Thus, by the Theorem of Green, we have

THEOREM 1.2. *In a compact orientable Riemannian space* M, *we have*

$$\int_{M} (g^{ji} \nabla_{j} \nabla_{i} f) d\sigma = 0 \tag{1.6}$$

*for any scalar field* $f$.

Since we have

$$\frac{1}{2} g^{ji} \nabla_{j} \nabla_{i} f^{2} = f(g^{ji} \nabla_{j} \nabla_{i} f) + g^{ji} (\nabla_{j} f)(\nabla_{i} f),$$

applying Theorem 1.2, we have

THEOREM 1.3. *In a compact orientable Riemannian space* M, *we have*

$$\int_M [f(g^{ji}\nabla_j\nabla_i f)+g^{ji}(\nabla_j f)(\nabla_i f)]d\sigma=0 \tag{1.7}$$

*for any scalar field* $f(\xi)$.

If a scalar function $f(\xi)$ satisfies

$$g^{ji}\nabla_j\nabla_i f \geqq 0$$

in the whole space, then, $d\sigma$ having a definite sign, Theorem 1.2 gives

$$g^{ji}\nabla_j\nabla_i f=0$$

and consequently Theorem 1.3 gives

$$g^{ji}(\nabla_j f)(\nabla_i f)=0,$$

from which $\qquad \nabla_i f=0, \quad f=\text{constant},$

the form $g^{ji}w_j w_i$ being positive definite. Thus we have

THEOREM 1.4†. *If, in a compact orientable Riemannian space* M, *we have*

$$g^{ji}\nabla_j\nabla_i f \geqq 0$$

*for a scalar field* $f(\xi)$, *then*

$$f=\text{constant}$$

*in the whole space.*

If a scalar function $f(\xi)$ which is not constant satisfies

$$g^{ji}\nabla_j\nabla_i f=cf \tag{1.8}$$

$c$ being a constant, the constant $c$ cannot be zero. Substituting this in (1.7), we find

$$\int_M [cf^2+g^{ji}(\nabla_j f)(\nabla_i f)]d\sigma=0.$$

Thus, if $c>0$, then we should have $f=0$, and we have

† The theorem is true also for a compact space, see E. Hopf [1].

THEOREM 1.5. *If, in a compact orientable Riemannian space* M, *a scalar function* $f(\xi)$ *which is not constant satisfies*

$$g^{ji}\nabla_j\nabla_i f = cf$$

*c being a constant, then the constant c is negative.*

We now apply the Laplacian $g^{ji}\nabla_j\nabla_i$ to the square of the length of a vector field $v^h$:

$$\frac{1}{2}\, g^{ji}\nabla_j\nabla_i(v^h v_h) = (g^{ji}\nabla_j\nabla_i v^h)v_h + (\nabla^j v^i)(\nabla_j v_i).$$

Thus, applying Theorem 1.2, we obtain

THEOREM 1.6. *In a compact orientable Riemannian space* M, we have

$$\int_M [g^{ji}(\nabla_j\nabla_i v^h)v_h + (\nabla^j v^i)(\nabla_j v_i)]d\sigma = 0 \tag{1.9}$$

*for an arbitrary vector field* $v^h$.

Suppose that the second covariant derivative $\nabla_j\nabla_i v^h$ of a vector $v^h$ vanishes, then, $(\nabla^j v^i)\,(\nabla_j v_i)$ being positive definite, we have, from (1.9),

$$\nabla_i v^h = 0,$$

that is, the first covariant derivative vanishes. Thus we have

THEOREM 1.7. *If, in a compact orientable Riemannian space* M, *the second covariant derivative of a vector vanishes, then the first covariant derivative of the vector vanishes too.*

If a vector field $v^h(\xi)$ which is not zero satisfies

$$g^{ji}\nabla_j\nabla_i v^h = cv^h \tag{1.10}$$

$c$ being a constant, substituting (1.10) into (1.9), we find

$$\int_M [cv^h v_h + (\nabla^j v^i)\,(\nabla_j v_i)]d\sigma = 0. \tag{1.11}$$

Thus, $v^h$ being different from zero, $c$ cannot be positive. Hence we have

THEOREM 1.8. *If, in a compact orientable Riemannian space* M, *a vector field* $v^h(\xi)$ *other than zero vector satisfies*

$$g^{ji}\nabla_j\nabla_i v^h = cv^h$$

*c being a constant, then c is non-positive.*

Consider a vector field $v^h(\xi)$ in a compact orientable Riemannian space M and calculate the divergence of the vector field

$$v^i(\nabla_i v^j) - v^j(\nabla_i v^i),$$

then we get

$$\nabla_j[v^i(\nabla_i v^j) - v^j(\nabla_i v^i)] = K_{ji}v^j v^i + (\nabla^j v^i)(\nabla_i v_j) - (\nabla_i v^i)^2,$$

from which

THEOREM 1.9. *In a compact orientable Riemannian space* M, *we have*

$$\int_M [K_{ji}v^j v^i + (\nabla^j v^i)(\nabla_i v_j) - (\nabla_i v^i)^2]d\sigma = 0 \tag{1.12}$$

*for an arbitrary vector field* $v^h(\xi)$.

Forming the difference (1.9) – (1.12), we have

THEOREM 1.10. *In a compact orientable Riemannian space* M, *we have*

$$\int_M [(g^{ji}\nabla_j\nabla_i v^h - K_i{}^h v^i)v_h + \frac{1}{2}(\nabla^j v^i - \nabla^i v^j)(\nabla_j v_i - \nabla_i v_j)$$
$$+ (\nabla_i v^i)^2]d\sigma = 0 \tag{1.13}$$

*for an arbitrary vector field* $v^h(\xi)$.

Forming next the sum (1.9)+(1.12), we have

THEOREM 1.11. *In a compact orientable Riemannian space* M, *we have*

$$\int_M [(g^{ji}\nabla_j\nabla_i v^h + K_i{}^h v^i)v_h + \frac{1}{2}(\nabla^j v^i + \nabla^i v^j)(\nabla_j v_i + \nabla_i v_j)$$
$$- (\nabla_i v^i)^2]d\sigma = 0 \tag{1.14}$$

*for an arbitrary vector field* $v^h(\xi)$.

Take two skew-symmetric tensors $u_{i_1 i_2 \ldots i_p}$ of $p$-th order and $v_{j i_1 \ldots i}$ of $(p+1)$-st order and form the vector field

$$u_{i_1 i_2 \ldots i_p} v^{j i_1 \ldots i_p}.$$

Then the theorem of Green gives

$$\int \nabla_j (u_{i_1 i_2 \ldots i_p} v^{j i_1 \ldots i_p}) d\sigma = 0$$

or

$$\int_M \frac{1}{p+1} (\nabla_j u_{i_1 i_2 \ldots i_p} - \nabla_{i_1} u_{j i_2 \ldots i_p} - \ldots - \nabla_{i_p} u_{i_1 i_2 \ldots j}) v^{j i_1 \ldots i_p} d\sigma$$
$$+ \int_M u_{i_1 i_2 \ldots i_p} \nabla_j v^{j i_1 \ldots i_p} d\sigma = 0.$$

Thus if we use the notation

$$\int_M \frac{1}{p!} a_{i_1 \ldots i_p} b^{i_1 \ldots i_p} d\sigma = (a, b)$$

for the integral of the scalar product of two skew-symmetric tensors $a_{i_1 \ldots i_p}$ and $b_{i_1 \ldots i_p}$ of the same order $p$, we have from the above equation

$$(du, v) + (u, \delta v) = 0.$$

Hence we have

**Theorem 1.12.** *For two skew-symmetric tensors, $u$ of the $p$-th order and $v$ of the $(p+1)$-st order, in a compact orientable Riemannian space, we have*

$$(du, v) + (u, \delta v) = 0. \qquad (1.15)$$

Take a skew-symmetric tensor $w$ of the $p$-th order and put

$$u = w, \qquad v = dw,$$

in (1.15), then we find

$$(\delta dw, w) + (dw, dw) = 0.$$

Next put

$$u = \delta w \qquad vw =$$

in (1.15), then we find

$$(d\delta w,\ w)+(\delta w,\ \delta w)=0.$$

Adding these two equations, we obtain

$$(\Delta w,\ w)+(dw,\ dw)+(\delta w,\ \delta w)=0,$$

where we have put

$$\Delta w=\delta dw+d\delta w.$$

Hence we have

THEOREM 1.13. *For a skew-symmetric tensor w in a compact orientable Riemannian space* M, *we have*

$$(\delta dw,\ w)+(dw,\ dw)=0, \tag{1.16}$$

$$(d\delta w,\ w)+(\delta w,\ \delta w)=0, \tag{1.17}$$

*and*

$$(\Delta w,\ w)+(dw,\ dw)+(\delta w,\ \delta w)=0. \tag{1.18}$$

The formulae (1.12), (1.13), (1.14), (1.15), (1.16), (1.17) and (1.18) will be found extremely useful in the following discussions.

We denote, in the sequel, by M a compact orientable Riemannian space with positive definite metric.

## § 2. Harmonic vectors and tensors†

A harmonic vector is defined as a vector field $v^h(\xi)$ which satisfies

$$\nabla_j v_i-\nabla_i v_j=0 \quad \text{and} \quad \nabla_i v^i=0. \tag{2.1}$$

It is well known that in a compact orientable Riemannian space the number of linearly independent (with constant

† See S. Bochner [2], [5], [6]; R. Couty [1]; H. Guggenheimer [2]; W. V. D. Hodge [1]; A. Lichnerowicz [1], [2], [9]; I. Mogi [1]; G. de Rham [1], [2], [4]; G. de Rham and K. Kodaira [1]; D. C. Spencer [1]; Y. Tomonaga [1]; K. Yano and S. Bochner [1].

coefficients) harmonic vectors is equal to the one-dimensional Betti number of the space.

Suppose that a harmonic vector $v_i$ has the form

$$v_i = \nabla_i f,$$

then we have

$$g^{ji}\nabla_j\nabla_i f = 0$$

and hence, following Theorem 1.4,

$$v_i = 0.$$

Thus we obtain

THEOREM 2.1. *If, in a space* M, *a harmonic vector* $v_i$ *has the form* $v_i = \nabla_i f$, *it is identically zero.*

For a harmonic vector field $v^h(\xi)$, formula (1.12) gives

$$\int_{\mathrm{M}} [K_{ji}v^j v^i + (\nabla^j v^i)(\nabla_j v_i)]d\sigma = 0,$$

from which we have

THEOREM 2.2. *In* M *with Ricci curvature* $K_{ji}v^j v^i \geqq 0$, *a harmonic vector* $v_h$ *has a vanishing covariant derivative:*

$$\nabla_j v^h = 0,$$

*and in a space* M *with positive definite Ricci curvature* $K_{ji}v^j v^i > 0$ *a harmonic vector other than the zero vector does not exist. Consequently the one-dimensional Betti number of* M *vanishes.* (S. Bochner [2], S. B. Myers [1]).

For a harmonic vector field $v_h$, we have

$$\nabla_i v_h - \nabla_h v_i = 0 \quad \text{and} \quad g^{ji}\nabla_j v_i = 0.$$

Applying the operator $g^{ji}\nabla_j$ to the first of these equations, we find

$$g^{ji}\nabla_j\nabla_i v_h - K_h{}^a v_a = 0 \tag{2.2}$$

by virtue of the second equation.

Conversely, if (2.2) is satisfied in M, then equation (1.13) gives

$$\nabla_j v_i - \nabla_i v_j = 0 \quad \text{and} \quad \nabla_i v^i = 0.$$

Hence we have

THEOREM 2.3. *A necessary and sufficient condition for a vector field $v^h(\xi)$ in* M *to be harmonic is*

$$g^{ji}\nabla_j\nabla_i v^h - K_i{}^h v^i = 0.$$

A harmonic tensor is defined as a skew-symmetric tensor field $w$: $w_{i_1 i_2 \ldots i_p}$ which satisfies

$$dw = 0, \qquad \delta w = 0. \tag{2.3}$$

It is well known that in M the number of linearly independent (with constant coefficients) harmonic tensors of the $p$-th order is equal to the $p$-dimensional Betti number of M.

Suppose that a harmonic tensor $w$ of the $p$-th order has the form

$$w = du,$$

then we have, by the definition of a harmonic tensor

$$\delta du = 0.$$

Hence, putting $v = du$ in (1.15), we find

$$(du,\ du) = 0$$

from which

$$w = du = 0,$$

and we have

THEOREM 2.4. *If, in* M, *a harmonic tensor $w$ has the form $w = du$, then it identically vanishes.*

For a harmonic tensor $w$, we have

$$dw = 0, \qquad \delta w = 0$$

and consequently

$$\Delta w = \delta dw + d\delta w = 0.$$

Conversely, if this is satisfied, we have from (1.18)

$$(dw,\ dw) + (\delta w,\ \delta w) = 0,$$

from which

$$dw = 0, \qquad \delta w = 0$$

and $w$ is harmonic.

Hence

THEOREM 2.5. *A necessary and sufficient condition for a skew-symmetric $w$ in M to be harmonic is*

$$\Delta w = \delta d w + d \delta w = 0. \tag{2.4}$$

Calculating explicitly $\Delta w$, we find

$$\Delta w:\ g^{ji}\nabla_j\nabla_i w_{i_1 i_2 \ldots i_p} - \sum_s^{1 \ldots p} K_{i_s}{}^a w_{i_1 \ldots a \ldots i_p}$$
$$- \sum_{s<t}^{1 \ldots p} K^{ba}{}_{i_s i_t} w_{i_1 \ldots b \ldots a \ldots i_p}. \tag{2.5}$$

Hence we have

THEOREM 2.6. *A necessary and sufficient condition for a skew-symmetric tensor $w_{i_1 i_2 \ldots i_p}$ in M to be harmonic is*

$$g^{ji}\nabla_j\nabla_i w_{i_1 i_2 \ldots i_p} - \sum_s^{1 \ldots p} K_{i_s}{}^a w_{i_1 \ldots a \ldots i_p}$$
$$- \sum_{s<t}^{1 \ldots p} K^{ba}{}_{i_s i_t} w_{i_1 \ldots b \ldots a \ldots i_p} = 0. \tag{2.6}$$

(K. Yano [4]).

## § 3. Killing vectors†

A Killing vector is defined as a vector field $v^h(\xi)$ which satisfies

$$\nabla_j v_i + \nabla_i v_j = 0 \tag{3.1}$$

† See S. B. Myers and N. E. Steenrod [1].

and consequently

$$\nabla_i v^i = 0. \tag{3.2}$$

A Killing vector generates a local one-parameter group of motions.

Suppose that a Killing vector $v^h$ has the form

$$v_i = \nabla_i f,$$

then we have from (3.1)

$$\nabla_j \nabla_i f = 0$$

and hence, following Theorem 1.4,

$$v_i = \nabla_i f = 0.$$

Thus we obtain

THEOREM 3.1. *If, in* M, *a Killing vector* $v^h$ *has the form* $v_i = \nabla_i f$, *it is identically zero.*

For a Killing vector field $v^h(\xi)$, formula (1.12) gives

$$\int_M [K_{ji} v^j v^i - (\nabla^j v^i)(\nabla_j v_i)] d\sigma = 0,$$

from which we have

THEOREM 3.2. *In a space* M *with Ricci curvature* $K_{ji} v^j v^i \leqq 0$, *a Killing vector has a vanishing covariant derivative:*

$$\nabla_j v^h = 0.$$

*In a space* M *with negative definite Ricci curvature* $K_{ji} v^j v^i < 0$, *a Killing vector does not exist other than the zero vector. Consequently* M *does not admit a local one parameter group of motions.* (Bochner [2]).

Suppose that M has vanishing Ricci curvature $K_{ji} = 0$ and admits a transitive group of motions, then, following the above theorem, the generators $v_a^h(\xi)(a = 1, 2, \ldots, r \geqq n)$ of this group should have vanishing covariant derivatives:

$$\nabla_j v_a^h = 0$$

from which

$$K_{kji}{}^h v^i_a = 0.$$

The rank of the matrix $(v^i_a)$ being $n$, this implies

$$K_{kji}{}^h = 0.$$

Hence we have

THEOREM 3.3. *If* M *has vanishing Ricci curvature and admits a transitive group of motions, then it is locally Euclidean.*

For a Killing vector field $v^h$, we have

$$\underset{v}{\mathfrak{L}} g_{ji} = \nabla_j v_i + \nabla_i v_j = 0,$$

and consequently

$$g^{ji}\left(\underset{v}{\mathfrak{L}}\left\{ {h \atop j\ i} \right\}\right) = g \quad \nabla_j \nabla_i v^h + K_i{}^h v^i = 0 \tag{3.3}$$

by virtue of (3.14) and (3.18) of Chapter I.

Conversely if (3.3) and $\nabla_i v^i = 0$ are satisfied in M, then equation (1.14) gives

$$\nabla_j v_i + \nabla_i v_j = 0.$$

Hence we have

THEOREM 3.4. *A necessary and sufficient condition for a vector field* $v^h(\xi)$ *in* M *to be a Killing vector is*

$$g^{ji}\nabla_j \nabla_i v^h + K_i{}^h v^i = 0 \quad \text{and} \quad \nabla_i v^i = 0. \tag{3.4}$$

(K. Yano [3]).

Assume that M admits a harmonic vector $w_i$ and a Killing vector $v^h$, then we have

$$g^{ji}\nabla_j \nabla_i w_a = K_a{}^i w_i \quad \text{and} \quad g^{ji}\nabla_j \nabla_i v^a = -K_j{}^a v^j.$$

Calculating the Laplacian of their inner product, we find

$$\begin{aligned} g^{ji}\nabla_j \nabla_i (w_a v^a) &= (g^{ji}\nabla_j \nabla_i w_a) v^a + 2(\nabla_j w_i)(\nabla^j v^i) + w_a (g^{ji}\nabla_j \nabla_i v^a) \\ &= K_{ji} w^j v^i + 0 - K_{ji} w^j v^i = 0 \end{aligned}$$

and hence

$$w_a v^a = \text{constant}$$

by virtue of Theorem **1.4**. Thus we have

THEOREM 3.5. *If* M *admits a harmonic vector* $w_i$ *and a Killing vector* $v^h$, *then their inner product is constant.* (S. Bochner [8]).

Assume that M admits a harmonic tensor $w_{i_1 i_2 \ldots, i_p}$ of the $p$-th order and a Killing vector $v^h$. Since we see from (3.17) of Chapter I that covariant differentiation $\nabla_i$ and Lie derivation $\underset{v}{\mathfrak{L}}$ with respect to a Killing vector are commutative, we have

$$\nabla_j\Big(\underset{v}{\mathfrak{L}}\, w_{i_1 i_2 \ldots i_p}\Big) - \nabla_{i_1}\Big(\underset{v}{\mathfrak{L}}\, w_{j i_2 \ldots i_p}\Big) - \cdots - \nabla_{i_p}\Big(\underset{v}{\mathfrak{L}}\, w_{i_1 i_2 \ldots j}\Big)$$
$$= \underset{v}{\mathfrak{L}}\, (\nabla_j w_{i_1 i_2 \ldots i_p} - \nabla_{i_1} w_{j i_2 \ldots i_p} - \cdots - \nabla_{i_p} w_{i_1 i_2 \ldots j}) = 0$$

and

$$g^{ji}\nabla_j\Big(\underset{v}{\mathfrak{L}}\, w_{i i_2 \ldots i_p}\Big) = \underset{v}{\mathfrak{L}}\, (g^{ji}\nabla_j w_{i i_2 \ldots i_p}) = 0$$

by virtue of the fact that $w_{i_1 i_2 \ldots i_p}$ is harmonic and $\underset{v}{\mathfrak{L}}\, g^{ji} = 0$.

These two equations show that the Lie derivative $\underset{v}{\mathfrak{L}}\, w_{i_1 i_2 \ldots i_p}$ of a harmonic tensor $w_{i_1 i_2 \ldots i_p}$ is also harmonic.

On the other hand, we have

$$\underset{v}{\mathfrak{L}}\, w_{i_1 i_2 \ldots i_p} = v^a \nabla_a w_{i_1 i_2 \ldots i_p} + w_{a i_2 \ldots i_p}\nabla_{i_1} v^a + \cdots + w_{i_1 i_2 \ldots a}\nabla_{i_p} v^a.$$

But, $w_{i_1 i_2 \ldots i_p}$ being harmonic, we have

$$\nabla_a w_{i_1 i_2 \ldots i_p} = \nabla_{i_1} w_{a i_2 \ldots i_p} + \nabla_{i_2} w_{i_1 a \ldots i_p} + \cdots + \nabla_{i_p} w_{i_1 i_2 \ldots a}$$

and consequently

$$\underset{v}{\mathfrak{L}}\, w_{i_1 i_2 \ldots i_p} = v^a(\nabla_{i_1} w_{a i_2 \ldots i_p} + \nabla_{i_2} w_{i_1 a \ldots i_p} + \cdots + \nabla_{i_p} w_{i_1 i_2 \ldots a})$$
$$+ w_{a i_2 \ldots i_p}\nabla_{i_1} v^a + \cdots + w_{i_1 i_2 \ldots a}\nabla_{i_p} v^a$$
$$= \nabla_{i_1}(v^a w_{a i_2 \ldots i_p}) - \nabla_{i_2}(v^a w_{a i_1 \ldots i_p}) - \cdots - \nabla_{i_p}(v^a w_{a i_2 \ldots i_{p-1} i_1}),$$

which shows that the harmonic tensor $\underset{v}{\mathfrak{L}}\, w_{i_1 i_2 \ldots i_p}$ is the rotation of

$$v^a w_{a i_2 \ldots i_p}.$$

Hence, by Theorem 2.4, $\underset{v}{\mathfrak{L}}\, w_{i_1 i_2 \ldots i_p}$ vanishes identically. Thus we have

THEOREM 3.6. *The Lie derivative of a harmonic tensor with respect to a Killing vector vanishes identically in* M. (K. Yano [3]).

If we apply this theorem to a harmonic vector $w_i$, we have

$$\underset{v}{\mathfrak{L}}\, w_i = v^a \nabla_a w_i + w_a \nabla_i v^a = v^a \nabla_i w_a + w_a \nabla_i v^a = \nabla_i (w_a v^a) = 0,$$

and consequently

$$w_a v^a = \text{constant}.$$

This also proves Theorem 3.5.

## § 4. Affine and projective Killing vectors†

The condition for a vector field $v^h(\xi)$ to be an affine Killing vector is given by

$$\underset{v}{\mathfrak{L}} \left\{ {h \atop j\ i} \right\} = \nabla_j \nabla_i v^h + K_{kji}{}^h v^k = 0 \tag{4.1}$$

from which, contracting with $g^{ji}$, we have

$$g^{ji} \nabla_j \nabla_i v^h + K_i{}^h v^i = 0, \tag{4.2}$$

and, contracting with respect to $h$ and $i$,

$$\nabla_j (\nabla_i v^i) = 0.$$

† See R. Couty [3]; A. Lichnerowicz [2]; K. Nomizu [I]; T. Sumitomo [1]; K. Yano [11]; K. Yano and T. Nagano [1].

Thus $\nabla_i v^i$ is a constant. But since we have

$$\int_M (\nabla_i v^i) d\sigma = 0$$

in M, this constant must be zero:

$$\nabla_i v^i = 0. \tag{4.3}$$

According to Theorem 3.4, equations (4.2) and (4.3) show that $v^h$ is a Killing vector. Hence

THEOREM 4.1. *An infinitesimal affine motion is a motion in* M. (K. Yano [3]).†

The condition for a vector field $v^h(\xi)$ to be a projective Killing vector is given by

$$\underset{v}{\mathfrak{L}} \left\{ \begin{matrix} h \\ j \ i \end{matrix} \right\} = \nabla_j \nabla_i v^h + K_{kji}{}^h v^k = A_j^h p_i + A_i^h p_j \tag{4.4}$$

$p_i$ being a certain vector field, from which, contracting $g^{ji}$ we have

$$g^{ji} \nabla_j \nabla_i v^h + K_i{}^h v^i = 2p^h, \tag{4.5}$$

and, contracting with respect to $h$ and $i$,

$$\nabla_j (\nabla_a v^a) = (n+1) p_j. \tag{4.6}$$

Equations (4.5) and (4.6) give

$$g^{ji} \nabla_j \nabla_i v^h - K_i{}^h v^i = -2K_i{}^h v^i + \frac{2}{n+1} \nabla^h (\nabla_a v^a). \tag{4.7}$$

Substituting (4.7) into (1.13), we find

$$\int_M \Big[ -2K_{ji} v^j v^i + \frac{2}{n+1} v^h \nabla_h (\nabla_a v^a) + \frac{1}{2} (\nabla^j v^i - \nabla^i v^j)(\nabla_j v_i - \nabla_i v_j) + (\nabla_i v^i)^2 \Big] d\sigma = 0.$$

---

† See also J. Hano [1], S. Ishihara and M. Obata [1]; S. Kobayashi [1]; A. Lichnerowicz [16]; K. Nomizu [2]; K. Yano and T. Nagano [3].

Since we have, on the other hand,

$$0=\int_{\mathrm{M}}\nabla_h(v^h\nabla_a v^a)d\sigma=\int(\nabla_a v^a)^2 d\sigma+\int_{\mathrm{M}} v^h\nabla_h(\nabla_a v^a)d\sigma,$$

the above equation may be put in the form

$$\int_{\mathrm{M}}\Big[-2K_{ji}v^j v^i+\frac{1}{2}(\nabla^j v^i-\nabla^i v^j)(\nabla_j v_i-\nabla_i v_j)$$

$$+\frac{n-1}{n+1}(\nabla_i v^i)^2\Big]d\sigma=0. \tag{4.8}$$

Hence, in a space M with Ricci curvature $K_{ji}v^j v^i \leqq 0$, we have

$$\nabla_j v_i-\nabla_i v_j=0, \qquad \nabla_i v^i=0,$$

that is, $v^h$ is a harmonic vector. Thus, from (4.6) we find

$$p_i=0.$$

Consequently the projective motion is an affine motion. According to Theorem 4.1, the projective motion is then a motion, and consequently we have

$$\nabla_j v_i+\nabla_i v_j=0,$$

from which

$$\nabla_j v_i=0.$$

In an M with Ricci curvature $K_{ji}v^j v^i<0$, we have immediately $v^h=0$. Hence

THEOREM 4.2. *In a space* M *with Ricci curvature* $K_{ji}v^j v^i \leqq 0$, *a projective Killing vector has a vanishing covariant derivative:*

$$\nabla_j v^h=0,$$

*and if the Ricci curvature satifies* $K_{ji}v^j v^i<0$, *a projective Killing vector does not exist other than the zero vector. Con-*

*sequently* M *does not admit a local one-parameter group of projective motions.* (R. Couty [2], [3]).

Suppose that M has vanishing Ricci curvature $K_{ji}=0$ and admits a transitive group of projective motions, then according to the above theorem, the generators $v^i_{\alpha}$ ($\alpha = 1, 2, \ldots, r \geqq n$) of the group satisfy

$$\nabla_j v^h_{\alpha} = 0$$

from which

$$K_{kji}{}^h = 0.$$

Thus

THEOREM 4.3. *If* M *has vanishing Ricci curvature and admits a transitive group of projective motions, then it is locally Euclidean.*

## § 5. Conformal and homothetic Killing vectors†

The condition for a vector field $v^h(\xi)$ to be a conformal Killing vector is given by

$$\underset{v}{\mathfrak{L}}\, g_{ji} = \nabla_j v_i + \nabla_i v_j = 2\Phi g_{ji}. \tag{5.1}$$

A conformal Killing vector generates a local one-parameter group of conformal motions.

Since (5.1) gives

$$\Phi = \frac{1}{n} \nabla_i v^i \tag{5.2}$$

and

$$\nabla_i v_j = \frac{2}{n} (\nabla_a v^a) g_{ji} - \nabla_j v_i,$$

† See R. Couty [3]; M. S. Knebelman and K. Yano [1]; A. Lichnerowicz [13] T. Nagano [1]; I. Sato [1]; T. Sumitomo [1]; K. Yano [1], [2], [3], [11]; K. Yano and T. Nagano [1], [2]

substituting this into (1.12), we find

$$\int_M \left[ K_{ji}v^j v^i - (\nabla^j v^i)(\nabla_j v_i) - \frac{n-2}{n}(\nabla_i v^i)^2 \right] d\sigma = 0.$$

From this we have

THEOREM 5.1. *In a space* M *with Ricci curvature* $K_{ji}v^j v^i \leqq 0$, *a conformal Killing vector has a vanishing covariant derivative:*

$$\nabla_j v^h = 0$$

*and if* $K_{ji}v^j v^i < 0$, *a conformal Killing vector does not exist other than the zero vector. Consequently* M *does not admit a local one-parameter group of conformal motions.* (K. Yano [3]).

In the exactly same way as in the cases of motions and projective motions, we can prove

THEOREM 5.2. *If* M *has vanishing Ricci curvature and admits a transitive group of conformal motions, then it is locally Euclidean.*

From the equation

$$\underset{v}{\mathfrak{L}}\, g_{ji} = 2\Phi g_{ji}$$

we get

$$\underset{v}{\mathfrak{L}} \left\{ {h \atop j\ i} \right\} = A_j^h \Phi_i + A_i^h \Phi_j - \Phi^h g_{ji} \tag{5.3}$$

by virtue of (3.18) of Chapter I, where

$$\Phi_i = \nabla_i \Phi \quad \text{and} \quad \Phi^h = g^{ha}\Phi_a.$$

Contracting $g^{ji}$ to (5.3) we find

$$g^{ji}\nabla_j \nabla_i v^h + K_i{}^h v^i = -(n-2)\Phi^h$$

or

$$g^{ji}\nabla_j \nabla_i v^h + K_i{}^h v^i + \frac{n-2}{n} \nabla^h \nabla_i v^i = 0, \tag{5.4}$$

by virtue of (5.2).

Now, equation (1.14) can be written in the form

$$\int_M \Big[ (g^{ji}\nabla_j\nabla_i v^h + K_i{}^h v^i)v_h + \frac{1}{2}\left(\nabla^j v^i + \nabla^i v^j - \frac{2}{n}\, g^{ji}\nabla_b v^b\right)\left(\nabla_j v_i + \nabla_i v_j - \frac{2}{n}\, g_{ji}\nabla_a v^a\right) - \frac{n-2}{n}\,(\nabla_j v^j)(\nabla_i v^i) \Big] d\sigma = 0.$$

But we have on the other hand

$$0 = \int_M \nabla_j [v^j(\nabla_i v^i)] d\sigma = \int_M (\nabla_j v^j)(\nabla_i v^i) d\sigma + \int_M v^j \nabla_j (\nabla_i v^i) d\sigma,$$

and consequently the above equation can also be put in the form

$$\int_M \Big[ \left( g^{ji}\nabla_j\nabla_i v^h + K_i{}^h v^i + \frac{n-2}{n}\,\nabla^h\nabla_i v^i \right) v_h + \frac{1}{2}\left(\nabla^j v^i + \nabla^i v^j - \frac{2}{n}\, g^{ji}\nabla_b v^b\right)\left(\nabla_j v_i + \nabla_i v_j - \frac{2}{n}\, g_{ji}\nabla_a v^a\right) \Big] d\sigma = 0. \tag{5.5}$$

Thus we have

**Theorem 5.3.** *A necessary and sufficient condition for a vector field $v^h(\xi)$ in* M *to be a conformal Killing vector is*

$$g^{ji}\nabla_j\nabla_i v^h + K_i{}^h v^i + \frac{n-2}{n}\,\nabla^h\nabla_i v^i = 0. \tag{5.6}$$

(I. Sato [1]; A. Lichnerowicz [19]).

For a conformal Killing vector $v^h$, we have

$$\nabla_j \Phi_i = \underset{v}{\mathfrak{L}}\, C_{ji}, \tag{5.7}$$

where

$$C_{ji} = -\frac{K_{ji}}{n-2} + \frac{K g_{ji}}{2(n-1)(n-2)}$$

(See (3.39) and (3.40) of Chapter I).

From (5.7), we have

$$g^{ji}\nabla_j\Phi_i = g^{ji} \underset{v}{\mathfrak{L}} C_{ji} = \underset{v}{\mathfrak{L}} (g^{ji}C_{ji}) - (\underset{v}{\mathfrak{L}} g^{ji})C_{ji} = \underset{v}{\mathfrak{L}} (g^{ji}C_{ji}) + 2\Phi g^{ji}C_{ji},$$

from which

$$g^{ji}\nabla_j\nabla_i\Phi = -\frac{1}{2(n-1)} (\underset{v}{\mathfrak{L}} K + 2\Phi K). \tag{5.8}$$

Hence, if we assume that

$$K = \text{constant},$$

then we get

$$g^{ji}\nabla_j\nabla_i\Phi = -\frac{K}{n-1} \Phi. \tag{5.9}$$

Substituting this into the identity

$$\int_M [\Phi(g^{ji}\nabla_j\nabla_i\Phi) + g^{ji}(\nabla_j\Phi)(\nabla_i\Phi)]d\sigma = 0$$

(See (1.7)), we find

$$\int_M \left[ -\frac{K}{n-1} \Phi^2 + g^{ji}(\nabla_j\Phi)(\nabla_i\Phi) \right] d\sigma = 0.$$

Hence if $K<0$, then $\Phi=0$ and the conformal transformation is a motion. If $K=0$, then $\Phi=\text{constant}$ and the conformal transformation is homothetic. But a homothetic transformation is an affine transformation and consequently, in M, it is a motion according to Theorem 4.1. Thus we have

THEOREM 5.4. *If* M *has a constant curvature scalar* $K \leqq 0$, *then a conformal transformation is a motion.*

The condition for a vector field $v^h(\xi)$ to be a homothetic conformal Killing vector is given by

$$\underset{v}{\mathfrak{L}} g_{ji} = \nabla_j v_i + \nabla_i v_j = 2c g_{ji} \tag{5.10}$$

where $c$ is a constant. Since a homothetic motion is an

affine motion, it is a motion according to Theorem 4.1. Hence a proper homothetic transformation does not exist in an M.

Suppose that $v^h$ is a conformal Killing vector and $w_{i_p i_{p-1} \ldots i_1}$ is a skew-symmetric tensor, then we have

$$
\begin{aligned}
&\underset{v}{\mathfrak{L}} \nabla_j w_{i_p i_{p-1} \ldots i_1} - \nabla_j \underset{v}{\mathfrak{L}} w_{i_p i_{p-1} \ldots i_1} \\
&= \left( \underset{v}{\mathfrak{L}} \left\{ {t \atop j\ i_p} \right\} \right) w_{t i_{p-1} \ldots i_1} + \left( \underset{v}{\mathfrak{L}} \left\{ {t \atop j\ i_{p-1}} \right\} \right) w_{i_p t i_{p-2} \ldots i_1} + \ldots \\
&\qquad \ldots + \left( \underset{v}{\mathfrak{L}} \left\{ {t \atop j\ i_1} \right\} \right) w_{i_p i_{p-1} \ldots i_2 t} \\
&= \Phi_{i_p} w_{j i_{p-1} \ldots i_1} + \Phi_j w_{i_p i_{p-1} \ldots i_1} - g_{j i_p} \Phi^t w_{t i_{p-1} \ldots i_1} \\
&+ \Phi_{i_{p-1}} w_{i_p j \ldots i_1} + \Phi_j w_{i_p i_{p-1} \ldots i_1} - g_{j i_{p-1}} \Phi^t w_{i_p t \ldots i_1} \\
&+ \ldots \\
&+ \Phi_{i_1} w_{i_p i_{p-1} \ldots j} + \Phi_j w_{i_p i_{p-1} \ldots i_1} - g_{j i_1} \Phi^t w_{i_p i_{p-1} \ldots t}, \qquad (5.11)
\end{aligned}
$$

from which

$$
\begin{aligned}
&\underset{v}{\mathfrak{L}} [\nabla_j w_{i_p i_{p-1} \ldots i_1} - \nabla_{i_p} w_{j i_{p-1} \ldots i_1} - \ldots - \nabla_{i_1} w_{i_p \ldots i_2 j}] \\
&= \nabla_j \underset{v}{\mathfrak{L}} w_{i_p i_{p-1} \ldots i_1} - \nabla_{i_p} \underset{v}{\mathfrak{L}} w_{j i_{p-1} \ldots i_1} - \ldots - \nabla_{i_1} \underset{v}{\mathfrak{L}} w_{i_p \ldots i_2 j}
\end{aligned}
$$

and consequently

THEOREM 5.5. *The Lie derivative of a closed skew-symmetric tensor with respect to a conformal Killing vector is closed.*

Transvecting (5.11) with $g^{ji_p}$ we find

$$
g^{ji} \underset{a}{\mathfrak{L}} \nabla_j w_{i i_{p-1} \ldots i_1} - g^{ji} \nabla_j \underset{v}{\mathfrak{L}} w_{i i_{p-1} \ldots i_1} = (2p-n) \Phi^t w_{t i_{p-1} \ldots i_1}
$$

or

$$
\begin{aligned}
\underset{v}{\mathfrak{L}} g^{ji} \nabla_j w_{i i_{p-1} \ldots i_1} + 2\Phi g^{ji} \nabla_j w_{i i_{p-1} \ldots i_1} - g^{ji} \nabla_j \underset{v}{\mathfrak{L}} w_{i i_{p-1} \ldots i_1} \\
= (2p-n) \Phi^t w_{t i_{p-1} \ldots i_1},
\end{aligned}
$$

from which

THEOREM 5.6. *The Lie derivative of a closed skew-symmetric tensor of order $p = n/2$ in a Riemannian space of even di-*

*mension $n$ with respect to a conformal Killing vector is also closed.*

From Theorems 5.5 and 5.6, we get

THEOREM 5.7. *The Lie derivative of a harmonic tensor of order $p=n/2$ in a Riemannian space with even dimension $n$ with respect to a conformal Killing vector is harmonic.*

But the Lie derivative of a harmonic tensor is the rotation of another tensor. Thus

THEOREM 5.8. *The Lie derivative of a harmonic tensor of order $p=n/2$ in M with even dimension $n$ with respect to a conformal Killing vector is identically zero.* (S. I. Goldberg [4], [5]; K. Yano [16], [17]).

## § 6. Symmetric spaces†

A Riemannian space is called a symmetric space when it satisfies

$$\nabla_l K_{kji}{}^h = 0. \tag{6.1}$$

From this we get

$$\nabla_l K_{ji} = 0, \qquad \nabla_l K = 0. \tag{6.2}$$

We define in general the tensor $H$ as

$$\begin{aligned} H_{mlkjih} &= -(\nabla_m \nabla_l K_{kjih} - \nabla_l \nabla_m K_{kjih}) \qquad (6.3) \\ &= K_{mlk}{}^a K_{ajih} + K_{mlj}{}^a K_{kaih} + K_{mli}{}^a K_{kjah} + K_{mlh}{}^a K_{kjia}. \end{aligned}$$

For a symmetric space we have

$$H_{mlkjih} = 0. \tag{6.4}$$

Now put

$$f = K_{kjih} K^{kjih},$$

then we have

$$\frac{1}{2} \Delta f = (g \quad \nabla_m \nabla_l K_{kjih}) K^{kjih} + (\nabla_l K_{kjih})(\nabla^l K^{kjih}).$$

† See also B. Kostant [1]; K. Nomizu [3], [5].

On the other hand

$$\begin{aligned}(g^{ml}\nabla_m\nabla_l K_{kjih})K^{kjih} &= -g^{ml}[\nabla_m(\nabla_k K_{jlih}+\nabla_j K_{lkih})]K^{kjih}\\ &= -g^{ml}(-H_{mkjlih}+\nabla_k\nabla_m K_{ihjl}-H_{mjlkih}+\nabla_j\nabla_m K_{ihlk})K^{kjih}\\ &= [H_{kjlih}{}^{l}+g^{m}\nabla_k(\nabla_i K_{hmjl}+\nabla_h K_{mijl})-H_{jklih}{}^{l}\\ &\qquad +g^{ml}\nabla_j(\nabla_i K_{hmlk}+\nabla_h K_{milk})]K^{kjih}\\ &= 2H_{kjlih}{}^{l}K^{kjih}-(\nabla_k\nabla_i K_{hj}-\nabla_k\nabla_h K_{ij}-\nabla_j\nabla_i K_{hk}+\nabla_j\nabla_h K_{ik})K^{kjih}\end{aligned}$$

and hence

$$(g^{ml}\nabla_m\nabla_l K_{kjih})K^{kjih}=2H_{kjlih}{}^{l}K^{kjih}-4(\nabla_k\nabla_i K_{hj})K^{kjih}.$$

Consequently, we have

$$\frac{1}{2}\Delta f=2H_{kjlih}{}^{l}K^{kjih}-4(\nabla_k\nabla_i K_{hj})K^{kjih}+(\nabla_l K_{kjih})(\nabla K^{kjih}).$$

Thus, in an M satisfying

$$2H_{kjlih}{}^{l}K^{kjih}\geqq 0, \qquad \nabla_k K_{ji}=0,$$

we have $\frac{1}{2}\Delta f\geqq 0$ and consequently $\frac{1}{2}\Delta f=0$, from which

$$\nabla_l K_{kjih}=0.$$

Hence we have

THEOREM 6.1. *A space* M *for which*

$$2H_{kjlih}{}^{l}K^{kjih}\geqq 0, \qquad \nabla_k K_{ji}=0$$

*is symmetric.*

## § 7. Theorems on Betti numbers

We know already that a necessary and sufficient condition for a skew-symmetric tensor $w_{i_1 i_2 \dots i_p}$ in a compact orientable Riemannian space to be harmonic is

$$\begin{aligned}g^{ji}\nabla_j\nabla_i w_{i_1 i_2 \dots i_p}-\textstyle\sum_{s=1}^{p}K_{i_s}{}^{l}w_{i_1\dots l\dots i_p}\\ -\textstyle\sum_{t<s}^{1\dots p}K^{ba}{}_{i_t i_s}w_{i_1\dots b\dots a\dots i_p}=0.\end{aligned} \qquad (7.1)$$

Now, substituting this into

$$\frac{1}{2}\int_M g^{ji}\nabla_j\nabla_i(w_{i_1\ldots i_p}w^{i_1\cdots i_p})d\sigma = \int_M (g^{ji}\nabla_j\nabla_i w_{i_1\ldots i_p})w^{i_1\cdots i_p}d\sigma$$
$$+\int_M (\nabla^j w^{i_1\cdots i_p})(\nabla_j w_{i_1\ldots i_p})d\sigma = 0,$$

we find

$$\int_M p\cdot F(w,\ w)\,d\sigma + \int_M (\nabla^j w^{i_1\cdots i_p})(\nabla_j w_{i_1\ldots i_p})d\sigma = 0, \qquad (7.2)$$

where

$$F(w,\ w) = K_{ji}w^{ji_2\cdots i_p}w^i{}_{i_2\ldots i_p}$$
$$+\frac{(p-1)}{2}K_{kjih}w^{kji_3\cdots i_p}w^{ih}{}_{i_3\ldots i_p}. \qquad (7.3)$$

This formula may be considered to be valid even when $p=1$.

From (7.2), we see that if $F(w,\ w)$ is positive definite, then we should have $w_{i_1\ldots i_p}=0$, and hence we have,

THEOREM 7.1. *If, in a compact orientable Riemannian space, the form $F(w,\ w)$ is positive definite, then we have*

$$B_p = 0 \qquad (p=1,\ \ldots,\ n-1)$$

*$B_p$ being p-dimensional Betti numbers.* (A. Lichnerowicz [1], [2], [9],; K. Yano [4]; K. Yano and S. Bochner [1]).

Suppose that our Riemannian space is of constant curvature, then we have

$$K_{kjih} = \frac{K}{n(n-1)}(g_{kh}g_{ji} - g_{jh}g_{ki}) \qquad (7.4)$$

and

$$K_{ji} = \frac{K}{n}g_{ji} \qquad (7.5)$$

and consequently

$$F(w,\ w) = \frac{n-p}{n(n-1)}Kw^{i_1\cdots i_p}w_{i_1\ldots i_p}, \qquad (7.6)$$

which is positive definite for $K>0$, hence

THEOREM 7.2. *For a compact orientable Riemannian space of positive constant curvature, we have*

$$B_p=0 \qquad (p=1,\ 2,\ \ldots,\ n-1).$$

Suppose next that our Riemannian space is conformally flat, then we have

$$K_{kjih}=\frac{1}{n-2}\,(g_{kh}K_{ji}-g_{jh}K_{ki}+K_{kh}g_{ji}-K_{jh}g_{ki})$$

$$-\frac{K}{(n-1)(n-2)}\,(g_{kh}g_{ji}-g_{jh}g_{ki}). \tag{7.7}$$

For this case, we have

$$F(w,\ w)=\frac{n-2p}{n-2}K_{ji}w^{ji_2\cdots i_p}w^{i}{}_{i_2\cdots i_p}$$

$$+\frac{p-1}{(n-1)(n-2)}K\cdot w^{i_1i_2\cdots i_p}w_{i_1i_2\cdots i_p}. \tag{7.8}$$

Now we assume that $K_{ji}v^jv^i$ is positive definite and denote by $L$ the smallest (positive) proper value of the matrix $(K_{ji})$, then we have

$$K_{ji}v^jv^i\geqq Lg_{ji}v^jv^i \tag{7.9}$$

and

$$K=g^{ji}K_{ji}\geqq nL>0. \tag{7.10}$$

Therefore from (7.8), we have

$$F(w,\ w)\geqq\frac{n-2p}{n-2}\cdot Lw^{i_1\cdots i_p}w_{i_1\cdots i_p} \tag{7.11}$$

$$+\frac{n(p-1)}{(n-1)(n-2)}Lw^{i_1\cdots i_p}w_{i_1\cdots i_p}=\frac{n-p}{n-1}Lw^{i_1\cdots i_p}w_{i_1\cdots i_p}.$$

which is positive definite, hence

THEOREM 7.3. *For a compact orientable conformally Euclidean Riemannian space, with positive definite Ricci curvature, we have*

$$B_p=0 \qquad (p=1,\ 2,\ \ldots,\ n-1).$$

We next assume that the curvature tensor $K_{kjih}$ of our Riemannian space satisfies

$$0<\frac{1}{2}k\leqq -\frac{K_{kjih}w^{kj}w^{ih}}{w_{ji}w^{ji}}\leqq k \tag{7.12}$$

for a certain positive number $k$ and for any skew-symmetric tensor $w^{ji}$.

Taking two mutually orthogonal unit vectors $u^h$ and $v^h$, and putting

$$w^{ji}=v^j u^i - v^i u^j, \tag{7.13}$$

we find, from (7.12),

$$0<\frac{1}{4}k\leqq -K_{kjih}v^k u^j v^i u^h \leqq \frac{1}{2}k. \tag{7.14}$$

We now put $v^h_{(1)}=u^h$ and take $n-1$ unit vectors $v^h_{(2)}, \ldots, v^h_{(n)}$ which are orthogonal to $v^h_{(1)}$ and to each other, then we have

$$\sum_{s=1}^{n} v^j_{(s)} v^i_{(s)} = g^{ji}. \tag{7.15}$$

On the other hand, we have, from (7.14)

$$0<\frac{1}{4}k\leqq -K_{kjih}v^k_{(a)}u^j v^i_{(a)}u^h\leqq \frac{1}{2}k$$

from which

$$0<\frac{n-1}{4}k\leqq K_{ji}u^j u^i\leqq \frac{n-1}{2}k, \tag{7.16}$$

by virtue of (7.15).

Thus, we have

$$F(w, w) = K_{ji} w^{ji_2 \ldots i_p} w^{i}{}_{i_2 \ldots i_p} + \frac{p-1}{2} K_{kjih} w^{kji_3 \ldots i_p} w^{ih}{}_{i_3 \ldots i_p}$$

$$\geqq \frac{n-1}{4} k w^{i_1 \ldots i_p} w_{i_1 \ldots i_p} - \frac{p-1}{2} k w^{i_1 \ldots i_p} w_{i_1 \ldots i_p}$$

$$= \frac{n-2p+1}{4} k w^{i_1 \ldots i_p} w_{i_1 \ldots i_p}$$

which is positive definite for $n \geqq 2p$. Hence

THEOREM 7.4. *If, in a compact orientable Riemannian space, the curvature tensor satisfies*

$$0 < \frac{1}{2} k \leqq -\frac{K_{kjih} w^{kj} w^{ih}}{w_{ji} w^{ji}} \leqq k$$

*for a certain positive constant* $k$ *and for any bivector* $w^{ji}$, *then we have*

$$B_p = 0, \qquad (p = 1, 2, \ldots, n-1)\dagger.$$

† Cf. H. E. Rauch [1].

CHAPTER III

# COMPLEX MANIFOLDS

## § 1. Complex manifolds†

We consider a $2n$-dimensional real manifold $M_{2n}$ of class $C^\infty$ covered by a system of coordinate neighbourhoods $(\xi^h)$, where Latin indices $h, i, j, \ldots$ run over the range $1, 2, \ldots, n, \bar{1}, \bar{2}, \ldots, \bar{n}$ and Greek indices $\varkappa, \lambda, \mu, \nu, \ldots$ over the range $1, 2, \ldots, n$. We can introduce in each coordinate neighbourhood $(\xi^h)$ complex coordinates $(z^\varkappa)$ defined by

$$z^\varkappa = \xi^\varkappa + \sqrt{-1}\,\xi^{\bar{\varkappa}}. \tag{1.1}$$

We call $\xi^h$ real coordinates and $z^\varkappa$ complex coordinates of a point with respect to these systems of coordinates respectively.

If we can cover the whole manifold $M_{2n}$ by a system of complex coordinate neighbourhoods $(z^\varkappa)$ in such a way that, in the intersection of two complex coordinate neighbourhoods $(z^\varkappa)$ and $(z^{\varkappa'})$, we have

$$z^{\varkappa'} = f^{\varkappa'}(z^\lambda), \qquad |A_\varkappa^{\varkappa'}| \neq 0, \tag{1.2}$$

where $f^{\varkappa'}(z^\lambda)$ are regular complex functions of complex variables $z^1, z^2, \ldots, z^n$ and $A_\varkappa^{\varkappa'} = \partial_\varkappa z^{\varkappa'}$, we say that the manifold $M_{2n}$ admits a complex structure defined by the existence of such a system of complex coordinate neigh-

† See S. Bochner [4], [7], [9]; H. Cartan [1]; S. S. Chern [4], [5]; A. Frölicher [1]; P. R. Garabedian and D. C. Spencer [1], [2]; H. Hopf [1], [2]; K. Yano [10]. K. Yano and S. Bochner [1].

bourhoods and we call $M_{2n}$ a *complex manifold*. We often denote such $M_{2n}$ by $C_n$.

When we write equations (1.2) in the form

$$\xi^{h'} = \xi^{h'}(\xi), \tag{1.3}$$

the $\xi^{h'}$ are real analytic functions of $\xi^h$ and the Jacobian is then given by

$$|A_h^{h'}| = |A_\varkappa^{\varkappa'}|\,|\bar{A}_\varkappa^{\varkappa'}| > 0, \tag{1.4}$$

where $A_h^{h'} = \partial_h \xi^{h'}$ and the bar denotes complex conjugate, so that a complex manifold is of class $C^\omega$ and is orientable.

We call the complex conjugate manifold a manifold $\bar{C}_n$ which can be put in a homeomorphic correspondence with the complex manifold $C_n$ in such a way that, if the local complex coordinate neighbourhoods $(z^\varkappa)$ of $C_n$ correspond to the local complex coordinate neighbourhoods $(z^{\bar{\varkappa}})$ of $\bar{C}_n$, then we have the relation

$$\bar{z}^\varkappa = z^{\bar{\varkappa}} \tag{1.5}$$

between the complex coordinates of the corresponding points, $\bar{z}^\varkappa$ being the complex conjugates of $z^\varkappa$.

If we have a real analytic function $f(\xi)$ of real coordinates $\xi^h$, we develop this function in series of $\xi^h$ and replace $\xi^\varkappa$ by $\frac{1}{2}(z^\varkappa + \bar{z}^\varkappa)$ and $\xi^{\bar{\varkappa}}$ by $\frac{1}{2i}(z^\varkappa - \bar{z}^\varkappa)$. If we replace again $\bar{z}^\varkappa$ by $z^{\bar{\varkappa}}$, we get an analytic function $f(z^\lambda, z^{\bar{\lambda}})$ of $2n$ independent complex variables $z^\lambda$ and $z^{\bar{\lambda}}$. Hence we can always regard a real analytic function $f(\xi)$ on the complex manifold $M_{2n}$ as an analytic function on the product manifold $C_n \times \bar{C}_n$.

To define the partial derivatives of a function $f(z, \bar{z})$ in $C_n$, we first regard $f(z, \bar{z})$ as a function of $z^\lambda$ and $z^{\bar{\lambda}}$ and take partial derivatives

$$\frac{\partial f}{\partial z^\lambda}, \qquad \frac{\partial f}{\partial z^{\bar{\lambda}}}$$

and then we put

$$z^{\bar{\lambda}}=\bar{z}^{\lambda}.$$

We denote these partial derivatives by

$$\partial_{\lambda}f, \qquad \partial_{\bar{\lambda}}f$$

respectively.

A transformation of complex coordinates in the intersection of two complex coordinate neighbourhoods in $C_n \times \overline{C}_n$ is of the form

$$z^{\varkappa'}=f^{\varkappa'}(z^1, \ldots, z^n), \qquad z^{\bar{\varkappa}'}=\bar{f}^{\varkappa'}(z^{\bar{1}}, \ldots, z^{\bar{n}}), \tag{1.6}$$

where $\bar{f}^{\varkappa'}$ are complex conjugates of the functions $f^{\varkappa'}$.

The equations $z^{\bar{\varkappa}}=\bar{z}^{\varkappa}$ may be regarded as those defining the original manifold $M_{2n}$ in $C_n \times \overline{C}_n$.

## § 2. Vectors and tensors in a complex manifold

A contravariant vector in the product space $C_n \times \overline{C}_n$ has components of the form

$$v^h=(v^{\varkappa}, v^{\bar{\varkappa}}) \tag{2.1}$$

and their law of transformation under (1.6) is

$$v^{\varkappa'}=A_{\varkappa}^{\varkappa'}v^{\varkappa}, \qquad v^{\bar{\varkappa}'}=A_{\bar{\varkappa}}^{\bar{\varkappa}'}v^{\bar{\varkappa}}, \tag{2.2}$$

where

$$A_{\varkappa}^{\varkappa'}=\partial_{\varkappa}z^{\varkappa'}, \qquad A_{\bar{\varkappa}}^{\bar{\varkappa}'}=\partial_{\bar{\varkappa}}z^{\bar{\varkappa}'}.$$

When the vector is considered at a point of the subspace defined by

$$z^{\bar{\varkappa}}=\bar{z}^{\varkappa}, \tag{2.3}$$

we call the vector a contravariant vector in the original complex manifold $M_{2n}$. In this case the law of transformation (2.2) can be written as

$$v^{\varkappa'}=A_{\varkappa}^{\varkappa'}v^{\varkappa}, \qquad v^{\bar{\varkappa}'}=\bar{A}_{\varkappa}^{\varkappa'}v^{\bar{\varkappa}}. \tag{2.4}$$

Thus, if $(v^{\varkappa}, v^{\bar{\varkappa}})$ are components of a contravariant vector in $M_{2n}$, then

$$(v^{\varkappa}, 0), \qquad (0, v^{\bar{\varkappa}}), \qquad (\overline{v^{\bar{\varkappa}}}, \overline{v^{\varkappa}})$$

are also components of contravariant vectors in $M_{2n}$.

We regard the contravariant vector $v^h = (v^{\varkappa}, v^{\bar{\varkappa}})$ in $M_{2n}$ as a contravariant vector in $C_n \times \overline{C}_n$ defined on the subspace (2.3) and denote by $\pi$ and $\bar{\pi}$ the tangent planes to $C_n$ and $\overline{C}_n$ respectively.

Then the vector $(v^{\varkappa}, 0)$ is the projection of $(v^{\varkappa}, v^{\bar{\varkappa}})$ on the plane $\pi$ and $(0, v^{\bar{\varkappa}})$ is the projection of $(v^{\varkappa}, v^{\bar{\varkappa}})$ on the plane $\bar{\pi}$.

Now by the natural correspondence between $C_n$ and $\overline{C}_n$, to a vector $(v^{\varkappa}, 0)$ tangent to $C_n$ corresponds a vector $(0, \overline{v^{\varkappa}})$ tangent to $\overline{C}_n$ and to a vector $(0, v^{\bar{\varkappa}})$ tangent to $\overline{C}_n$ corresponds a vector $(\overline{v^{\bar{\varkappa}}}, 0)$ tangent to $C_n$.

Thus if we have a vector $(v^{\varkappa}, v^{\bar{\varkappa}})$ in $M_{2n}$ we decompose it as

$$(v^{\varkappa}, v^{\bar{\varkappa}}) = (v^{\varkappa}, 0) + (0, v^{\bar{\varkappa}})$$

and consider vectors

$$(0, \overline{v^{\varkappa}}), \qquad (\overline{v^{\bar{\varkappa}}}, 0)$$

corresponding to $(v^{\varkappa}, 0)$ and $(0, v^{\bar{\varkappa}})$ respectively, then their sum is given by

$$(\overline{v^{\bar{\varkappa}}}, \overline{v^{\varkappa}}).$$

We call the vector $(\overline{v^{\bar{\varkappa}}}, \overline{v^{\varkappa}})$ the vector *conjugate* to the original vector $(v^{\varkappa}, v^{\bar{\varkappa}})$. When the original vector and its conjugate coincide, we call the vector a *self-conjugate* vector. Thus for a self-conjugate vector, we have

$$v^{\bar{\varkappa}} = \overline{v^{\varkappa}}. \tag{2.5}$$

A self-conjugate vector is tangent to the subspace defined by $z^{\bar{\varkappa}} = \bar{z}^{\varkappa}$.

The transition from complex coordinates $(z^{\kappa}, z^{\bar{\kappa}})$ to real coordinates $\xi^h$ is given by

$$\xi^{\kappa} = \frac{1}{2}(z^{\kappa} + z^{\bar{\kappa}}), \qquad \xi^{\bar{\kappa}} = \frac{1}{2i}(z^{\kappa} - z^{\bar{\kappa}})$$

and consequently the components of a vector $v^h$ in these real coordinates are given by

$$\frac{\partial \xi^h}{\partial z^{\lambda}} v^{\lambda} + \frac{\partial \xi^h}{\partial z^{\bar{\lambda}}} v^{\bar{\lambda}},$$

that is,

$$\frac{1}{2}(v^{\kappa} + v^{\bar{\kappa}}) \quad \text{and} \quad \frac{1}{2i}(v^{\kappa} - v^{\bar{\kappa}}). \tag{2.6}$$

Thus a self-conjugate vector has real components in a real coordinate system and hence we sometimes call it a *real vector*.

Similar considerations may be applied to a covariant vector and also to a general tensor.

A covariant vector in the product space $C_n \times \overline{C}_n$ has components of the form

$$w_i = (w_{\lambda}, w_{\bar{\lambda}}) \tag{2.7}$$

and their law of transformation under (1.6) is

$$w_{\lambda'} = A_{\lambda'}^{\lambda} w_{\lambda}, \qquad w_{\bar{\lambda}'} = A_{\bar{\lambda}'}^{\bar{\lambda}} w_{\bar{\lambda}}. \tag{2.8}$$

When the vector is considered at a point of the subspace (2.3) we call the vector a covariant vector in the original complex manifold $M_{2n}$. In this case the law of transformation (2.8) can be written as

$$w_{\lambda'} = A_{\lambda'}^{\lambda} w_{\lambda}, \qquad w_{\bar{\lambda}'} = \bar{A}_{\lambda'}^{\lambda} w_{\bar{\lambda}}. \tag{2.9}$$

Thus if $(w_{\lambda}, w_{\bar{\lambda}})$ are components of a covariant vector in $M_{2n}$, then

$$(w_{\lambda}, 0), \qquad (0, w_{\bar{\lambda}}), \qquad (\bar{w}_{\bar{\lambda}}, \bar{w}_{\lambda})$$

are also components of covariant vector in $M_{2n}$. We call a covariant vector satisfying

$$(w_\lambda,\ w_{\bar\lambda}) = (\bar w_{\bar\lambda},\ \bar w_\lambda)$$

that is,

$$w_{\bar\lambda} = \bar w_\lambda$$

a *self-conjugate* covariant vector.

The components of the vector in a real coordinate system are given by

$$w_\lambda + w_{\bar\lambda} \quad \text{and} \quad i(w_\lambda - w_{\bar\lambda}). \tag{2.10}$$

Thus a *self-conjugate* covariant vector has real components in a real coordinate system. So we call such a covariant vector a *real* covariant vector.

A mixed tensor in the product space $C_n \times \bar C_n$ has components of the form

$$T_i{}^h = \begin{pmatrix} T_\lambda{}^\varkappa, & T_\lambda{}^{\bar\varkappa} \\ T_{\bar\lambda}{}^\varkappa, & T_{\bar\lambda}{}^{\bar\varkappa} \end{pmatrix} \tag{2.11}$$

and its law of transformation under (1.6) is

$$\begin{aligned} T_\lambda{}^{\varkappa'} &= A_\varkappa^{\varkappa'} A_{\lambda'}^\lambda T_\lambda{}^\varkappa, \quad & T_{\lambda'}{}^{\bar\varkappa'} &= A_{\bar\varkappa}^{\bar\varkappa'} A_{\lambda'}^\lambda T_\lambda{}^{\bar\varkappa}, \\ T_{\lambda'}{}^{\varkappa'} &= A_\varkappa^{\varkappa'} A_{\bar\lambda'}^{\bar\lambda} T_{\bar\lambda}{}^\varkappa, \quad & T_{\bar\lambda'}{}^{\bar\varkappa'} &= A_{\bar\varkappa}^{\bar\varkappa'} A_{\bar\lambda'}^{\bar\lambda} T_{\bar\lambda}{}^{\bar\varkappa}. \end{aligned} \tag{2.12}$$

When such a tensor is considered at a point of the subspace (2.3) we call the tensor a mixed tensor of type (1.1) in the original complex manifold $M_{2n}$. In this case the law of transformation (2.12) can be written as

$$\begin{aligned} T_{\lambda'}{}^{\varkappa'} &= A_{\varkappa'}^{\varkappa'} A_{\lambda'}^\lambda T_\lambda{}^\varkappa, \quad & T_{\lambda'}{}^{\bar\varkappa'} &= \overline{A_{\bar\varkappa}^{\varkappa'}} A_{\lambda'}^\lambda T_\lambda{}^{\bar\varkappa}, \\ T\ {}_{'}{}^{\varkappa'} &= A_\varkappa^{\varkappa'} \overline{A_{\lambda'}^\lambda} T_{\bar\lambda}{}^\varkappa, \quad & T_{\bar\lambda'}{}^{\bar\varkappa'} &= \bar A_{\bar\varkappa}^{\bar\varkappa'} \bar A_{\bar\lambda'}^{\bar\lambda} T_{\bar\lambda}{}^{\bar\varkappa}. \end{aligned} \tag{2.13}$$

Thus if $T_i{}^h$ are components of a tensor of type (1.1) in $M_{2n}$, then

$$\begin{pmatrix} T_\lambda{}^\varkappa & 0 \\ 0 & 0 \end{pmatrix}, \quad \begin{pmatrix} 0 & T_\lambda{}^{\bar\varkappa} \\ 0 & 0 \end{pmatrix}, \quad \ldots, \quad \begin{pmatrix} T_\lambda{}^\varkappa T_\lambda{}^{\bar\varkappa} \\ 0 \quad 0 \end{pmatrix}, \quad \ldots$$

and

$$\begin{pmatrix} \overline{T_{\bar\lambda}^{\ \bar\kappa}} & \overline{T_{\bar\lambda}^{\ \kappa}} \\ \overline{T_{\lambda}^{\ \bar\kappa}} & \overline{T_{\lambda}^{\ \kappa}} \end{pmatrix}$$

are all components of a mixed tensor of type (1.1). We call a tensor satisfying

$$\overline{T_i^{\ h}} = T_{\bar i}^{\ \bar h}$$

a self-conjugate mixed tensor, where

$$\bar i = \bar\kappa \quad \text{when} \quad i = \kappa \quad \text{and} \quad \bar i = \kappa \quad \text{when} \quad i = \bar\kappa .$$

The components of the tensor in a real coordinate system are given by

$$\begin{pmatrix} \frac{1}{2}(T_\lambda^{\ \kappa} + T_{\bar\lambda}^{\ \bar\kappa}) + \frac{1}{2}(T_\lambda^{\ \bar\kappa} + T_{\bar\lambda}^{\ \kappa}), & \frac{1}{2i}(T_\lambda^{\ \kappa} - T_{\bar\lambda}^{\ \bar\kappa}) - \frac{1}{2i}(T_\lambda^{\ \bar\kappa} - T_{\bar\lambda}^{\ \kappa}) \\ \frac{i}{2}(T_\lambda^{\ \kappa} - T_{\bar\lambda}^{\ \bar\kappa}) + \frac{i}{2}(T_\lambda^{\ \bar\kappa} - T_{\bar\lambda}^{\ \kappa}), & \frac{1}{2}(T_\lambda^{\ \kappa} + T_{\bar\lambda}^{\ \bar\kappa}) - \frac{1}{2}(T_\lambda^{\ \bar\kappa} + T_{\bar\lambda}^{\ \kappa}) \end{pmatrix}.$$

Thus a self-conjugate tensor has real components in a real coordinate system. So we call such a tensor a *real* tensor.

## § 3. The tensor $F_i^{\ h}$.†

We know that the transformation law of the components $(v^\kappa, v^{\bar\kappa})$ of a contravariant vector in $M_{2n}$ is

$$v^{\kappa'} = A_\kappa^{\kappa'} v^\kappa, \qquad v^{\bar\kappa'} = \bar A_\kappa^{\kappa'} v^{\bar\kappa}. \tag{3.1}$$

These equations show that if $(v^\kappa, v^{\bar\kappa})$ are components of a contravariant vector, then $(iv^\kappa, -iv^{\bar\kappa})$ are also components of a contravariant vector.

† See A. Weil [1].

Moreover, if the vector $(v^{\kappa}, v^{\bar{\kappa}})$ is self-conjugate, then

$$v^{\bar{\kappa}}=\overline{v^{\kappa}} \quad \text{and consequently} \quad -iv^{\bar{\kappa}}=(\overline{iv^{\kappa}}),$$

which shows that $(iv^{\kappa}, -iv^{\bar{\kappa}})$ is also self-conjugate.

The transition from $(v^{\kappa}, v^{\bar{\kappa}})$ to $(iv^{\kappa}, -iv^{\bar{\kappa}})$ is described as a transition from $v^h$ to $F_i{}^h v^i$, where

$$F_i{}^h=\begin{pmatrix} i\delta_\lambda^\kappa & 0 \\ 0 & -i\delta_{\bar{\lambda}}^{\bar{\kappa}} \end{pmatrix}. \tag{3.2}$$

The matrix $F_i{}^h$ carries an arbitrary self-conjugate vector $v^h$ to a self-conjugate vector $F_i{}^h v^i$ and consequently, by the quotient law, $F_i{}^h$ are components of a self-conjugate mixed tensor. Indeed the tensor $F_i{}^h$ has the components

$$F_i{}^h=\begin{pmatrix} 0 & \delta_\lambda^\kappa \\ -\delta_\lambda^\kappa & 0 \end{pmatrix} \tag{3.3}$$

in the real coordinate system $(\xi^h)$. It satisfies

$$F_j^i F_i{}^h=-A_j^h. \tag{3.4}$$

If we put

$$P_i{}^h=\begin{pmatrix} \delta_\lambda^\kappa & 0 \\ 0 & 0 \end{pmatrix}, \qquad Q_i{}^h=\begin{pmatrix} 0 & 0 \\ 0 & \delta_{\bar{\lambda}}^{\bar{\kappa}} \end{pmatrix} \tag{3.5}$$

in a complex coordinate system, $P_i{}^h$ is an operator which projects any vector

$$v^h=(v^\kappa, v^{\bar{\kappa}}) \quad \text{into} \quad P_i{}^h v^i=(v^\kappa, 0)$$

and $Q_i{}^h$ an operator which projects

$$v^h=(v^\kappa, v^{\bar{\kappa}}) \quad \text{into} \quad Q_i{}^h v^i=(0, v^{\bar{\kappa}}).$$

We also have

$$P_i{}^h w_h=(w_\lambda, 0), \quad Q_i{}^h w_h=(0, w_{\bar{\lambda}})$$

for a covariant vector $w_i$.

These operators may be expressed in terms of $A_i^h$ and $F_i{}^h$ as follows:

$$\begin{aligned} P_i{}^h &= \frac{1}{2}(A_i^h - iF_i{}^h), \\ Q_i{}^h &= \frac{1}{2}(A_i^h + iF_i{}^h), \end{aligned} \tag{3.6}$$

from which

$$\begin{aligned} A_i^h &= P_i{}^h + Q_i{}^h, \\ F_i{}^h &= i(P_i{}^h - Q_i{}^h). \end{aligned} \tag{3.7}$$

## § 4. Pure and hybrid tensors

If a tensor $T_i{}^h$ has components of the type

$$T_i{}^h = \begin{pmatrix} T_\lambda{}^\varkappa & 0 \\ 0 & T_{\bar\lambda}{}^{\bar\varkappa} \end{pmatrix}, \tag{4.1}$$

we say that $T_i{}^h$ is *pure* and if it has components of the type

$$T_i{}^h = \begin{pmatrix} 0 & T_\lambda{}^{\bar\varkappa} \\ T_{\bar\lambda}{}^\varkappa & 0 \end{pmatrix} \tag{4.2}$$

we say that $T_i{}^h$ is *hybrid*. Similarly if a tensor $T_{ji}$ has components of the type

$$T_{ji} = \begin{pmatrix} T_{\mu\lambda} & 0 \\ 0 & T_{\bar\mu\bar\lambda} \end{pmatrix}, \tag{4.3}$$

we say that $T_{ji}$ is *pure* and if it has components of the type

$$T_{ji} = \begin{pmatrix} 0 & T_{\mu\bar\lambda} \\ T_{\bar\mu\lambda} & 0 \end{pmatrix}, \tag{4.4}$$

we say that $T_{ji}$ is *hybrid*.

Now take a general tensor

$$T_i^{\ h} = \begin{pmatrix} T_\lambda^{\ \kappa} & T_\lambda^{\ \bar\kappa} \\ T_{\bar\lambda}^{\ \kappa} & T_{\bar\lambda}^{\ \bar\kappa} \end{pmatrix} \tag{4.5}$$

and form the tensors $P_i^{\ b} P_a^{\ h} T_b^{\ a}$ and $Q_i^{\ b} Q_a^{\ h} T_b^{\ a}$, then we find

$$P_i^{\ b} P_a^{\ h} T_b^{\ a} = \begin{pmatrix} T_\lambda^{\ \kappa} & 0 \\ 0 & 0 \end{pmatrix}, \quad Q_i^{\ b} Q_a^{\ h} T_b^{\ a} = \begin{pmatrix} 0 & 0 \\ 0 & T_{\bar\lambda}^{\ \bar\kappa} \end{pmatrix}$$

and consequently

$$(P_i^{\ b} P_a^{\ h} + Q_i^{\ b} Q_a^{\ h}) T_b^{\ a} = \begin{pmatrix} T_\lambda^{\ \kappa} & 0 \\ 0 & T_{\bar\lambda}^{\ \bar\kappa} \end{pmatrix}.$$

On the other hand, from (3.6), we have

$$P_i^{\ b} P_a^{\ h} + Q_i^{\ b} Q_a^{\ h} = \frac{1}{4}(A_i^b - iF_i^{\ b})(A_a^h - iF_a^{\ h}) + \frac{1}{4}(A_i^b + iF_i^{\ b})(A_a^h + iF_a^{\ h})$$

$$= \frac{1}{2}(A_i^b A_a^h - F_i^{\ b} F_a^{\ h}),$$

and consequently, if we put

$$O_{ia}^{bh} = \frac{1}{2}(A_i^b A_a^h - F_i^{\ b} F_a^{\ h}), \tag{4.6}$$

we have

$$O_{ia}^{bh} T_b^{\ a} = \begin{pmatrix} T_\lambda^{\ \kappa} & 0 \\ 0 & T_{\bar\lambda}^{\ \bar\kappa} \end{pmatrix}. \tag{4.7}$$

Thus the operator $O_{ia}^{bh}$ applied to a general tensor $T_i^{\ h}$ gives a pure tensor given by (4.7). We call (4.7) the *pure part* of $T_i^{\ h}$.

We next form, from a general tensor (4.5), the tensors $P_i^{\ b} Q_a^{\ h} T_b^{\ a}$ and $Q_i^{\ b} P_a^{\ h} T_b^{\ a}$, then

$$P_i^{\ b} Q_a^{\ h} T_b^{\ a} = \begin{pmatrix} 0 & T_\lambda^{\ \bar\kappa} \\ 0 & 0 \end{pmatrix} \qquad Q_i^{\ b} P_a^{\ h} T_b^{\ a} = \begin{pmatrix} 0 & 0 \\ T_{\bar\lambda}^{\ \kappa} & 0 \end{pmatrix}$$

and consequently

$$(P_i^{\ b}Q_a^{\ h}+Q_i^{\ b}P_a^{\ h})T_b^{\ a}=\begin{pmatrix}0 & T_\lambda^{\ \bar{\varkappa}}\\ T_{\bar{\lambda}}^{\ \varkappa} & 0\end{pmatrix}.$$

On the other hand, from (3.6), we have

$$P_i^{\ b}Q_a^{\ h}+Q_i^{\ b}P_a^{\ h}=\frac{1}{4}(A_i^b-iF_i^{\ b})(A_a^h+iF_a^{\ h})$$

$$+\frac{1}{4}(A_i^b+iF_i^{\ b})(A_a^h-iF_a^{\ h})=\frac{1}{2}(A^b\,A^h{}_i+F_i^{\ b}F_a^{\ h});$$

and consequently, if we put

$$*O_{ia}^{bh}T_b^{\ a}=\frac{1}{4}(A_i^bA_a^h+F_i^{\ b}F_a^{\ h})T_b^{\ a}. \qquad (4.8)$$

we have

$$*O_{ia}^{bh}T_b^{\ a}=\begin{pmatrix}0 & T_\lambda^{\ \bar{\varkappa}}\\ T_{\bar{\lambda}}^{\ \varkappa} & 0\end{pmatrix}. \qquad (4.9)$$

Thus the operator $*O_{ia}^{bh}$ applied to a general mixed tensor $T_i^{\ h}$ gives a hybrid tensor given by (4.9). We call (4.9) the hybrid part of $T_i^{\ h}$.

Take next a general covariant tensor $T_{ji}$, then

$$O_{ji}^{cb}T_{cb}=\begin{pmatrix}T_{\mu\lambda} & 0\\ 0 & T_{\bar{\mu}\bar{\lambda}}\end{pmatrix}. \qquad (4.10)$$

Thus the operator $O_{ji}^{cb}$ applied to a general covariant tensor $T_{ji}$ gives a pure tensor (4.10). We call (4.10) the *pure part* of $T_{ji}$.

Applying next the operator $*O_{ji}^{cb}$ to $T_{ji}$ we find

$$*O_{ji}^{cb}T_{cb}=\begin{pmatrix}0 & T_{\mu\bar{\lambda}}\\ T_{\bar{\mu}\lambda} & 0\end{pmatrix}. \qquad (4.11)$$

Thus the operator $*O_{ji}^{cb}$ applied to a general covariant tensor $T_{ji}$ gives a hybrid tensor (4.11). We call (4.11) the *hybrid part* of $T_{ji}$. A similar result is obtained also for a tensor $T^{ih}$.

It is easily seen that the operator $O$ and $*O$ satisfy the following relations:

$$O+{}^*O=A,\ O\cdot O=O,\ O\cdot{}^*O=0,\ {}^*O\cdot O=0,\ {}^*O\cdot{}^*O={}^*O. \tag{4.12}$$

Thus the condition for $T$ to be pure is given by

$$OT=T \quad \text{or} \quad {}^*OT=0 \tag{4.13}$$

and that for $T$ to be hybrid is given by

$$^*OT=T \quad \text{or} \quad OT=0. \tag{4.14}$$

## § 5. Linear connexions in $\mathrm{M}_{2n}$

Now let us suppose that a linear connexion $\Gamma_{ji}^{h}(z)$ is given in the manifold $\mathrm{C}_n\times\overline{\mathrm{C}}_n$. Under a transformation of coordinates

$$z^{\varkappa'}=z^{\varkappa'}(z^{1},\ \ldots,\ z^{n}),\qquad z^{\bar\varkappa'}=z^{\bar\varkappa'}(z^{\bar 1},\ \ldots,\ z^{\bar n}), \tag{5.1}$$

the linear connexion $\Gamma_{ji}^{h}$ has the following transformation law:

$$\Gamma_{j'i'}^{h'}=\frac{\partial z^{h'}}{\partial z^{h}}\left(\frac{\partial z^{j}}{\partial z^{j'}}\,\frac{\partial z^{i}}{\partial z^{i'}}\,\Gamma_{ji}^{h}+\frac{\partial^{2}z^{h}}{\partial z^{j'}\partial z^{i'}}\right). \tag{5.2}$$

Since the transformation of coordinates is of the form (5.1), we can see from (5.2) that the quantities

$$\Gamma_{\mu\bar\lambda}^{\varkappa},\quad \Gamma_{\bar\mu\lambda}^{\varkappa},\quad \Gamma_{\bar\mu\bar\lambda}^{\varkappa},\quad \Gamma_{\mu\lambda}^{\bar\varkappa},\quad \Gamma_{\mu\bar\lambda}^{\bar\varkappa},\quad \Gamma_{\bar\mu\lambda}^{\bar\varkappa}$$

are all components of tensors.

Take a contravariant vector $v^h$, then its covariant differential is given by

$$\begin{aligned}\delta v^{\varkappa}&=dv^{\varkappa}+\Gamma_{\mu\lambda}^{\varkappa}dz^{\mu}v^{\lambda}+\Gamma_{\mu\bar\lambda}^{\varkappa}dz^{\mu}v^{\bar\lambda}+\Gamma_{\bar\mu\lambda}^{\varkappa}dz^{\bar\mu}v^{\lambda}+\Gamma_{\bar\mu\bar\lambda}^{\varkappa}dz^{\bar\mu}v^{\bar\lambda},\\ \delta v^{\bar\varkappa}&=dv^{\bar\varkappa}+\Gamma_{\mu\lambda}^{\bar\varkappa}dz^{\mu}v^{\lambda}+\Gamma_{\mu\bar\lambda}^{\bar\varkappa}dz^{\mu}v^{\bar\lambda}+\Gamma_{\bar\mu\lambda}^{\bar\varkappa}dz^{\bar\mu}v^{\lambda}+\Gamma_{\bar\mu\bar\lambda}^{\bar\varkappa}dz^{\bar\mu}v^{\bar\lambda},\end{aligned} \tag{5.3}$$

and its covariant derivative by

$$\begin{cases}\nabla_{\mu}v^{\varkappa}=\partial_{\mu}v^{\varkappa}+\Gamma_{\mu\lambda}^{\varkappa}v^{\lambda}+\Gamma_{\mu\bar\lambda}^{\varkappa}v^{\bar\lambda},\ \nabla_{\mu}v^{\bar\varkappa}=\partial_{\mu}v^{\bar\varkappa}+\Gamma_{\mu\lambda}^{\bar\varkappa}v^{\lambda}+\Gamma_{\mu\bar\lambda}^{\bar\varkappa}v^{\bar\lambda},\\ \nabla_{\bar\mu}v^{\varkappa}=\partial_{\bar\mu}v^{\varkappa}+\Gamma_{\bar\mu\lambda}^{\varkappa}v^{\lambda}+\Gamma_{\bar\mu\bar\lambda}^{\varkappa}v^{\bar\lambda},\ \nabla_{\bar\mu}v^{\bar\varkappa}=\partial_{\bar\mu}v^{\bar\varkappa}+\Gamma_{\bar\mu\lambda}^{\bar\varkappa}v^{\lambda}+\Gamma_{\bar\mu\bar\lambda}^{\bar\varkappa}v^{\bar\lambda}.\end{cases} \tag{5.4}$$

The linear connexion in $M_{2n}$ is defined as a linear connexion which carries a vector of $M_{2n}$, that is, a vector tangent to the subspace $z^{\bar{\varkappa}}=\overline{z^{\varkappa}}$ to a vector which is still tangent to the subspace. It is easily seen that the necessary and sufficient condition for $\Gamma^h_{ji}$ to define a linear connexion in $M_{2n}$ is given by

$$\overline{\Gamma^h_{ji}}=\Gamma^{\bar{h}}_{\bar{j}\bar{i}} \tag{5.5}$$

where the same remarks as stated near the end of § 2 apply. In the sequel we assume (5.5).

When a contravariant vector $v^h=(v^{\varkappa},\ 0)$ tangent to $C_n$ is displaced along $C_n$, we have

$$\begin{cases} \delta v^{\varkappa}=dv^{\varkappa}+\Gamma^{\varkappa}_{\mu\lambda}dz^{\mu}v^{\lambda}, \\ \delta v^{\bar{\varkappa}}= \qquad\quad \Gamma^{\bar{\varkappa}}_{\mu\lambda}dz^{\mu}v^{\lambda}, \end{cases}$$

from which

THEOREM 5.1. *An arbitrary contravariant vector tangent to* $C_n$, *displaced parallelly along* $C_n$, *is still tangent to* $C_n$ *if and only if*

$$\Gamma^{\bar{\varkappa}}_{\mu\lambda}=0. \tag{5.6}$$

When a contravariant vector $v^h=(v^{\varkappa},\ 0)$ tangent to $C_n$ is displaced along $\overline{C}_n$, we have

$$\begin{cases} \delta v^{\varkappa}=dv^{\varkappa}+\Gamma^{\varkappa}_{\bar{\mu}\lambda}dz^{\bar{\mu}}v^{\lambda}, \\ \delta v^{\bar{\varkappa}}= \qquad\quad \Gamma^{\bar{\varkappa}}_{\bar{\mu}\lambda}dz^{\bar{\mu}}v^{\lambda}, \end{cases}$$

from which

THEOREM 5.2. *An arbitrary contravariant vector tangent to* $C_n$, *displaced parallelly along* $\overline{C}_n$, *is still tangent to* $C_n$ *if and only if*

$$\Gamma^{\bar{\varkappa}}_{\bar{\mu}\lambda}=0. \tag{5.7}$$

Combining the above two theorems, we have

THEOREM 5.3. *An arbitrary contravariant vector tangent to* $C_n$, *displaced parallelly in any direction, is still tangent to* $C_n$, *if and only if*

$$\Gamma^{\bar{\varkappa}}_{\mu\lambda}=0, \qquad \Gamma^{\bar{\varkappa}}_{\bar{\mu}\lambda}=0. \tag{5.8}$$

In this case, the $C_n$'s are parallel.

In a similar way, we can prove

THEOREM 5.4. *An arbitrary contravariant vector tangent to* $\overline{C}_n$, *displaced parallelly along* $C_n$, *is still tangent to* $\overline{C}_n$, *if and only if*

$$\Gamma^{\varkappa}_{\mu\bar{\lambda}}=0. \tag{5.9}$$

THEOREM 5.5. *An arbitrary contravariant vector tangent to* $\overline{C}_n$, *displaced parallelly along* $\overline{C}_n$, *is still tangent to* $\overline{C}_n$, *if and only if*

$$\Gamma^{\varkappa}_{\bar{\mu}\bar{\lambda}}=0. \tag{5.10}$$

THEOREM 5.6. *An arbitrary contravariant vector tangent to* $\overline{C}_n$, *displaced parallelly in any direction, is still tangent to* $\overline{C}_n$, *if and only if*

$$\Gamma^{\varkappa}_{\mu\bar{\lambda}}=0, \qquad \Gamma^{\varkappa}_{\bar{\mu}\bar{\lambda}}=0. \tag{5.11}$$

In this case, the $\overline{C}_n$'s are parallel.

## § 6. Hermite and Kähler metrics†

We now assume that there is given a self-conjugate positive definite Riemannian metric

$$ds^2=g_{ji}dz^j dz^i \tag{6.1}$$

in our complex manifold.

If the fundamental tensor $g_{ji}$ is hybrid, that is, if it has the form

$$g_{ji}=\begin{pmatrix} 0 & g_{\mu\bar{\lambda}} \\ g_{\bar{\mu}\lambda} & 0 \end{pmatrix}, \tag{6.2}$$

† See M. Apte [1]; M. Apte and A. Lichnerowicz [1]; S. Bochner [3], [4], [8], [9], [10]; W. M. Boothby [1]; E. Calabi [1], [3]; S. S. Chern [1]; A. Frölicher [1]; P. R. Garabedian and D. C. Spencer [1], [2]; H. Guggenheimer [3]; H. Hombu [1]; E. Kähler [1]; A. Lichnerowicz [17], [15], [21]; G. de Rham [4]; K. Yano and S. Bochner [1].

then we call such a metric a *Hermite metric* and the complex space with a Hermitian metric a *Hermite manifold.* The fact that $g_{ji}$ is hybrid may be written also as

$$O_{ji}^{cb} g_{cb} = 0 \quad \text{or} \quad F_j^{\ c} F_i^{\ b} g_{cb} = g_{ji}. \tag{6.3}$$

Since the covariant fundamental tensor $g_{ji}$ is hybrid, the contravariant fundamental tensor $g^{ih}$ is also hybrid, that is,

$$g^{ih} = \begin{pmatrix} 0 & g^{\lambda\bar{\varkappa}} \\ g^{\bar{\lambda}\varkappa} & 0 \end{pmatrix} \tag{6.4}$$

or

$$O_{ba}^{ih} g^{ba} = 0 \quad \text{or} \quad F_b^{\ i} F_a^{\ h} g^{ba} = g^{ih}. \tag{6.5}$$

We now put

$$F_j^{\ t} g_{ti} = F_{ji}, \tag{6.6}$$

then $F_{ji}$ has the components

$$F_{ji} = \begin{pmatrix} 0 & i g_{\mu\bar{\lambda}} \\ -i g_{\bar{\mu}\lambda} & 0 \end{pmatrix} \tag{6.7}$$

by virtue of (3.2) and (6.2) and is skew-symmetric. Since $F_i^{\ h}$ and $g_{ji}$ are both self-conjugate, $F_{ji}$ is also self-conjugate.

Since $F_{ji}$ is skew-symmetric, we can form a differential form

$$F_{ji} dz^j \wedge dz^i = 2 i g_{\mu\bar{\lambda}} dz^\mu \wedge dz^{\bar{\lambda}}. \tag{6.8}$$

The differential of this form is given by

$$i(\partial_\nu g_{\mu\bar{\lambda}} - \partial_\mu g_{\nu\bar{\lambda}}) dz^\nu \wedge dz^\mu \wedge dz^{\bar{\lambda}} - i(\partial_{\bar{\nu}} g_{\bar{\mu}\lambda} - \partial_{\bar{\mu}} g_{\bar{\nu}\lambda}) dz^{\bar{\nu}} \wedge dz^{\bar{\mu}} \wedge dz^{\lambda}. \tag{6.9}$$

We next form the Christoffel symbols

$$\left\{ {h \atop j\ i} \right\} = \frac{1}{2} g^{hr} (\partial_j g_{ir} + \partial_i g_{jr} - \partial_r g_{ji}) \tag{6.10}$$

constructed from $g_{ji}$; then we find

$$\left\{ {\varkappa \atop \mu\lambda} \right\} = \left\{ {\varkappa \atop \lambda\mu} \right\} = \frac{1}{2} g^{\varkappa\bar{\varrho}}(\partial_\mu g_{\lambda\bar{\varrho}} + \partial_\lambda g_{\mu\bar{\varrho}}), \quad \text{conj.} \tag{6.11}$$

$$\left\{ {\varkappa \atop \mu\bar{\lambda}} \right\} = \left\{ {\varkappa \atop \bar{\lambda}\mu} \right\} = \frac{1}{2} g^{\varkappa\bar{\varrho}}(\partial_{\bar{\lambda}} g_{\mu\bar{\varrho}} - \partial_{\bar{\varrho}} g_{\mu\bar{\lambda}}), \quad \text{conj.}$$

$$\left\{ {\varkappa \atop \bar{\mu}\bar{\lambda}} \right\} = \left\{ {\varkappa \atop \bar{\lambda}\bar{\mu}} \right\} = 0, \quad \text{conj.}$$

where conj. means that there exists a formula which is the complex conjugate of the formula written at the left.

Thus, following Theorems 5.1 and 5.5, we have

THEOREM 6.1. *In a Hermite space, an arbitrary contravariant vector tangent to* $C_n(\overline{C}_n)$, *displaced parallelly along* $C_n(\overline{C}_n)$ *according to Levi–Civita's parallelism, is still tangent to* $C_n(\overline{C}_n)$.

Following Theorems 4.3 and 4.6, a necessary and sufficient condition that $C_n$ and $\overline{C}_n$ are always parallel is

$$\left\{ {\bar{\varkappa} \atop \mu\lambda} \right\} = 0, \qquad \left\{ {\bar{\varkappa} \atop \bar{\lambda}\mu} \right\} = 0, \qquad \left\{ {\varkappa \atop \mu\bar{\lambda}} \right\} = 0, \qquad \left\{ {\varkappa \atop \bar{\lambda}\bar{\mu}} \right\} = 0.$$

When the Levi–Civita parallelism defined by a Hermite metric satisfies these conditions, we call the metric a *Kähler metric* and a complex space with a Kähler metric a *Kähler space*. (E. Kähler [1]; J. A. Schouten [1], [3]; J. A. Schouten and D. van Dantzig [1], [2]). Thus a Kähler metric is characterized by

$$\partial_{\bar{\lambda}} g_{\mu\bar{\varrho}} - \partial_{\bar{\varrho}} g_{\mu\bar{\lambda}} = 0, \quad \text{conj.}$$

or

$$\partial_\mu g_{\lambda\bar{\varkappa}} = \partial_\lambda g_{\mu\bar{\varkappa}}, \qquad \text{conj.} \tag{6.12}$$

Equation (6.12) shows that there exists a function $\varphi$ such that

$$g_{\mu\bar{\lambda}} = \partial_\mu \partial_{\bar{\lambda}} \varphi. \tag{6.13}$$

From (6.9), we have

THEOREM 6.2. *A necessary and sufficient condition that a Hermite space be Kählerian is that the differential of the form* $F_{ji} dz^j \wedge dz^i$ *vanishes identically.*

Let us now compute the covariant derivative $\nabla_j F_i^{\ h}$ of the tensor $F_i^h$ in a Hermite space. Then we have

$$\nabla_\mu F_\lambda^{\ \varkappa}=0, \quad \nabla_\mu F_\lambda^{\ \bar\varkappa}=2i\Gamma_{\mu\lambda}^{\bar\varkappa}, \quad \nabla_\mu F_{\bar\lambda}^{\ \varkappa}=-2i\Gamma_{\mu\bar\lambda}^{\varkappa}, \quad \nabla_\mu F_{\bar\lambda}^{\ \bar\varkappa}=0, \quad \text{conj}.$$

and consequently we have

THEOREM 6.3. *A necessary and sufficient condition that a Hermite space be Kählerian is that*

$$\nabla_j F_i^{\ h}=0 \tag{6.14}$$

*or*

$$\nabla_j F_{ih}=0. \tag{6.15}$$

In a Kähler space, the Christoffel symbols take the form

$$\left\{ {\varkappa \atop \mu\lambda} \right\} = g^{\varkappa\bar\varrho}(\partial_\mu g_{\lambda\bar\varrho}), \qquad \left\{ {\bar\varkappa \atop \bar\mu\bar\lambda} \right\} = g^{\bar\varkappa\varrho}(\partial_{\bar\mu} g_{\bar\lambda\varrho}) \tag{6.16}$$

all the others being zero, and the covariant derivative of a contravariant vector $v^h$ is given by

$$\begin{aligned} &\nabla_\mu v^\varkappa = \partial_\mu v^\varkappa + \left\{ {\varkappa \atop \mu\lambda} \right\} v^\lambda, \qquad &&\nabla_\mu v^{\bar\varkappa} = \partial_\mu v^{\bar\varkappa}, \\ &\nabla_{\bar\mu} v^\varkappa = \partial_{\bar\mu} v^\varkappa, &&\nabla_{\bar\mu} v^{\bar\varkappa} = \partial_{\bar\mu} v^{\bar\varkappa} + \left\{ {\bar\varkappa \atop \bar\mu\bar\lambda} \right\} v^{\bar\lambda}, \end{aligned} \tag{6.17}$$

and that of a covariant vector $w_i$ by

$$\begin{aligned} &\nabla_\mu w_\lambda = \partial_\mu w_\lambda - \left\{ {\varkappa \atop \mu\lambda} \right\} w_\varkappa, \qquad &&\nabla_\mu w_{\bar\lambda} = \partial_\mu w_{\bar\lambda}, \\ &\nabla_{\bar\mu} w_\lambda = \partial_{\bar\mu} w_\lambda, &&\nabla_{\bar\mu} w_{\bar\lambda} = \partial_{\bar\mu} w_{\bar\lambda} - \left\{ {\bar\varkappa \atop \bar\mu\bar\lambda} \right\} w_{\bar\varkappa}. \end{aligned} \tag{6.18}$$

Suppose that two Kähler spaces $'M_{2n}$ and $M_{2n}$ are in projective correspondence preserving complex structure. Then taking coordinate systems in respective spaces such that the corresponding points have the same coordinates, we have

$$'\left\{ {h \atop j\ i} \right\} = \left\{ {h \atop j\ i} \right\} + p_j A_i^h + p_i A_j^h. \tag{6.19}$$

Put $h=\varkappa$, $i=\lambda$, $j=\bar{\mu}$ in (6.19), then the spaces being both Kählerian, we have

$$0=0+p_{\bar{\mu}}A_{\lambda}^{\varkappa}+p_{\lambda}A_{\bar{\mu}}^{\varkappa},$$

from which

$$p_{\bar{\mu}}=0 \quad \text{and consequently} \quad p_{\mu}=0.$$

Hence

THEOREM 6.4. *A projective correspondence between two Kähler spaces preserving complex structure is necessarily affine.* (N. Coburn [1])†.

Suppose that two Kähler spaces $'M_{2n}$ and $M_{2n}$ are in conformal correspondence *preserving complex structures.* Then taking coordinate systems in respective spaces such that the corresponding points have the same coordinates, we have

$$'\begin{Bmatrix} h \\ j\ i \end{Bmatrix}=\begin{Bmatrix} h \\ j\ i \end{Bmatrix}+\varrho_j A_i^h+\varrho_i A_j^h-\varrho^h g_{ji}, \tag{6.20}$$

$\varrho_j$ being the gradient of a real function $\varrho$. Put $h=\varkappa$, $i=\lambda$, $j=\bar{\mu}$ in (6.20), then the spaces being both Kähler, we have

$$0=0+\varrho_{\bar{\mu}}A_{\lambda}^{\varkappa}+\varrho_{\lambda}A_{\bar{\mu}}^{\varkappa}-\varrho^{\varkappa}g_{\bar{\mu}\lambda}$$

from which, by contraction with respect to $\varkappa$ and $\lambda$,

$$\varrho_{\bar{\mu}}=0 \quad \text{and consequently} \quad \varrho_{\mu}=0.$$

Thus the function $\varrho$ is a constant. Hence

THEOREM 6.5. *A conformal correspondence between two Kähler spaces is necessarily homothetic.* (N. Coburn [2]).

## § 7. Curvature tensors in a Kähler space

We now consider the curvature tensor

$$K_{kji}{}^h=\partial_k\begin{Bmatrix} h \\ ji \end{Bmatrix}-\partial_j\begin{Bmatrix} h \\ ki \end{Bmatrix}+\begin{Bmatrix} h \\ kt \end{Bmatrix}\begin{Bmatrix} t \\ ji \end{Bmatrix}-\begin{Bmatrix} h \\ jt \end{Bmatrix}\begin{Bmatrix} t \\ ki \end{Bmatrix}. \tag{7.1}$$

† See also S. I. Goldberg [3]; K. Yano [6].

Since the Christoffel symbols have the form (6.16), we have from (7.1)

$$K_{kj\lambda}{}^{\bar{\kappa}}=0, \qquad K_{kj\bar{\lambda}}{}^{\kappa}=0; \tag{7.2}$$

that is, the curvature tensor $K_{kji}{}^{h}$ is pure in $h$ and $i$. Consequently, for the covariant components

$$K_{kjih}=K_{kji}{}^{t}g_{th} \tag{7.3}$$

of the curvature tensor, we have

$$K_{kj\lambda\kappa}=0, \qquad K_{kj\bar{\lambda}\bar{\kappa}}=0; \tag{7.4}$$

that is, the curvature tensor $K_{kjih}$ is hybrid in $i$ and $h$. Since we have

$$K_{kjih}=K_{ihkj}$$

the curvature tensor $K_{kjih}$ is also hybrid in $k$ and $j$.

Thus the only non-zero components of the curvature tensor $K_{kjih}$ are those of the type

$$K_{\nu\bar{\mu}\lambda\bar{\kappa}}, \qquad K_{\bar{\nu}\mu\lambda\bar{\kappa}}, \qquad K_{\nu\bar{\mu}\bar{\lambda}\kappa}, \qquad K_{\bar{\nu}\mu\bar{\lambda}\kappa} \tag{7.5}$$

and the non-zero components of the curvature tensor $K_{kji}{}^{h}$ are those of the type

$$K_{\nu\bar{\mu}\lambda}{}^{\kappa}, \qquad K_{\bar{\nu}\mu\lambda}{}^{\kappa}, \qquad K_{\nu\bar{\mu}\bar{\lambda}}{}^{\bar{\kappa}}, \qquad K_{\bar{\nu}\mu\bar{\lambda}}{}^{\bar{\kappa}}. \tag{7.6}$$

As we see from (7.1), these components are given by

$$K_{\bar{\nu}\mu\lambda}{}^{\kappa}=\partial_{\bar{\nu}}\left\{ {\kappa \atop \mu\lambda} \right\}, \qquad K_{\nu\bar{\mu}\bar{\lambda}}{}^{\bar{\kappa}}=\partial_{\nu}\left\{ {\bar{\kappa} \atop \bar{\mu}\bar{\lambda}} \right\}, \tag{7.7}$$

from which we have

THEOREM 7.1. *If the Christoffel symbols* $\left\{ {\kappa \atop \mu\lambda} \right\}$ *of a Kähler space are analytic functions of complex coordinates* $(z^{\kappa})$, *then the space is flat.*

Equations (7.7) show that

$$K_{\bar{\nu}\mu\lambda}{}^{\kappa}=K_{\bar{\nu}\lambda\mu}{}^{\kappa}, \tag{7.8}$$

$$K_{\nu\bar{\mu}\lambda}{}^{\kappa}=K_{\lambda\bar{\mu}\nu}{}^{\kappa} \tag{7.9}$$

These equations may be obtained also from the first Bianchi identity

$$K_{kji}{}^h + K_{jik}{}^h + K_{ikj}{}^h = 0. \tag{7.10}$$

We now calculate the covariant components $K_{\nu\bar{\mu}\lambda\bar{\kappa}}$ of the curvature tensor obtaining

$$K_{\nu\bar{\mu}\lambda\bar{\kappa}} = K_{\nu\bar{\mu}\lambda}{}^{\omega} g_{\omega\bar{\kappa}} = -\left(\partial_{\bar{\mu}} \left\{ {\omega \atop \nu\lambda} \right\}\right) g_{\omega\bar{\kappa}} = -\left\{\partial_{\bar{\mu}}(g^{\omega\bar{\varrho}}\partial_\nu g_{\lambda\bar{\varrho}})\right\} g_{\omega\bar{\kappa}}$$
$$= -\partial_{\bar{\mu}}\partial_\nu g_{\lambda\bar{\kappa}} + g^{\omega\bar{\varrho}}(\partial_\nu g_{\lambda\bar{\varrho}})(\partial_{\bar{\mu}} g_{\omega\bar{\kappa}})$$

and hence

$$K_{\nu\bar{\mu}\lambda\bar{\kappa}} = -\partial_\nu\partial_{\bar{\mu}}\partial_\lambda\partial_{\bar{\kappa}}\varphi + g^{\bar{\varrho}\omega}(\partial_\nu\partial_\lambda\partial_{\bar{\varrho}}\varphi)(\partial_{\bar{\mu}}\partial_{\bar{\kappa}}\partial_\omega\varphi) \tag{7.11}$$

from which we have

$$K_{\nu\bar{\mu}\lambda\bar{\kappa}} = K_{\lambda\bar{\mu}\nu\bar{\kappa}} = K_{\nu\bar{\kappa}\lambda\bar{\mu}} = K_{\lambda\bar{\kappa}\nu\bar{\mu}}. \tag{7.12}$$

For the Ricci tensor $K_{ji}$, we have

$$K_{\mu\lambda} = K_{t\mu\lambda}{}^t = K_{\nu\mu\lambda}{}^{\nu} + K_{\bar{\nu}\mu\lambda}{}^{\bar{\nu}} = 0$$

and consequently $K_{\bar{\mu}\bar{\lambda}} = 0$. Thus $K_{ji}$ is hybrid in $j$ and $i$.

On the other hand,

$$K_{\mu\bar{\lambda}} = K_{t\mu\bar{\lambda}}{}^t = K_{\nu\mu\bar{\lambda}}{}^{\nu} + K_{\bar{\nu}\mu\bar{\lambda}}{}^{\bar{\nu}} = -\partial_\mu \left\{ {\bar{\nu} \atop \bar{\nu}\bar{\lambda}} \right\}.$$

Since

$$\left\{ {\bar{\nu} \atop \bar{\nu}\bar{\lambda}} \right\} = \partial_{\bar{\lambda}} \log \sqrt{\mathfrak{g}},$$

we find

$$K_{\mu\bar{\lambda}} = -\partial_\mu\partial_{\bar{\lambda}} \log \sqrt{\mathfrak{g}} \tag{7.13}$$

where

$$\mathfrak{g} = |g_{ji}| = |g_{\mu\bar{\lambda}}|^2. \tag{7.14}$$

Now if we put

$$H_{kj} = \frac{1}{2} K_{kjih} F^{ih} \tag{7.15}$$

we have

$$H_{\nu\bar\mu} = -iK_{\nu\bar\mu\lambda\bar\varkappa}g^{\lambda\bar\varkappa} = -iK_{\lambda\bar\mu\nu\bar\varkappa}g^{\lambda\bar\varkappa}$$

and hence

$$H_{\nu\bar\mu} = -iK_{\nu\bar\mu}. \tag{7.16}$$

We suppose now that a Kähler space is of constant curvature:

$$K_{kjih} = k(g_{kh}g_{ji} - g_{jh}g_{ki}),$$

where $k$ is a constant. Put $k=\nu$, $j=\mu$, $i=\bar\lambda$, $h=\bar\varkappa$, then we have

$$0 = K_{\nu\mu\bar\lambda\bar\varkappa} = k(g_{\nu\bar\varkappa}g_{\mu\bar\lambda} - g_{\mu\bar\varkappa}g_{\nu\bar\lambda}).$$

Transvecting this with $g^{\nu\bar\varkappa}$, we find

$$(n-1)\ kg_{\mu\bar\lambda} = 0, \quad \text{from which} \quad k = 0.$$

Thus we have

THEOREM 7.2. *A Kähler space of constant curvature is flat.*

We next suppose that a Kähler space is conformally flat:

$$K_{kjih} + g_{kh}C_{ji} - g_{jh}C_{ki} + C_{kh}g_{ji} - C_{jh}g_{ki} = 0,$$

where

$$C_{ji} = -\frac{K_{ji}}{m-2} + \frac{Kg_{ji}}{2(m-1)(m-2)} \qquad (m = 2n).$$

Put $k=\nu$, $j=\mu$, $i=\bar\lambda$, $h=\bar\varkappa$ in the above equation, then we have

$$g_{\nu\bar\varkappa}C_{\mu} - - g_{\mu\bar\varkappa}C_{\nu\bar\varkappa\bar\lambda} + C_{\nu\bar\varkappa}g_{\mu\bar\lambda} - C_{\mu\bar\varkappa}g_{\nu\bar\lambda} = 0.$$

Transvecting this with $g^{\nu\bar\varkappa}$, we find

$$nC_{\mu\bar\lambda} - C_{\mu\bar\lambda} + \frac{1}{2}Cg_{\mu\bar\lambda} - C_{\mu\bar\lambda} = 0 \qquad (C = g^{ji}C_{ji})$$

from which, transvecting again with $g^{\mu\bar\lambda}$

$$(n-1)C = 0 \quad \text{and consequently} \quad C = 0,$$

and consequently, if $n > 2$,

$$C_{\mu\bar\lambda} = 0.$$

Since $C_{ji}$ is hybrid, this shows that $C_{ji} = 0$ and conseqently $K_{kji}{}^{h} = 0$. Thus we have

THEOREM 7.3. *A conformally flat Kähler space is flat.*

CHAPTER IV

# KÄHLER SPACES

## § 1. Kähler spaces in real coordinate systems

Kähler space is an even-dimensional space with a mixed tensor $F_i{}^h$ and with a Riemannian metric $g_{ji}$ which satisfy the following conditions:

$$F_j{}^i F_i{}^h = -A_j^h, \tag{1.1}$$

$$F_j{}^t F_i{}^s g_{ts} = g_{ji} \tag{1.2}$$

and

$$\nabla_j F_i{}^h = 0, \tag{1.3}$$

$\nabla_j$ being the operator of covariant differentiation with respect to the Christoffel symbols $\left\{ {h \atop j\ i} \right\}$ formed with $g_{ji}$. If we put

$$F_{ji} = F_j{}^h g_{hi}, \tag{1.4}$$

then from (1.1) and (1.2), we can see that

$$F_{ji} = -F_{ij} \tag{1.5}$$

and from (1.3), we get

$$\nabla_k F_{ji} = 0. \tag{1.6}$$

Now, from (1.3) and the Ricci identity

$$\nabla_k \nabla_j F_i{}^h - \nabla_j \nabla_k F_i{}^h = K_{kjs}{}^h F_i{}^s - K_{kji}{}^s F_s{}^h,$$

we get

$$K_{kji}{}^s F_s{}^h = K_{kjs}{}^h F_i{}^s, \tag{1.7}$$

from which

$$K_{kji}{}^{h}+K_{kjs}{}^{r}F_{i}{}^{s}F_{r}{}^{h}=0 \quad \text{or} \quad {}^{*}O_{ir}^{sh}K_{kjs}{}^{r}=0 \tag{1.8}$$

which shows that $K_{kji}{}^{h}$ is pure in $h$ and $i$.

From (1.7), we find

$$K_{kjis}F_{h}{}^{s}+K_{kjsh}F_{i}{}^{s}=0, \tag{1.9}$$

from which

$$K_{kjih}-K_{kjsr}F_{i}{}^{s}F_{h}{}^{r}=0 \quad \text{or} \quad O_{ih}^{sr}K_{kjsr}=0, \tag{1.10}$$

which shows that $K_{kjih}$ is hybrid in $i$ and $h$. The $K_{kjih}$ is hybrid also in $k$ and $j$.

From (1.7), by transvection with $g^{ji}$, we obtain

$$K_{k}{}^{s}F_{s}{}^{h}=K_{ksr}{}^{h}F^{sr}=\frac{1}{2}(K_{ksr}{}^{h}-K_{krs}{}^{h})F^{sr}$$

or

$$K_{k}{}^{s}F_{s}{}^{h}=-\frac{1}{2}K_{srk}{}^{h}F^{sr}, \tag{1.11}$$

from which

$$K_{k}{}^{s}F_{sh}+K_{h}{}^{s}F_{sk}=0$$

or

$$K_{k}{}^{s}F_{s}{}^{h}-K_{s}{}^{h}F_{k}{}^{s}=0, \tag{1.12}$$

from which again

$$K_{k}{}^{h}+K_{s}{}^{r}F_{k}{}^{s}F_{r}{}^{h}=0 \quad \text{or} \quad {}^{*}O_{kr}^{sh}K_{s}{}^{r}=0, \tag{1.13}$$

which shows that $K_{k}{}^{h}$ is pure in $h$ and $k$.

From (1.12) and (1.13) we obtain respectively

$$K_{ks}F_{h}{}^{s}+K_{sh}F_{k}{}^{s}=0 \tag{1.14}$$

and

$$K_{kh}-K_{sr}F_{k}{}^{s}F_{h}{}^{r}=0 \quad \text{or} \quad O_{kh}^{sr}K_{sr}=0, \tag{1.15}$$

which shows that $K_{kh}$ is hybrid in $k$ and $h$.

We defined $H_{kj}$ by

$$H_{kj}=\frac{1}{2}K_{kjih}F^{ih}. \tag{1.16}$$

From (1.11), we obtain

$$K_{ks}F_j{}^s = H_{kj}, \quad H_{ks}F_j{}^s = -K_{kj}, \tag{1.17}$$

from which

$$K = -H_{kj}F^{kj}. \tag{1.18}$$

From (1.17), we obtain

THEOREM 1.1. *The tensor $H_{kj}$ is proportional to $F_{kj}$ if and only if the space is an Einstein space.*

From the Bianchi identity

$$\nabla_l K_{kjih} + \nabla_k K_{jlih} + \nabla_j K_{lkih} = 0,$$

we deduce

$$\nabla_l H_{kj} + \nabla_k H_{jl} + \nabla_j H_{lk} = 0. \tag{1.19}$$

On the other hand, we have

$$\nabla^k H_{kj} = \nabla^k (K_{ks}) F_j{}^s$$

or

$$\nabla H_{kj} = \frac{1}{2}(\nabla_s K) F_j{}^s \tag{1.20}$$

by virtue of the identity

$$\nabla^k K_{ks} = \frac{1}{2}\nabla_s K.$$

From (1.18), (1.19) and (1.20), we have

THEOREM 1.2. *The tensor $H_{kj}$ is harmonic if and only if the curvature scalar is constant and is effective (that is, it satisfies $F^{ji}H_{ji} = 0$) if and only if the curvature scalar is zero.*

In particular

THEOREM 1.3. *In a homogeneous Kähler space, the tensor $H_{kj}$ is harmonic.*

Suppose that we have a Hermite space, that is, a space with a skew-symmetric tensor $F_{ji}$ and with a Riemannian metric $g_{ji}$ satisfying

$$F_j{}^t F_i{}^s g_{ts} = g_{ji}.$$

By a conformal transformation

$$\bar{g}_{ji} = \varrho^2 g_{ji}, \tag{1.21}$$

the tensor $F_i{}^h$ remains unchanged, but the tensor $F_{ji}$ is transformed into $\bar{F}_{ji}$ by (1.22)

$$\bar{F}_{ji} = \varrho^2 \bar{F}_{ji}. \tag{1.22}$$

We suppose that the transformed Hermite space is a Kähler space, hence

$$\bar{F}_{kji} = \partial_k \bar{F}_{ji} + \partial_j \bar{F}_{ik} + \partial_i \bar{F}_{kj} = 0.$$

Then we have from (1.22) and this equation

$$\varrho_k F_{ji} + \varrho_j F_{ik} + \varrho_i F_{kj} + \frac{1}{2} F_{kji} = 0, \tag{1.23}$$

where

$$\varrho_k = \partial_k \log \varrho$$

and

$$F_{kji} = \partial_k F_{ji} + \partial_j F_{ik} + \partial_i F_{kj}.$$

Transvecting (1.23) with $F^{ji}$, we find

$$2(n-1)\varrho_k + F_k = 0$$

from which

$$\varrho_k = -\frac{1}{2(n-1)} F_k, \tag{1.24}$$

where

$$F_k = \frac{1}{2} F_{kji} F^{ji}. \tag{1.25}$$

Substituting (1.24) into (1.23), we find

$$F_{kji} - \frac{1}{2(n-1)} [F_k F_{ji} + F_j F_{ik} + F_i F_{kj}] = 0. \tag{1.26}$$

Thus (1.23) and the set of (1.24) and (1.26) are equivalent to each other.

From (1.26), we find

$$0=\partial_l F_{kji}-\partial_k F_{lji}-\partial_j F_{kli}-\partial_i F_{kjl}$$
$$=\frac{1}{2(n-1)}[(\partial_l F_k-\partial_k F_l)F_{ji}-(\partial_l F_j-\partial_j F_l)F_{ki}-(\partial_l F_i-\partial_i F_l)F_{jk}$$
$$+F_{lk}(\partial_j F_i-\partial_i F_j)-F_{lj}(\partial_k F_i-\partial_i F_k)-F_{li}(\partial_j F_k-\partial_k F_j)],$$

from which, by transvection with $F^{ji}$,

$$(2n-4)(\partial_l F_k-\partial_k F_l)+F_{lk}(\partial_j F_i-\partial_i F_j)F^{ji}=0,$$

from which, by transvection with $F^{lk}$,

$$4(n-1)(\partial_l F_k-\partial_k F_l)F^{lk}=0$$

and consequently

$$(2n-4)(\partial_l F_k-\partial_k F_l)=0.$$

Thus, for $2n\neq 4$, (1.26) implies

$$\partial_l F_k-\partial_k F_l=0. \tag{1.27}$$

Thus the integrability conditions of (1.24) are automatically satisfied.

For the case $2n=4$, (1.26) becomes an identity, so, as the integrability conditions of (1.24), we should have (1.27). Thus we have

THEOREM 1.4. *A necessary and sufficient condition for a Hermite space to be conformal to a Kähler space is that, for* $2n>4$,

$$F_{kji}-\frac{1}{2(n-1)}[F_k F_{ji}+F_j F_{ik}+F_i F_{kj}]=0,$$

*and, for* $2n=4$,

$$\partial_k F_l-\partial_l F_k=0,$$

*where* $F_k=\frac{1}{2}F_{kji}F^{ji}$. (W. J. Westlake [2], K. Yano [1]).

## § 2. Kähler spaces of constant holomorphic curvature†

Take a vector $u^h$ at a point in a Kähler space and consider the transform $F_i{}^h u^i$ of $u^h$ by $F_i{}^h$. Then $u^h$ and $F_i{}^h u^i$ are mutually orthogonal and consequently they are linearly independent. The plane element determined by $u^h$ and $F_i{}^h u^i$ is called a *holomorphic section.*

The sectional curvature $k$ of the space at this point with respect to this holomorphic section is given by

$$k = -\frac{K_{mjlh} F_k{}^m u^k u^j F_i{}^l u^i u^h}{g_{kj} u^k u^j g_{ih} u^i u^h}. \tag{2.1}$$

We call this the *holomorphic sectional curvature* with respect to the vector $u^h$. If the holomorphic sectional curvature is always constant with respect to any vector at every point of the space, then we call the space a *space of constant holomorphic sectional curvature.* The constant $k$ here may depend on the point of the space.

For a space of constant holomorphic sectional curvature,

$$K_{mjlh} F_q{}^m u^q u^j F_p{}^l u^p u^h = -k g_{qj} u^q u^j g_{ph} u^p u^h$$

must be satisfied for any $u^h$, from which we obtain

$$K_{mjlh} F_q{}^m F_p{}^l + K_{mplh} F_j{}^m F_q{}^l + K_{mqlh} F_p{}^m F_j{}^l$$
$$= -k(g_{qj} g_{ph} + g_{jp} g_{qh} + g_{pq} g_{jh})$$

by virtue of the symmetry of $K_{mjlh} F_q{}^m F_p{}^l$ with respect to $q$, $j$ and $p$, $h$. Transvecting the above equation with $F_k{}^q$ $F_i{}^p$, we find

$$K_{kjih} - K_{jikh} - K_{iqlh} F_k{}^q F_j{}^l = -k(F_{kj} F_{ih} - F_{ji} F_{kh} + g_{ki} g_{jh}).$$

Taking the skew-symmetric part of this equation with respect to $k$ and $j$ and taking account of

$$K_{iqlh} F_k{}^q F_j{}^l - K_{iqlh} F_j{}^q F_k{}^l = (K_{iqlh} - K_{ilqh}) F_k{}^q F_j{}^l$$
$$= -K_{qlih} F_k{}^q F_j{}^l = -K_{kjih},$$

† See N. S. Hawley [1]; K. Yano and I. Mogi [1], [2].

we obtain

$$2K_{kjih}-K_{jikh}+K_{kijh}+K_{kjih}=-k[2F_{kj}F_{ih}+(g_{ki}g_{jh}-g_{ji}g_{kh})\\ -(F_{ji}F_{kh}-F_{ki}F_{jh})]$$

or

$$K_{kjih}=\frac{k}{4}[(g_{kh}g_{ji}-g_{jh}g_{ki})+(F_{kh}F_{ji}-F_{jh}F_{ki})-2F_{kj}F_{ih}].$$

Using the Bianchi identity, we can easily prove that $k$ is an absolute constant. Thus we have

THEOREM 2.1. *If a Kähler space has a constant holomorphic sectional curvature $k$ at each point, then the curvature tensor of the space is of the form*

$$K_{kjih}=\frac{k}{4}[(g_{kh}g_{ji}-g_{jh}g_{ki})+(F_{kh}F_{ji}-F_{jh}F_{ki})-2F_{kj}F_{ih}] \quad (2.2)$$

*and the scalar $k$ is an absolute constant.* (T. Fukami [1], [2]; K. Yano and I. Mogi [1], [2]).

Now we consider the general sectional curvature $\varkappa$ with respect to a general section determined by two mutually orthogonal unit vectors $u^h$ and $v^h$ in this space. We have

$$\varkappa=-K_{kjih}v^ku^jv^iu^h=\frac{k}{4}[1+3(F_{ji}v^ju^i)^2].$$

But $F_{ji}v^j$ being a unit vector, $F_{ji}v^ju^i$ represents $\cos\theta$, $\theta$ being the angle between $F_{ji}v^j$ and $u^i$. Thus

$$\varkappa=\frac{k}{4}(1+3\cos^2\theta) \quad (2.3)$$

and consequently we have

THEOREM 2.2. *Denote by $\varkappa$ a general sectional curvature of a Kähler space of constant holomorphic sectional curvature $k\neq 0$; then we have*

$$\frac{k}{4}\leqq\varkappa\leqq k \qquad \textit{for} \quad k>0$$

*and*

$$\frac{k}{4} \geqq \varkappa \geqq k \qquad \text{for} \quad k<0.$$

The equality $\frac{k}{4}=\varkappa$ occurs when $F_{ji}v^j$ and $u^i$ are orthogonal, that is, when $F_{ji}v^ju^i=0$ and the equality $k=\varkappa$ occurs when $F_{ji}v^j$ and $u^i$ coincide, that is, when the section is holomorphic.

From (2.2) we have

THEOREM 2.3. *A Kähler space of constant holomorphic sectional curvature is an Einstein space:*

$$K_{ji}=\frac{n+1}{2}kg_{ji}. \tag{2.4}$$

## § 3. Kähler spaces admitting holomorphic free mobility

Suppose that a space admits a group of motions which carry any two vectors $u^h(P)$ and $F_i^{\ h}(P)u^i(P)$ at a point P to any two other vectors $w^h(Q)$ and $F_i^{\ h}(Q)w^i(Q)$ at any point Q. We then say that the space admits a *holomorphic free mobility.*

If this is the case, there exists a motion which carries any two vectors $u^h(P)$ and $F_i^{\ h}(P)u^i(P)$ into any two vectors $w^h(P)$ and $F_i^{\ h}(P)w^i(P)$ at the same point. Thus there exists a motion which carries any holomorphic section into any holomorphic section at the same point. This means that the so-called holomorphic sectional curvature of the space at the point is a constant, and hence the curvature tensor must be of the form

$$K_{kjih}=\frac{k}{4}[(g_{kh}g_{ji}-g_{jh}g_{ki})+(F_{kh}F_{ji}-F_{jh}F_{ki})-2F_{kj}F_{ih}], \tag{3.1}$$

where $k$ is a constant.

Suppose now that the curvature tensor of our Kähler space has this form. We are going to prove that the space admits holomorphic free mobility.

Consider an infinitesimal transformation defined by a vector $v^h$. The condition that this is a motion is given by

$$\underset{v}{\mathfrak{L}} g_{ji} = \nabla_j v_i + \nabla_i v_j = 0 \tag{3.2}$$

and the condition that, if this motion carrries a direction $u^h(P)$ at P to a direction $w^h(Q)$ at Q, then it carries the direction $F_i^{\ h}(P)u^i(P)$ at P to the direction $F_i^{\ h}(Q)w^i(Q)$ at the point Q, is expressed by

$$\underset{v}{\mathfrak{L}} F_i^{\ h} = -F_i^{\ t}\nabla_t v^h + F_t^{\ h}\nabla_i v^t = 0. \tag{3.3}$$

From (3.2), we derive

$$\underset{v}{\mathfrak{L}} \left\{ {h \atop j\ i} \right\} = \nabla_j \nabla_i v^h + K_{kji}{}^h v^k = 0. \tag{3.4}$$

Thus, to see if there exists such a motion, we must examine the integrability conditions of the system of partial differential equations

$$\left.\begin{aligned} \nabla_i v^h &= v_i^{\ h} \\ \nabla_j v_i^{\ h} &= -K_{kji}{}^h v^k \end{aligned}\right\} \tag{3.5}$$

with additional conditions

$$\begin{gathered} v_{ji} + v_{ij} = 0 \\ F_i^{\ t} v_{th} + F_h^{\ t} v_{it} = 0, \end{gathered} \tag{3.6}$$

where $v_{ji} = v_j^{\ h} g_{ih}$.

Now, the integrability conditions of the system (3.5) are given by

$$\underset{v}{\mathfrak{L}} K_{kji}{}^h = v^l \nabla_l K_{kji}{}^h - K_{kji}{}^l v_l^{\ h} + K_{lji}{}^h v_k^{\ l} + K_{kli}{}^h v_j^{\ l} + K_{kjl}{}^h v_i^{\ l} = 0. \tag{3.7}$$

But we know that the curvature tensor $K_{kji}{}^h$ has the form (3.1) and consequently it is not so difficult to see that this integrability condition is identically satisfied for the values of $v_i{}^h$ which satisfy (3.6).

Moreover we have

$$\nabla_k(v_{ji}+v_{ij})=0;$$

that is,

$$\partial_k(v_{ji}+v_{ij})-\left\{{l \atop k\ j}\right\}(v_{li}+v_{il})-\left\{{l \atop k\ i}\right\}(v_{jl}+v_{lj})=0$$

and

$$\nabla_j(F_i{}^t v_{th}+F_h{}^t v_{it})=0;$$

that is,

$$\partial_j(F_i{}^t v_{th}+F_h{}^t v_{it})-\left\{{l \atop j\ i}\right\}(F_l{}^t v_{th}+F_h{}^t v_{lt})-\left\{{l \atop j\ h}\right\}(F_i{}^t v_{tl}+F_l{}^t v_{it})=0$$

by virtue of (3.5).

Thus if the conditions

$$v_{ji}+v_{ij}=0 \quad \text{and} \quad F_i{}^t v_{th}+F_h{}^t v_{it}=0$$

are satisfied by their initial values, they are always satisfied by the solutions. Thus the system (3.5) with conditions (3.6) is completely integrable.

Since we can choose the initial values of $v^h$ arbitrarily, the group is transitive, and, since we can choose $v_{ji}$ in such a way that they satisfy (3.6), the isotropy group is a real representation of a unitary group, and consequently we see that the space admits holomorphic free mobility. Thus we have

THEOREM 3.1. *A necessary and sufficient condition that a Kähler space admits holomorphic free mobility is that the space be a space of constant holomorphic sectional curvature.* (K. Yano and I. Mogi [1], [2]).

## § 4. Harmonic tensors in a Kähler space

W. V. D. Hodge [1] proved the following theorem.

THEOREM 4.1. *In a compact Kähler manifold of real dimension* $2n$, *we have*

$$B_0 \leqq B_2 \leqq B_4 \leqq, \ldots \leqq B_{2[n/2]},$$
$$B_1 \leqq B_3 \leqq B_5 \leqq, \ldots \leqq B_{2[n/2]+1}$$

*and the number* $B_{p+2} - B_p$ $(0 < p+2 < n)$ *is the number of linearly independent (with constant coefficients) self-conjugate effective harmonic tensors of order* $p+2$.

*Thus if there are no self-conjugate effective harmonic tensors, then*

$$B_0 = B_2 = B_4 = \ldots = B_{2[n/2]},$$
$$B_1 = B_3 = B_5 = \ldots = B_{2[n/2]+1}.$$

We are going to apply this theorem to a Kähler space and to evaluate the Betti numbers of a Kähler space of constant holomorphic sectional curvature.

We find from (7.2) of Chapter II that, in a compact orientable Riemannian space, there exists no harmonic tensor field of order $p$ which satisfies

$$F(w, w) = K_{ji} w^{ji_2 \ldots i_p} w^i{}_{i_2 \ldots i_p} + \frac{p-1}{2} K_{kjih} w^{kji_3 \ldots i_p} w^{ih}{}_{i_3 \ldots i_p} \geqq 0, \tag{4.1}$$

unless we have

$$\nabla_j w_{i_1 i_2 \ldots i_p} = 0.$$

In particular if the form $F(w, w)$ is positive definite, then there does not exist a harmonic tensor field of order $p$.

We assume moreover that the harmonic tensor $w_{i_1 i_2 \ldots i_p}$ is effective, that is

$$F^{ji} w_{jii_3 \ldots i_p} = 0, \tag{4.2}$$

and evaluate the form $F(w, w)$ for a Kähler space of constant holomorphic sectional curvature.

For a Kähler space of constant holomorphic curvature, we have

$$K_{kjih}=\frac{k}{4}\,[(g_{kh}g_{ji}-g_{jh}g_{ki})+(F_{kh}F_{ji}-F_{jh}F_{ki})-2F_{kj}F_{ih}] \quad (4.3)$$

from which

$$K_{ji}=\frac{n+1}{2}\,kg_{ji}. \quad (4.4)$$

We first find that, if $p=1$ and $k>0$, $F(w, w)$ is positive.

Substituting (4.3) and (4.4) into (4.1), we find

$$F(w,\ w)=\frac{n+1}{2}\,kw^{i_1i_2\dots i_p}w_{i_1i_2\dots i_p}$$

$$-\frac{p-1}{4}\,k(w^{i_1i_2\dots i_p}w_{i_1i_2\dots i_p}+F_m{}^kF_l{}^jw^{mli_3\dots i_p}w_{kji_3\dots i_p})$$

$$=\frac{n+1}{2}\,kw^{i_1i_2\dots i_p}w_{i_1i_2\dots i_p}-\frac{p-1}{2}\,k(*O_{ml}^{kj}w^{mli_3\dots i_p})w_{kji_3\dots i_p}.$$

On the other hand, we have

$$\begin{aligned} &w^{i_1i_2i_3\dots i_p}w_{i_1i_2i_3\dots i_p}\\ &\quad=(O_{t\,s}^{i_1i_2}+*O_{t\,s}^{i_1i_2})w^{tsi_3\dots i_p}(O_{i_1i_2}^{ml}+*O_{i_1i_2}^{ml})w_{mli_3\dots i_p}\\ &\quad=(O_{t\,s}^{i_1i_2}w^{tsi_3\dots i_p})(O_{i_1i_2}^{ml}w_{mli_3\dots i_p})\\ &\qquad\qquad+(*O_{t\,s}^{i_1i_2}w^{tsi_3\dots i_p})(*O_{i_1i_2}^{ml}w_{mli_3\dots i_p}) \end{aligned}$$

and

$$*O_{ml}^{kj}w^{mli_3\dots i_p}w_{kji_3\dots i_p}=(*O_{ml}^{kj}w^{mli_3\dots i_p})(*O_{kj}^{ts}w_{tsi_3\dots i_p}),$$

and consequently, we obtain

$$F(w,\ w)=\frac{n+1}{2}\,k(O_{t\,s}^{i_1i_2}w^{tsi_3\dots i_p})(O_{i_1i_2}^{ml}w_{mli_3\dots i_p})$$

$$+\frac{n-p+2}{2}\,k(*O_{ml}^{i_1i_2}w^{mli_3\dots i_p})(*O_{i_1i_2}^{ts}w_{tsi_3\dots i_p}),$$

which is positive for $k>0$ and $1\leqq p\leqq n$. Thus we have

THEOREM 4.2. *In a compact Kähler space of constant holomorphic sectional curvature $k>0$, we have*

$$B_{2l}=1, \qquad B_{2l+1}=0, \qquad (0\leqq 2l,\ 2l+1\leqq 2n).$$

## § 5. Covariant analytic vectors

Consider a self-conjugate covariant vector field $w_i = (w_\lambda, w_{\bar\lambda})$ in a Kähler space. If the components $w_\lambda$ are functions of $z^\nu$ only and the components $w_{\bar\lambda}$ are functions of $z^{\bar\nu}$ only, then we call the vector a *covariant analytic vector* in a Kähler space. The condition for $w_i$ to be covariant analytic is expressed by

$$\partial_{\bar\mu} w_\lambda = 0, \qquad \partial_\mu w_{\bar\lambda} = 0$$

or

$$\nabla_{\bar\mu} w_\lambda = 0, \qquad \nabla_\mu w_{\bar\lambda} = 0, \tag{5.1}$$

which show that $\nabla_j w_i$ is pure in $j$ and $i$. Consequently the condition (5.1) may also be written as

$$^*O_{ji}^{ts}\nabla_t w_s = 0, \tag{5.2}$$

from which

$$F_j^{\ s}\nabla_s w_i - F_i^{\ s}\nabla_j w_s = 0. \tag{5.3}$$

Applying the operator $F^{jt}\nabla_t$ to this equation, we find

$$g^{ts}\nabla_t\nabla_s w_i - F^{jt}F_i^{\ s}\nabla_t\nabla_j w_s = 0,$$

$$g^{ts}\nabla_t\nabla_s w_i + \frac{1}{2}F^{tj}F_i^{\ s}(\nabla_t\nabla_j w_s - \nabla_j\nabla_t w_s) = 0,$$

$$g^{ts}\nabla_t\nabla_s w_i - \frac{1}{2}F^{tj}F_i^{\ s}K_{tjs}^{\ \ \ h}w_h = 0,$$

$$g^{ts}\nabla_t\nabla_s w_i - K_i^{\ h}w_h = 0$$

by virtue of (1.16) and (1.17). This is a necessary condition for $w_i$ to be covariant analytic.

On the other hand we have

$$\frac{1}{2}(F^{js}\nabla_s w^i - F^{is}\nabla^j w_s)(F_j{}^r\nabla_r w_i - F_i{}^r\nabla_j w_r)$$
$$= (\nabla^j w^i)(\nabla_j w_i) - F^{js}F^{ir}(\nabla_s w_i)(\nabla_j w_r)$$

and

$$\nabla_j[(\nabla^j w_i)w^i - (F^{js}F^{ir}\nabla_s w_i)w_r]$$
$$= (g^{ts}\nabla_t\nabla_s w_i)w^i + (\nabla^j w^i)(\nabla_j w_i) - F^{js}F^{ir}(\nabla_j\nabla_s w_i)w_r$$
$$- F^{js}F^{ir}(\nabla_s w_i)(\nabla_j w_r)$$
$$= (g^{ts}\nabla_t\nabla_s w_i - K_i{}^h w_h)w^i + (\nabla^j w^i)(\nabla_j w_i) - F^{js}F^{ir}(\nabla_s w_i)(\nabla_j w_r).$$

Thus assuming the space to be compact, we apply Green's theorem and we get

$$\int_M \Big[(g^{ts}\nabla_t\nabla_s w_i - K_i{}^h w_h)w^i + \frac{1}{2}(F^{js}\nabla_s w^i - F^{is}\nabla^j w_s)$$
$$(F_j{}^r\nabla_r w_i - F_i{}^r\nabla_j w_r)\Big]\,d\sigma = 0, \quad (5.4)$$

(K. Yano [13]), from which we have

THEOREM 5.1. *A necessary and sufficient condition for a vector $w_i$ in a compact Kähler space to be covariant analytic is*

$$g^{ts}\nabla_t\nabla_s w_i - K_i{}^h w_h = 0. \quad (5.5)$$

Since (5.5) is a necessary and sufficient condition that the vector $w_i$ be harmonic, we have

THEOREM 5.2. *A necessary and sufficient condition for a vector $w_i$ in a compact Kähler space to be covariant analytic is that the vector $w_i$ be harmonic*

Thus applying Bochner's theorem, (see Theorem 2.2 of Chapter II) we can state

THEOREM 5.3. *In a compact Kähler space with positive definite Ricci curvature, there exists no covariant analytic vector field other than the zero vector.*

*If the Ricci curvature is only positive semi-definite, a covariant analytic vector field has vanishing covariant derivative.*

## § 6. Contravariant analytic vectors†

Consider a self-conjugate vector field $v^h = (v^\varkappa, v^{\bar{\varkappa}})$ in a Kähler space. If the components $v^\varkappa$ are functions of $z^\lambda$ only and the components $v^{\bar{\varkappa}}$ are functions of $z^{\bar{\lambda}}$ only, then we call the vector a *contravariant analytic vector* in a Kähler space. The condition for $v^h$ to be contravariant analytic is expressed by

$$\partial_{\bar{\mu}} v^\varkappa = 0, \qquad \partial_\mu v^{\bar{\varkappa}} = 0$$

or

$$\nabla_{\bar{\mu}} v^\varkappa = 0, \qquad \nabla_\mu v^{\bar{\varkappa}} = 0, \tag{6.1}$$

which show that $\nabla_j v^h$ is pure in $j$ and $h$. Consequently the condition (6.1) may also be written as

$$*O_{ir}^{sh} \nabla_s v^r = 0 \tag{6.2}$$

or

$$O_{ih}^{sr} \nabla_s v_r = 0. \tag{6.3}$$

These equations may be written respectively as

$$\underset{v}{\mathfrak{L}} F_i{}^h = -F_i{}^s \nabla_s v^h + F_s{}^h \nabla_i v^s = 0 \tag{6.4}$$

or

$$F_i{}^s \nabla_s v_h + F_h{}^s \nabla_i v_s = 0. \tag{6.5}$$

From (6.4), we get

$$-F_i{}^s \nabla_s (F_t{}^h v^t) - \nabla_i v^h = 0$$

or

$$-F_i{}^s \nabla_s (F_t{}^h v^t) + F_s{}^h (\nabla_i F_t{}^s v^t) = 0,$$

which shows that $F_r{}^h v^r$ is also contravariant analytic. Moreover if $u^h$ and $v^h$ are both contravariant analytic, then from

$$\underset{v}{\mathfrak{L}} \underset{u}{\mathfrak{L}} F_i{}^h - \underset{u}{\mathfrak{L}} \underset{v}{\mathfrak{L}} F_i{}^h = \underset{[vu]}{\mathfrak{L}} F_i{}^h,$$

we see that $[v, u]$ is also contravariant analytic. Thus we have

† See S. Sasaki and K. Yano [1].

THEOREM 6.1. *When $u^h$ and $v^h$ are both contravariant analytic, the vectors*

$$F_i^{\ h}u^i, \quad F_i^{\ h}v^i, \quad \underset{v}{\mathfrak{L}}\, u^h, \quad \underset{Fv}{\mathfrak{L}}\, u^h, \quad \underset{v}{\mathfrak{L}}\, F_i^{\ h}u^i, \quad \underset{Fv}{\mathfrak{L}}\, F_i^{\ h}u^i$$

*are all contravariant analytic.*

Applying the operator $F^{ti}\nabla_t$ to (6.4), we find

$$g^{ts}\nabla_t\nabla_s v^h + F^{ti}F_s^{\ h}\nabla_t\nabla_i v^s = 0,$$

$$g^{ts}\nabla_t\nabla_s v^h + \frac{1}{2}F^{ti}F_s^{\ h}(\nabla_t\nabla_i v^s - \nabla_i\nabla_t v^s) = 0,$$

$$g^{ts}\nabla_t\nabla_s v^h + \frac{1}{2}F^{ti}F_s^{\ h}K_{tir}^{\ \ \ s}v^r = 0,$$

$$g^{ts}\nabla_t\nabla_s v^h + K_i^{\ h}v^i = 0.$$

This is a necessary condition for $v^h$ to be contravariant analytic. Thus we have from Theorem 3.4 of Chapter II,

THEOREM 6.2. *In a compact Kähler space a contravariant analytic vector $v^h$ satisfying $\nabla_i v^i = 0$ is a Killing vector.*

For a contravariant analytic vector $v^h$, we have

$$\begin{aligned} g^{ji}\nabla_j\nabla_i(v^h v_h) &= 2[(g^{ji}\nabla_j\nabla_i v^h)v_h + (\nabla^j v^i)(\nabla_j v_i)] \\ &= 2[-K_{ji}v^j v^i + (\nabla^j v^i)(\nabla_j v_i)], \end{aligned}$$

from which

THEOREM 6.3. *If a compact Kähler space has negative definite Ricci curvature, then there exists no contravariant analytic vector field other than the zero vector.*

*If the Ricci curvature is only negative semi-definite, then a contravariant analytic field has vanishing covariant derivative.*

Suppose that a contravariant analytic vector field $v^h$ satisfies $F^{ji}\nabla_j v_i = 0$, then a contravariant analytic vector field $F_i^{\ h}v^i$ satisfies $\nabla_h(F_i^{\ h}v^i) = 0$, and consequently is a Killing vector according to Theorem 6.2. Thus we have

THEOREM 6.4. *If a contravariant analytic vector field $v^h$ satisfies $F^{ji}\nabla_j v_i = 0$, then $F_i^{\ h}v^i$ is a Killing vector.*

On the other hand, we have

$$\frac{1}{2}(F^{js}\nabla_s v^i - F_s{}^i\nabla^j v^s)(F_j{}^r\nabla_r v_i - F_{ri}\nabla_j v^r)$$
$$= (\nabla^j v^i)(\nabla_j v_i) - F^{js}F^{ri}(\nabla_s v_i)(\nabla_j v_r)$$

and

$$\nabla_j[(\nabla^j v^i)v_i - F^{js}F^{ri}(\nabla_s v_i)v_r] = (g^{ts}\nabla_t\nabla_s v^i)v_i + (\nabla^j v^i)(\nabla_j v_i)$$
$$- F^{js}F^{ri}(\nabla_j\nabla_s v_i)v_r - F^{js}F^{ri}(\nabla_s v_i)(\nabla_j v_r)$$
$$= (g^{ts}\nabla_t\nabla_s v^i + K_r{}^i v^r)v_i + (\nabla^j v^i)(\nabla_j v_i) - F^{js}F^{ri}(\nabla_s v_i)(\nabla_j v_r)$$
$$= (g^{ts}\nabla_t\nabla_s v^h + K_i{}^h v^i)v_h + \frac{1}{2}(F^{js}\nabla_s v^i - F_s{}^i\nabla^j v^s)(F_j{}^r\nabla_r v_i - F_{ri}\nabla_j v^r).$$

Thus assuming the space to be compact, we apply Green's theorem and we obtain

$$\int\left[(g^{ts}\nabla_t\nabla_s v^h + K_i{}^h v^i)v_h + \frac{1}{2}(F^{js}\nabla_s v^i - F_s{}^i\nabla^j v^s)(F_j{}^r\nabla_r v_i - F_{ri}\nabla_j v^r)\right]d\sigma = 0. \tag{6.6}$$

(K. Yano [13]). From this we have

THEOREM 6.5. *A necessary and sufficient condition for a vector $v^h$ in a compact Kähler space to be contravariant analytic is*

$$g^{ts}\nabla_t\nabla_s v^h + K_i{}^h v^i = 0.$$

(K. Yano [13]).† This gives

THEOREM 6.6. *A Killing vector in a compact Kähler space is contravariant analytic.*††

This can also be proved in the following way. Since the Lie derivative of a harmonic tensor with respect to a Killing

---

† See also A. Lichnerowicz [18], [19]; K. Yano and T. Nagano [4].

†† On motions in a Kähler space, see, M. Apte [2]; J. Hano [2]; S. Kobayashi and K. Nomizu [1]; A. Lichnerowicz [10], [11], [12], [14] [15], [20], [21]; Y. Matsushima [2]; Y. Matsushima and J. Hano [1]; K. Nomizu [7]; J. A. Schouten and K. Yano [4]; K. Yano [5] and on, motions in a Hermite space, see S. Ishihara [1]; J. Koszul [1].

vector $v^h$ vanishes and since $F_{ji}$ is a harmonic tensor, we have

$$\underset{v}{\mathfrak{L}} F_{ji}=0,$$

from which

$$\underset{v}{\mathfrak{L}} F_i{}^h=\underset{v}{\mathfrak{L}}(F_{is}g^{sh})=0,$$

thus $v^h$ is contravariant analytic.

A Killing vector $v^h$ in a compact Kähler space is contravariant analytic. Consequently $F_i{}^h v^i$ is also contravariant analytic. But for $F_i v_t$ we have

$$\begin{aligned}\nabla_j(F_i{}^t v_t)-\nabla_i(F_j{}^t v_t)\quad & F_i{}^t(\nabla_j v_t)-F_j{}^t(\nabla_i v_t)\\ &=-(F_i{}^t\nabla_t v_j+F_j{}^t\nabla_i v_t)=0,\end{aligned}$$

and hence $F_i{}^s v_s$ is closed. Thus we have

THEOREM 6.7. *If $v^h$ is a Killing vector in a compact Kähler space, then $F_i{}^h v^i$ is contravariant analytic and closed.*

Conversely suppose that a contravariant analytic vector $v^h$ is closed, then we have

$$F^{ji}\nabla_j v_i=\frac{1}{2}F^{ji}(\nabla_j v_i-\nabla_i v_j)=0,$$

hence by Theorem 6.4, $F_i{}^h v^i$ is a Killing vector. Thus

THEOREM 6.8. *If a contravariant analytic vector $v^h$ is closed, then $F_i{}^h v^i$ is a Killing vector.*

We now consider an equation of the form

$$\Delta f=g^{ji}\nabla_j\nabla_i f=\lambda f \qquad (\lambda=\text{constant}<0) \tag{6.7}$$

in a compact Kähler space, from which

$$\Delta f_h=g^{ji}\nabla_j\nabla_i f_h-K_h{}^i f_i=\lambda f_h \qquad (f_h=\nabla_h f).$$

Putting

$$v_h=F_h{}^i f_i$$

we have from the above equation

$$g^{ji}\nabla_j\nabla_i v_h-K_h^i v_i=\lambda v_h. \tag{6.8}$$

Substituting this into the integral formula

$$\int\left[(g^{ji}\nabla_j\nabla_i v_h + K_h{}^i v_i)v^h + \frac{1}{2}(\nabla^j v^i + \nabla^i v^j)(\nabla_j v_i + \nabla_i v_j) - (\nabla_i v^i)^2\right] d\sigma = 0,$$

and taking account of $\nabla_i v^i = \nabla_i(F^{ij}f_j) = F^{ij}(\nabla_i\nabla_j f) = 0$, we find

$$\int\left[(2K_{ji} + \lambda g_{ji})v^j v^i + \frac{1}{2}(\nabla^j v^i + \nabla^i v^j)(\nabla_j v_i + \nabla_i v_j)\right] d\sigma = 0,$$

from which

THEOREM 6.9. *If, in a compact Kähler space, the form* $(2K_{ji} + \lambda g_{ji})v^j v^i$ ($\lambda = constant < 0$) *is positive definite, the equation* $\Delta f = \lambda f$ *has no solution other than zero. If* $(2K_{ji} + \lambda g_{ji})v^j v^i$ *is positive semi-definite, then* $v_h = F_h{}^i f_i$ *is a Killing vector for a solution of* $\Delta f = \lambda f$.

THEOREM 6.10. *If* $\frac{K}{n} + \lambda > 0$ *in a compact Kähler–Einstein space with* $K > 0$, *then the equation* $\Delta f = \lambda f$ *has no solution other than zero. Consequently if the equation* $\Delta f = \lambda f$ *admits a solution other than zero then*

$$\frac{K}{n} + \lambda \leqq 0, \quad \text{that is,} \quad \lambda \leqq -\frac{K}{n}.$$

THEOREM 6.11. *If, in a compact Kähler–Einstein space with* $K > 0$, *the equation* $\Delta f = -\frac{K}{n} f$ *admits a solution other than zero, then* $v_h = F_h{}^i f_i$ *is a Killing vector.*

Suppose now that a compact Kähler space admits a Killing vector $v^h$, then, following Theorem 6.7, $F_i{}^h v^i$ is closed:

$$\nabla_j(F_i{}^t v_t) - \nabla_i(F_j{}^t v_t) = 0.$$

Thus, if we have moreover

$$F^{ji}\nabla_j v_i = 0,$$

then $F_i^{\ t}v_t$ is a harmonic vector and consequently $v_i$ is harmonic too. Thus $v^i$ being a Killing vector and at the same time a harmonic vector, we have

$$\nabla_j v_i = 0.$$

Thus we have

THEOREM 6.12. *In a compact Kähler space which does not admit a parallel vector field, we have* $F^{ji}\nabla_j v_i \neq 0$ *for a non-vanishing Killing vector* $v^h$.

We now consider a compact Kähler–Einstein space with $K > 0$ and suppose that the space admits a Killing vector field $v^h$, then following Theorem 6.12 we have

$$f = \frac{n}{K} F^{ji}\nabla_j v_i \neq 0.$$

On the other hand, from

$$\nabla_j \nabla_i v^h + K_{kji}{}^h v^k = 0.$$

we find

$$f_k = \nabla_k f = \nabla_k \left( \frac{n}{K} F^{ji}\nabla_j v_i \right) = -F_k^{\ s} v_s$$

by virtue of $H_{kj} = -\dfrac{K}{2n} F_{kj}$, and consequently

$$f_i = -F_i^{\ s} v_s, \qquad v_i = F_i^{\ s} f_s,$$

from which

$$g^{ji}\nabla_j \nabla_i f = -\frac{K}{n} f.$$

Thus we have

THEOREM 6.13. *If a compact Kähler–Einstein space with* $K > 0$ *admits a Killing vector field* $v^h$, *then the equation* $g^{ji}\nabla_j\nabla_i f = -\dfrac{K}{n} f$ *admits a solution other than zero given by* $f = \dfrac{n}{K} F^{ji}\nabla_j v_i$ *and vice versa.*

Suppose that a compact Kähler–Einstein space with $K>0$ admits two Killing vector fields $v^h$ and $w^h$ to which correspond $f$ and $g$ respectively, then we have

$$F^{ji}\nabla_j[v,\ w]_i = F^{ji}\nabla_j \underset{v}{\mathfrak{L}}\, w_i = \underset{v}{\mathfrak{L}}\,(F^{ji}\nabla_j w_i)$$

$$= \frac{K}{n}\underset{v}{\mathfrak{L}}\, g = \frac{K}{n} v^i\nabla_i g = -\frac{K}{n} F^{si} f_s \nabla_i g = -\frac{K}{n} F^{ji}(\nabla_j f)(\nabla_i g).$$

Thus if we define $[f,\ g]$ by

$$[f,\ g] = -F^{ji}(\nabla_j f)(\nabla_i g),$$

we have

THEOREM 6.14. *If a compact Kähler–Einstein space with $K>0$ admits two Killing vectors $v^h$ and $w^h$ to which correspond $f$ and $g$ respectively, then $[v,\ w]^h$ and $[f,\ g]$ correspond to each other.*

Suppose that $v^h$ is a conformal Killing vector, then we have

$$g^{ji}\nabla_j\nabla_i v^h + K_i{}^h v^i + \frac{n-1}{n}\nabla^h(\nabla_s v^s) = 0. \tag{6.9}$$

Substituting this into (6.6) we find

$$\int\Big[-\frac{n-1}{n}\nabla^h(\nabla_t v^t)v_h + \frac{1}{2}(F^{jt}\nabla_t v^i - F_t{}^i\nabla^j v^t)(F_j{}^s\nabla_s v_i - F_{si}\nabla_j v^s)\Big]d\sigma = 0$$

or

$$\int\Big[\frac{n-1}{n}(\nabla_t v^t)^2 + \frac{1}{2}(F^{jt}\nabla_t v^i - F_t{}^i\nabla^j v^t)(F_j{}^s\nabla_s v_i - F_{si}\nabla_j v^s)\Big]d\sigma = 0,$$

from which we have

THEOREM 6.15. *A conformal Killing vector in a compact Kähler space is a Killing vector and consequently is contravariant analytic.*

We defined a skew-symmetric tensor $H_{kj}$ by

$$H_{kj} = \frac{1}{2}K_{kjih}F^{ih}. \tag{6.10}$$

This tensor is related to the Ricci tensor $K_{ji}$ by

$$H_{kj} = -K_k^{\ t} F_{tj} = +K_{kt} F_j^{\ t} \tag{6.11}$$

and is closed

$$\nabla_l H_{kj} + \nabla_k H_{jl} + \nabla_j H_{lk} = 0. \tag{6.12}$$

From (6.11), we have

THEOREM 6.16. *If an infinitesimal analytic transformation does not change the tensor $H_{kj}$, it does not change $K_{ji}$ and vice versa.*

Suppose that $w^h$ is a contravariant analytic vector, then we have

$$g^{ts}\nabla_t\nabla_s w_i + K_i^{\ t} w_t = 0$$

or

$$g^{ts}\nabla_t\nabla_s w_i - K_i^{\ t} w_t = 2H_{is}(F_t^{\ s} w^t),$$

from which, $H_{kj}$ being closed,

$$\begin{aligned} g^{ts}\nabla_t\nabla_s(\nabla_j w_i - \nabla_i w_j) - K_j^{\ t}(\nabla_t w_i - \nabla_i w_t) + K_i^{\ t}(\nabla_t w_j - \nabla_j w_t) \\ - K_{ji}{}^{ts}(\nabla_t w_s - \nabla_s w_t) = -2(F_t^{\ s} w^t)\nabla_s H_{ji} - 2\nabla_j(F_t^{\ s} w^t) H_{si} \\ - 2\nabla_i(F_t^{\ s} w^t) H_{js} \end{aligned}$$

or

$$\begin{aligned} g^{ts}\nabla_t\nabla_s(\nabla_j w_i - \nabla_i w_j) - K_j^{\ t}(\nabla_t w_i - \nabla_i w_t) - K_i^{\ t}(\nabla_j w_t - \nabla_t w_j) \\ - K_{ji}{}^{ts}(\nabla_t w_s - \nabla_s w_t) = -2 \underset{Fw}{\mathfrak{L}} H_{ji}. \end{aligned} \tag{6.13}$$

Thus if a contravariant analytic transformation $v^h$ does not change $H_{ji}$, we put

$$v^h = F_i^{\ h} w^i$$

and we see from (6.13) that

$$\nabla_j(F_i^{\ t} v_t) - \nabla_i(F_j^{\ t} v_t)$$

is a harmonic tensor. But it is a rotation of $F_i^{\ s} v_s$ and consequently identically zero,

$$\nabla_j(F_i^{\ t} v_t) - \nabla_i(F_j^{\ t} v_t) = 0,$$

from which, contracting $F^{ji}$,

$$\nabla_t v^t = 0,$$

and consequently $v^h$ is a Killing vector. Thus

THEOREM 6.17. *A contravariant analytic transformation of a compact Kähler space which does not change $H_{kj}$ or $K_{ji}$ is a motion.*

## § 7. A theorem of Matsushima

A necessary and sufficient condition for a vector $v^h$ in a compact Kähler–Einstein space to be a contravariant analytic vector is given by

$$g^{ji}\nabla_j\nabla_i v^h + \frac{K}{2n} v^h = 0. \tag{7.1}$$

From this we get

$$g^{ji}\nabla_j\nabla_i(\nabla_s v^s) + \frac{K}{n}(\nabla_s v^s) = 0, \tag{7.2}$$

from which

$$g^{ji}\nabla_j\nabla_i\nabla_h(\nabla_s v^s) + \frac{K}{2n}\nabla_h(\nabla_s v^s) = 0. \tag{7.3}$$

This equation shows that $\nabla_h(\nabla_s v^s)$ is a contravariant analytic vector field. If we put

$$p^h = v^h + \frac{n}{K}\nabla^h(\nabla_s v^s), \tag{7.4}$$

$p^h$ is contravariant analytic and we have

$$\nabla_h p^h = \nabla_h v^h + \frac{n}{K}\nabla_h\nabla^h(\nabla_s v^s) = 0$$

by virtue of (7.2). Thus $p^h$ is a Killing vector.

If we put

$$q^h = F_t{}^h \frac{n}{K} \nabla^t(\nabla_s v^s), \tag{7.5}$$

then $q^h$ is contravariant analytic and

$$\nabla_h q^h = 0$$

and consequently $q^h$ is also a Killing vector.

Thus, from (7.4) and (7.5), we find

$$v^h = p^h + F_i{}^h q^i, \tag{7.6}$$

where $p^h$ and $q^h$ are both Killing vectors.

This decomposition of a contravariant analytic vector is unique. Because suppose that we have another decomposition

$$v^h = 'p^h + F_i{}^h{}'q^i,$$

$'p^h$ and $'q^h$ being both Killing vectors, then we have

$$('p^h - p^h) + F_i{}^h('q^i - q^i) = 0,$$

from which

$$F^{ih}\nabla_i('q_h - q_h) = 0.$$

Since a Kähler–Einstein space with $K \neq 0$ does not admit a parallel vector field, applying Theorem 6.12, we have, from this equation,

$$'q^h = q^h$$

and consequently

$$'p^h = p^h.$$

Thus we have proved

THEOREM 7.1. *In a compact Kähler–Einstein space with $K \neq 0$, a contravariant analytic vector $v^h$ is uniquely decomposed in the form*

$$v^h = p^h + F_i{}^h q^i,$$

*$p^h$ and $q^h$ being both Killing vectors.* (Y. Matsushima [1]; K. Yano [12]).†

† See also S. Sawaki [5], [6].

## § 8. A generalization by Lichnerowicz

We first remark that the identities

$$F_j{}^s(\nabla_s v_i - \nabla_i v_s) + F_i{}^s(\nabla_j v_s - \nabla_s v_j) - [\nabla_j(F_i{}^s v_s) - \nabla_i(F_j{}^s v_s)]$$
$$= F_{ji}(\nabla_s v^s) - \nabla^s(F_{sj} v_i + F_{ji} v_s + F_{is} v_j) \qquad (8.1)$$

and

$$F_j{}^s(\nabla_i v_s + \nabla_s v_i) - F_i{}^s(\nabla_j v_s + \nabla_s v_j) + [\nabla_j(F_i{}^s v_s) - \nabla_i(F_j{}^s v_s)]$$
$$= F_{ji}(\nabla_s v^s) - \nabla^s(F_{sj} v_i + F_{ji} v_s + F_{is} v_j) \qquad (8.2)$$

hold good for an arbitrary vector field $v^h$.

On the other hand, we have

$$F_j{}^s(\nabla_s v_i - \nabla_i v_s) + F_i{}^s(\nabla_j v_s - \nabla_s v_j) \qquad (8.3)$$
$$= F_j{}^t[(\nabla_t v_i - \nabla_i v_t) - F_t{}^s F_i{}^r(\nabla_s v_r - \nabla_r v_s)] = 2F_j{}^t O_{ti}^{sr}(\nabla_s v_r - \nabla_r v_s)$$

and

$$F_j{}^s(\nabla_i v_s + \nabla_s v_i) - F_i{}^s(\nabla_j v_s + \nabla_s v_j)$$
$$= F_j{}^t[(\nabla_i v_t + \nabla_t v_i) + F_t{}^s F_i{}^r(\nabla_s v_r + \nabla_r v_s)] = 2F_j{}^t {}^*O_{ti}^{sr}(\nabla_s v_r + \nabla_r v_s), \qquad (8.4)$$

and consequently (8.1) and (8.2) can be written respectively as

$$2F_j{}^t O_{ti}^{sr}(\nabla_s v_r - \nabla_r v_s) - [\nabla_j(F_i{}^t v_t) - \nabla_i(F_j{}^t v_t)]$$
$$= F_{ji}(\nabla_t v^t) - \nabla^t(F_{tj} v_i + F_{ji} v_t + F_{it} v_j) \qquad (8.5)$$

and

$$2F_j{}^t {}^*O_{ti}^{sr}(\nabla_s v_r + \nabla_r v_s) + [\nabla_j(F_i{}^t v_t) - \nabla_i(F_j{}^t v_t)]$$
$$= F_{ji}(\nabla_t v^t) - \nabla^t(F_{tj} v_i + F_{ji} v_t + F_{it} v_j). \qquad (8.6)$$

Now the condition for $v^h$ to be covariant analytic is given by

$$^*O_{ji}^{ts} \nabla_t v_s = 0 \qquad (8.7)$$

or

$$S_{ji}(v) = F_j{}^t \nabla_i v_t - F_i{}^t \nabla_t v_j = 0. \qquad (8.8)$$

This $S_{ji}(v)$ satisfies

$$S_{ji}(v) - F_j{}^t(\nabla_i v_t - \nabla_t v_i) = F_{ji}(\nabla_t v^t) - \nabla^t(F_{tj} v_i + F_{ji} v_t + F_{it} v_j) \qquad (8.9)$$

and also

$$S_{ji}(v)+S_{ij}(v)=F_j^{\ t}(\nabla_i v_t-\nabla_t v_i)+F_i^{\ t}(\nabla_j v_t-\nabla_t v_j) \quad (8.10)$$

and

$$S_{ji}(v)-S_{ij}(v)=F_j^{\ t}(\nabla_i v_t+\nabla_t v_i)-F_i^{\ t}(\nabla_j v_t+\nabla_t v_j). \quad (8.11)$$

Hence (8.1) and (8.2) now become

$$\begin{aligned} S_{ji}(v)+S_{ij}(v)+2F_j^{\ t}(\nabla_t v_i-\nabla_i v_t)-[\nabla_j(F_i^{\ t}v_t)-\nabla_i(F_j^{\ t}v_t)] \\ =F_{ji}(\nabla_t v^t)-\nabla^t(F_{tj}v_i+F_{ji}v_t+F_{it}v_j) \end{aligned} \quad (8.12)$$

and

$$\begin{aligned} S_{ji}(v)-S_{ij}(v)+[\nabla_j(F_i^{\ t}v_t)-\nabla_i(F_j^{\ t}v_t)] \\ =F_{ji}(\nabla_t v^t)-\nabla^t(F_{tj}v_i+F_{ji}v_t+F_{it}v_j). \end{aligned} \quad (8.13)$$

The condition for $v^h$ to be contravariant analytic is given by

$$*O_{ir}^{sh}\nabla_s v^r=0 \quad \text{or} \quad O_{ji}^{ts}\nabla_t v_s=0 \quad (8.14)$$

or

$$T_{ji}(v)=F_j^{\ t}\nabla_t v_i+F_i^{\ t}\nabla_j v_t=0. \quad (8.15)$$

This $T_{ji}(v)$ satisfies

$$T_{ji}(v)-F_i^{\ t}(\nabla_j v_t+\nabla_t v_j)=F_{ji}(\nabla_t v^t)-\nabla^t(F_{tj}v_i+F_{ji}v_t+F_{it}v_j). \quad (8.16)$$

We also have

$$T_{ji}(v)+T_{ij}(v)=F_j^{\ t}(\nabla_t v_i+\nabla_i v_t)+F_i^{\ t}(\nabla_j v_t+\nabla_t v_j) \quad (8.17)$$

and

$$T_{ji}(v)-T_{ij}(v)=F_j^{\ t}(\nabla_t v-{}_i\nabla_i v_t)+F_i^{\ t}(\nabla_j v_t-\nabla_t v_j). \quad (8.18)$$

Hence (8.1) and (8.2) take respectively the forms

$$\begin{aligned} T_{ji}(v)-T_{ij}(v)-[\nabla_j(F_i^{\ t}v_t)-\nabla_i(F_j^{\ t}v_t)] \\ =F_{ji}(\nabla_t v^t)-\nabla^t(F_{tj}v_i+F_{ji}v_t+F_{it}v_j) \end{aligned} \quad (8.19)$$

and

$$\begin{aligned} -T_{ji}(v)-T_{ij}(v)+2F_j^{\ t}(\nabla_t v_i+\nabla_i v_t)+[\nabla_j(F_i^{\ t}v_t)-\nabla_i(F_j^{\ t}v_t)] \\ =F_{ji}(\nabla_t v^t)-\nabla^t(F_{tj}v_i+F_{ji}v_t+F_{it}v_j). \end{aligned} \quad (8.20)$$

Now following the theory of harmonic integrals, an arbitrary vector $v^h$ can be decomposed as

$$v^h = p^h + r^h, \tag{8.21}$$

where

$$\nabla_t p^t = 0 \quad \text{and} \quad r_i = \nabla_i r \tag{8.22}$$

for a certain scalar $r$.

Assume that $v^h$ is a contravariant analytic vector and substitute (8.21) into (8.18); then we have, taking account of (8.22),

$$F_j{}^t(\nabla_t p_i - \nabla_i p_t) + F_i{}^t(\nabla_j p_t - \nabla_t p_j) = 0 \tag{8.23}$$

that is,

$$T_{ji}(p) - T_{ij}(p) = 0. \tag{8.24}$$

Substituting this into (8.19), we find

$$\nabla_j(F_i{}^t p_t) - \nabla_i(F_j{}^t p_t) = \nabla^t(F_{tj} p_i + F_{ji} p_t + F_{it} p_j). \tag{8.25}$$

But the left hand side of this equation is the rotation of $F_i{}^t p_t$ and the right hand side is the divergence of $F_{tj} p_i + F_{ji} p_t + F_{it} p_j$, and consequently

$$\nabla_j(F_i{}^t p_t) - \nabla_i(F_j{}^t p_t) = 0 \tag{8.26}$$

$$\nabla^t(F_{tj} p_i + F_{ji} p_t + F_{it} p_j) = 0. \tag{8.27}$$

Forming (8.26) + (8.27) and taking account of (8.22), we find

$$F_j{}^t(\nabla_i p_t + \nabla_t p_i) - F_i{}^t(\nabla_j p_t + \nabla_t p_j) = 0. \tag{8.28}$$

Thus putting $v_t = p_t$ in (8.4), we have

$$*O_{ji}^{ts}(\nabla_t p_s + \nabla_s p_t) = 0. \tag{8.29}$$

We now assume that the curvature scalar $K$ of the space is constant, and calculate the divergence of the vector $K_i{}^t p_t$:

$$\nabla^i(K_i{}^t p_t) = \frac{1}{2}(\nabla^t K)p_t + \frac{1}{2}K^{ji}(\nabla_j p_i + \nabla_i p_j) = \frac{1}{2}K^{ji}(\nabla_j p_i + \nabla_i p_j).$$

But $K^{ji}$ is hybrid in $j$ and $i$ and $\nabla_j p_i + \nabla_i p_j$ is pure in $j$ and $i$ as is seen from (8.29). Thus we have

$$\nabla_h(K_i^{\ h} p^i) = 0, \tag{8.30}$$

and consequently

$$\nabla_h[g^{ji}\nabla_j\nabla_i p^h + K_i^{\ h} p^i] = 0. \tag{8.31}$$

Now substituting (8.21) into

$$g^{ji}\nabla_j\nabla_i v^h + K_i^{\ h} v^i = 0,$$

we find

$$(g^{ji}\nabla_j\nabla_i p^h + K_i^{\ h} p^i) + (g^{ji}\nabla_j\nabla_i r^h + K_i^{\ h} r^i) = 0,$$

from which

$$\nabla_h(g^{ji}\nabla_j\nabla_i r^h + K_i^{\ h} r^i) = 0 \tag{8.32}$$

by virtue of (8.31).

On the other hand, we have, in general,

$$\int \left[ (g^{ji}\nabla_j\nabla_i r^h + K_i^{\ h} r^i) r_h + \frac{1}{2} T^{ji}(r) T_{ji}(r) \right] d\sigma = 0$$

and consequently, by virtue of (8.22) and (8.32),

$$\int \left[ -\nabla_h \{(g^{ji}\nabla_j\nabla_i r^h + K_i^{\ h} r^i)\} r + \frac{1}{2} T^{ji}(r) T_{ji}(r) \right] d\sigma = 0$$

or

$$\int T^{ji}(r) T_{ji}(r) d\sigma = 0$$

from which

$$T_{ji}(r) = 0$$

which means that $r^h$ is a contravariant analytic vector.

Thus if we put

$$r^h = F_i^{\ h} q^i \quad \text{or} \quad q^h = -F_i^{\ h} r^i, \tag{8.33}$$

$q_h$ is also contravariant analytic. Moreover we have

$$\nabla_h q^h = F^{ih}\nabla_i r_h = 0$$

and consequently $q^h$ is a Killing vector.

Thus

$$p^h = v^h - F_i{}^h q^i$$

is also contravariant analytic and satisfies $\nabla_h p^h = 0$ and hence is a Killing vector. Thus we have

TEHOREM 8.1. *A contravariant analytic vector $v^h$ in a compact Kähler space with constant curvature scalar is decomposed in the form*

$$v^h = p^h + F_i{}^h q^i,$$

*where $p^h$ and $q^h$ are both Killing vectors.* (A. Lichnerowicz [22], [23], [24], [25]).

## § 9. Analytic tensors†

Consider a self-conjugate tensor field $T_{i_1 i_2 \ldots i_q}{}^{h_1 h_2 \ldots h_p}$ $(1 \leqq p, q)$ which we write $T^{h_1 h_2 \ldots h_p}_{i_1 i_2 \ldots i_q}$ and which is pure in all its indices, that is, which has components of the form

$$T^{h_1 h_2 \ldots h_p}_{i_1 i_2 \ldots i_q} = (T^{\kappa_1 \kappa_2 \ldots \kappa_p}_{\lambda_1 \lambda_2 \ldots \lambda_q},\ 0,\ 0,\ \ldots,\ 0,\ T^{\bar{\kappa}_1 \bar{\kappa}_2 \ldots \bar{\kappa}_p}_{\bar{\lambda}_1 \bar{\lambda}_2 \ldots \bar{\lambda}_q}). \tag{9.1}$$

The fact that $T^{h_1 h_2 \ldots h_p}_{i_1 i_2 \ldots i_q}$ is pure in all its indices is expressed by

$$*O^{sh}_{ir} T^{h_1 \ldots r \ldots h_p}_{i_1 \ldots s \ldots i_q} = 0 \tag{9.2}$$

or

$$F_j{}^s T^{h_1 \ldots h \ldots h_p}_{i_1 \ldots s \ldots i_q} - F_r{}^h T^{h_1 \ldots r \ldots h_p}_{i_1 \ldots j \ldots i_q} = 0 \tag{9.3}$$

the indices $s$, $r$, taking all the positions.

Now, if the components $T^{\kappa_1 \kappa_2 \ldots \kappa_p}_{\lambda_1 \lambda_2 \ldots \lambda_q}$ are functions of $z^\lambda$ only and the components $T^{\bar{\kappa}_1 \bar{\kappa}_2 \ldots \bar{\kappa}_p}_{\bar{\lambda}_1 \bar{\lambda}_2 \ldots \bar{\lambda}_q}$ are functions of $z^{\bar{\lambda}}$ only, then we call the tensor $T^{h_1 h_2 \ldots h_p}_{i_1 i_2 \ldots i_q}$ an *analytic tensor* in a Kähler space. The condition for a tensor $T^{h_1 h_2 \ldots h_p}_{i_1 i_2 \ldots i_q}$ in a Kähler space to be analytic is given by

$$\partial_{\bar{\mu}} T^{\kappa_1 \kappa_2 \ldots \kappa_p}_{\lambda_1 \lambda_2 \ldots \lambda_q} = 0, \quad \partial_\mu T^{\bar{\kappa}_1 \bar{\kappa}_2 \ldots \bar{\kappa}_p}_{\bar{\lambda}_1 \bar{\lambda}_2 \ldots \bar{\lambda}_q} = 0 \tag{9.4}$$

---

†As to the generalization, see H. Guggenheimer [4]; S. Kotō [3], [4], [5]; S. Sawaki [1], [3], [4]; S. Sawaki and S. Kotō [1] S. Tachibana [6].

or

$$\nabla_{\bar{\mu}} T^{\varkappa_1 \varkappa_2 \dots \varkappa_p}_{\lambda_1 \lambda_2 \dots \lambda_q} = 0, \quad \nabla_{\mu} T^{\bar{\varkappa}_1 \bar{\varkappa}_2 \dots \bar{\varkappa}_p}_{\bar{\lambda}_1 \bar{\lambda}_2 \dots \bar{\lambda}_q} = 0. \tag{9.5}$$

Since the only non-zero components of the covariant derivative $\nabla_j T^{h_1 h_2 \dots h_p}_{i_1 i_2 \dots i_q}$ are

$$\nabla_{\mu} T^{\varkappa_1 \varkappa_2 \dots \varkappa_p}_{\lambda_1 \lambda_2 \dots \lambda_q}, \quad \nabla_{\mu} T^{\bar{\varkappa}_1 \bar{\varkappa}_2 \dots \bar{\varkappa}_p}_{\bar{\lambda}_1 \bar{\lambda}_2 \dots \bar{\lambda}_q}, \quad \nabla_{\bar{\mu}} T^{\varkappa_1 \varkappa_2 \dots \varkappa_p}_{\lambda_1 \lambda_2 \dots \lambda_q}, \quad \nabla_{\bar{\mu}} T^{\bar{\varkappa}_1 \bar{\varkappa}_2 \dots \bar{\varkappa}_p}_{\bar{\lambda}_1 \bar{\lambda}_2 \dots \bar{\lambda}_q},$$

equation (9.5) shows that covariant derivative $\nabla_j T^{h_1 h_2 \dots h_p}_{i_1 i_2 \dots i_q}$ is pure in all its indices.

The condition (9.5) may also be written as

$$*O^{ts}_{ji} \nabla_t T^{h_1 h_2 \dots h_p}_{s i_2 \dots i_q} = 0, \tag{9.6}$$

from which

$$F_j{}^s \nabla_s T^{h_1 h_2 \dots h_p}_{i_1 i_2 \dots i_q} - F_{i_1}{}^s \nabla_j T^{h_1 h_2 \dots h_p}_{s i_2 \dots i_q} = 0. \tag{9.7}$$

Applying the operator $F^{jt}\nabla_t$ to this equation, we find

$$g^{ts} \nabla_t \nabla_s T^{h_1 h_2 \dots h_p}_{i_1 i_2 \dots i_q} - F^{jt} F_{i_1}{}^s \nabla_t \nabla_j T^{h_1 h_2 \dots h_p}_{s i_2 \dots i_q} = 0,$$

$$g^{ts} \nabla_t \nabla_s T^{h_1 h_2 \dots h_p}_{i_1 i_2 \dots i_q} + \frac{1}{2} F^{tj} F_{i_1}{}^s (\nabla_t \nabla_j T^{h_1 h_2 \dots h_p}_{s i_2 \dots i_q} - \nabla_j \nabla_t T^{h_1 h_2 \dots h_p}_{s i_2 \dots i_q}) = 0,$$

$$g^{ts} \nabla_t \nabla_s T^{h_1 h_2 \dots h_p}_{i_1 i_2 \dots i_q} + \frac{1}{2} F^{tj} F_{i_1}{}^s (K_{tjr}{}^{h_1} T^{r h_2 \dots h_p}_{s i_2 \dots i_q} + \dots + K_{tjr}{}^{h_p} T^{h_1 h_2 \dots r}_{s i_2 \dots i_q}$$

$$- K_{tjs}{}^r T^{h_1 h_2 \dots h_p}_{r i_2 \dots i_q} - K_{tji_2}{}^r T^{h_1 h_2 \dots h_p}_{s r i_3 \dots i_q} - \dots - K_{tji_q}{}^r T^{h_1 h_2 \dots h_p}_{s i_2 \dots r}) = 0$$

$$g^{ts} \nabla_t \nabla_s T^{h_1 h_2 \dots h_p}_{i_1 i_2 \dots i_q} + (H_r{}^{h_1} F_s{}^r T^{s \dots h_p}_{i_1 i_2 \dots i_q} + \dots + H_r{}^{h_p} F_s{}^r T^{h_1 \dots s}_{i_1 i_2 \dots i_q})$$
$$- (F_{i_1}{}^s H_s{}^r T^{h_1 h_2 \dots h_p}_{r i_2 \dots i_q} + F_{i_1}{}^s H_{i_2}{}^r T^{h_1 h_2 \dots h_p}_{s r i_3 \dots i_q} + \dots$$
$$\dots + F_{i_1}{}^s H_{i_p}{}^r T^{h_1 h_2 \dots h_p}_{s i_2 \dots r}) = 0$$

by virtue of (9.3). From the above equation, we obtain

$$g^{ts} \nabla_t \nabla_s T^{h_1 h_2 \dots h_p}_{i_1 i_2 \dots i_q} + K_s{}^{h_1} T^{s h_2 \dots h_p}_{i_1 i_2 \dots i_q} + \dots + K_s{}^{h_p} T^{h_1 \dots h_{p-1} s}_{i_1 i_2 \dots i_q}$$
$$- K_{i_1}{}^s T^{h_1 h_2 \dots h_p}_{s i_2 \dots i_q} - \dots - K_{i_q}{}^s T^{h_1 h_2 \dots h_p}_{i_1 \dots i_{q-1} s} = 0. \tag{9.8}$$

This is a necessary condition for $T^{h_1 h_2 \dots h_p}_{i_1 i_2 \dots i_q}$ to be analytic.

On the other hand, we have

$$\frac{1}{2}(F^{js}\nabla_s T^{i_1 i_2 \dots i_q h_1 h_2 \dots h_p} - F^{i_1 s}\nabla^j T_s{}^{i_2 \dots i_q h_1 \dots h_p})$$
$$\times (F_j{}^r \nabla_r T_{i_1 i_2 \dots i_q h_1 h_2 \dots h_p} - F_{i_1}{}^r \nabla_j T_{r i_2 \dots i_q h_1 \dots h_p})$$
$$= (\nabla^j T^{i_1 i_2 \dots i_q h_1 h_2 \dots h_p})(\nabla_j T_{i_1 i_2 \dots i_q h_1 h_2 \dots h_p})$$
$$+ F^{js}F^{ir}(\nabla_s T_r{}^{i_2 \dots i_q h_1 h_2 \dots h_p})(\nabla_j T_{i i_2 \dots i_q h_1 \dots h_p})$$

and

$$\nabla_j[(\nabla^j T_{i_1 \dots i_q h_1 \dots h_p}) T^{i_1 \dots i_q h_1 \dots h_p}$$
$$+ F^{js}F^{ir}(\nabla_s T_{r i_2 \dots i_q h_1 \dots h_p}) T_i{}^{i_2 \dots i_q h_1 \dots h_p}]$$
$$= (g^{ts}\nabla_t \nabla_s T_{i_1 \dots i_q h_1 \dots h_p}) T^{i_1 \dots i_q h_1 \dots h_p}$$
$$+ (\nabla^j T^{i_1 \dots i_q h_1 \dots h_p})(\nabla_j T_{i_1 \dots i_q h_1 \dots h_p})$$
$$+ F^{js}F^{ir}(\nabla_j \nabla_s T^{h_1 \dots h_p}_{r i_2 \dots i_q}) T^{i_2 \dots i_q}_{i h_1 \dots h_p}$$
$$+ F^{js}F^{ir}(\nabla_s T^{h_1 \dots h_p}_{r i_2 \dots i_q})(\nabla_j T^{i_2 \dots i_q}_{i h_1 \dots h_p})$$
$$= (g^{ts}\nabla_t \nabla_s T^{h_1 \dots h_p}_{i_1 i_2 \dots i_q} + K_s{}^{h_1} T^{s h_2 \dots h_p}_{i_1 \dots i_q} + \dots + K_s{}^{h_p} T^{h_1 \dots s}_{i_1 \dots i_q}$$
$$- K_{i_1}{}^s T^{h_1 \dots h_p}_{s i_2 \dots i_q} - \dots - K_{i_q}{}^s T^{h_1 \dots h_p}_{i_1 \dots s}) T^{i_1 \dots i_q}_{h_1 \dots h_p}$$
$$+ \frac{1}{2}(F^{js}\nabla_s T^{i_1 \dots i_q h_1 \dots h_p} - F^{i_1 s}\nabla^j T_s{}^{i_2 \dots i_q h_1 \dots h_p})$$
$$\times (F_j{}^r \nabla_r T_{i_1 \dots i_q h_1 \dots h_p} - F_{i_1}{}^r \nabla_j T_{r i_2 \dots i_q h_1 \dots h_p}).$$

Thus assuming the space to be compact, we apply Green's theorem and obtain

$$\int [(g^{ts}\nabla_t \nabla_s T^{h_1 \dots h_p}_{i_1 \dots i_q}$$
$$+ K_s{}^{h_1} T^{s h_2 \dots h_p}_{i_1 \dots i_q} + \dots + K_s{}^{h_p} T^{h_1 \dots h_{p-1} s}_{i_1 \dots i_l}$$
$$- K_{i_1}{}^s T^{h_1 \dots h_p}_{s i_2 \dots i_q} - \dots - K_{i_q}{}^s T^{h_1 \dots h_p}_{i_1 \dots i_{q-1} s} \times T^{i_1 \dots i_q}_{h_1 \dots h_p}$$
$$+ \frac{1}{2}(F^{js}\nabla_s T^{i_1 \dots i_q h_1 \dots h_p} - F^{i_1 s}\nabla^j T_s^{i_2 \dots i_q h_1 \dots h_p})$$
$$\times (F_j{}^r \nabla_r T_{i_1 \dots i_q h_1 \dots h_p} - F_{i_1}{}^r \nabla_j T_{r i_2 \dots i_q h_1 \dots h_p})] d\sigma = 0, \quad (9.9)$$

from which we have

THEOREM 9.1†. *A necessary and sufficient condition for a pure tensor* $T^{h_1 \ldots h_p}_{i_1 \ldots i_q}$ *in a compact Kähler space to be analytic is that*

$$g^{ts}\nabla_t\nabla_s T^{h_1 \ldots h_p}_{i_1 \ldots i_q} + K_s{}^{h_1} T^{sh_2 \ldots h_p}_{i_1 \ldots i_q} + \ldots + K_s{}^{h_p} T^{h_1 \ldots h_{p-1}s}_{i_1 \ldots i_q}$$

$$- K_{i_1}{}^s T^{h_1 \ldots h_p}_{si_2 \ldots i_q} - \ldots - K_{i_q}{}^s T^{h_1 \ldots h_p}_{i_1 \ldots i_{q-1}s} = 0. \qquad (9.10)$$

As a corollary to this theorem, we have

THEOREM 9.2. *A necessary and sufficient condition for a pure skew-symmetric covariant tensor* $w_{i_1 \ldots i_q}$ *in a compact Kähler space to be (covariant) analytic is that*

$$g^{ts}\nabla_t\nabla_s w_{i_1 \ldots i_q} - \sum_{a=1}^{q} K_{i_a}{}^r w_{i_1 \ldots r \ldots i_q} = 0. \qquad (9.11)$$

On the other hand a necessary and sufficient condition for a skew-symmetric covariant tensor $w_{i_1 \ldots i_q}$ in a compact Kähler space to be harmonic is

$$g^{ts}\nabla_t\nabla_s w_{i_1 \ldots i_q} - \sum_{a=1}^{q} K_{i_a}{}^r w_{i_1 \ldots r \ldots i_q}$$

$$- \sum_{b<a}^{1 \ldots q} K_{i_b i_a}{}^{sr} w_{i_1 \ldots s \ldots r \ldots i_q} = 0. \qquad (9.12)$$

But, if $w_{i_1 \ldots i_q}$ is pure, then $K_{kj}{}^{ih}$ being hybrid in $i$ and $h$, equation (9.12) reduces to

$$g^{ts}\nabla_t\nabla_s w_{i_1 \ldots i_q} - \sum_{a=1}^{q} K_{i_a}{}^r w_{i_1 \ldots r \ldots i_q} = 0.$$

Thus we have

THEOREM 9.3. *A necessary and sufficient condition for a pure skew-symmetric covariant tensor in a compact Kähler space to be (covariant) analytic is that the tensor be harmonic.*

Take now a harmonic tensor $w_{i_1 \ldots i_q}$ in a compact Kähler space, then it satisfies (9.12) and consequently its pure part also satisfies (9.12), that is, (9.11). Thus we have

THEOREM 9.4. *The pure part of a harmonic tensor in a compact Kähler space is analytic.*

---

† See also S. Sawaki and S. Koto [1]; S. Tachibana [6].

Now substituting (9.10) into

$$\int_M [(g^{ts}\nabla_t\nabla_s T^{h_1\ldots h_p}_{i_1\ldots i_q})T^{i_1,\ldots i_q}_{h_1\ldots h_p} + (\nabla^j T^{i_1\ldots i_q h_1\ldots h_p})(\nabla_j T_{i_1\ldots i_q h_1\ldots h_p})]d\sigma = 0,$$

we find

$$\begin{aligned}\int_M [(K_t{}^s T^{ti_2\ldots i_q h_1\ldots h_p} T_{si_2\ldots i_q h_1\ldots h_p} + \cdots \\ + K_t{}^s T^{i_1\ldots i_{q-1}th_1\ldots h_p} T_{i_1\ldots i_{q-1}sh_1\ldots h_p} \\ - K_t{}^s T^{i_1\ldots i_q th_2\ldots h_p} T_{i_1\ldots i_q sh_2\ldots h_p} - \cdots \\ - K_t{}^s T^{i_1\ldots i_q h_1\ldots h_{p-1}t} T_{i_1\ldots i_q h_1\ldots h_{p-1}s}) \\ + (\nabla^j T^{i_1\ldots, i_q h_1\ldots h_p})(\nabla_j T_{i_1\ldots i_q h_1\ldots h_p})]d\sigma = 0.\end{aligned}$$

Thus we have

THEOREM 9.5. *When the quadratic form*

$$\begin{aligned}G(T, T) = K_t{}^s T^{ti_2\ldots i_q h_1\ldots h_p} T_{si_2\ldots i_q h_1\ldots h_p} + \cdots \\ + K_t{}^s T^{i_1\ldots th_1\ldots h_p} T_{i_1\ldots sh_1\ldots h_p} \\ - K_t{}^s T^{i_1\ldots i_q t\ldots h_p} T_{i_1\ldots i_q sh_2\ldots h_p} - \cdots \\ - K_t{}^s T^{i_1\ldots i_q h_1\ldots h_{p-1}t} T_{i_1\ldots i_q h_1\ldots h_{p-1}s}\end{aligned} \qquad (9.13)$$

*is positive definite, then there does not exist an analytic tensor of the type* $T^{h_1\ldots h_p}_{i_1 i_2\ldots i_q}$ *other than the zero tensor. When* $G(T, T)$ *is positive semi-definite, an analytic tensor of the type* $T^{h_1\ldots h_p}_{i_1\ldots i_q}$ *has vanishing covariant derivative.*

If we denote, at every point of the Kähler space, by $M$ and $m$ the algebraically largest and smallest eigenvalues of the matrix $K_i{}^h$ respectively, then we have

$$G(T, T) \geqq (qm - pM) T^{i_1\ldots i_q h_1\ldots h_p} T_{i_1\ldots i_q h_1\ldots h_p}$$

from which

THEOREM 9.6. *If*

$$qm - pM \geqq 0 \qquad (9.14)$$

*then an analytic tensor of the type* $T^{h_1 \cdots h_p}_{i_1 \cdots i_q}$ *has vanishing covariant derivative. If*

$$qm - pM \geqq 0$$

*everywhere and*

$$qm - pM > 0$$

*somewhere, then there exists no analytic tensor of the type* $T^{h_1 h_2 \cdots h_p}_{i_1 i_2 \cdots i_q}$ *other than the zero tensor.* (S. Bochner [10]).

As a corollary to this theorem we have

THEOREM 9.7. *In a compact Kähler–Einstein space*

$$K_{ji} = \lambda g_{ji},$$

*for* $\lambda > 0$, *there exists no analytic tensor field of the type* $T^{h_1 \cdots h_p}_{i_1 i_2 \cdots i_q}(q > p)$, *and, for* $\lambda < 0$, *there exists no analytic tensor field of the type* $T^{h_1 h_2 \cdots h_p}_{i_1 i_2 \cdots i_q}$ $(q < p)$.

## § 10. Subspaces

Consider a Kähler space $K_{2n}$ which is covered by a system of complex coordinate neighbourhoods

$$z^{\varkappa} = \xi^{\varkappa} + i\xi^{\bar{\varkappa}}, \tag{10.1}$$

and a subspace $K_{2m}$ defined by

$$z^{\varkappa} = \xi^{\varkappa}(\eta^{a}) + i\xi^{\bar{\varkappa}}(\eta^{a}), \tag{10.2}$$

where $\eta^a$ are parameters on the subspace and

$$a,\ b,\ c,\ \ldots = 1,\ 2,\ \ldots,\ m;\ \bar{1},\ \bar{2},\ \ldots,\ \bar{m}.$$

Introduce a complex coordinate system $(u^{\alpha})$ defined by

$$u^{\alpha} = \eta^{\alpha} + i\eta^{\bar{\alpha}}, \tag{10.3}$$

where

$$\alpha,\ \beta,\ \ldots = 1,\ 2,\ \ldots,\ m.$$

If the equations (10.2) can be written as

$$z^{\kappa}=z^{\kappa}(u^{\alpha}) \tag{10.4}$$

we call such a subspace an *analytic subspace*.

For an analytic subspace, we have

$$\begin{aligned} F_i{}^{\kappa}\frac{\partial z^i}{\partial u^{\beta}} &= 'F_{\beta}{}^{a}\frac{\partial z^{\kappa}}{\partial u^{a}}, & F_i{}^{\kappa}\frac{\partial z^i}{\partial u^{\bar\beta}} &= 'F_{\bar\beta}{}^{a}\frac{\partial z^{\kappa}}{\partial u^{a}}, \\ F_i{}^{\bar\kappa}\frac{\partial z^i}{\partial u^{\beta}} &= 'F_{\beta}{}^{a}\frac{\partial z^{\bar\kappa}}{\partial u^{a}}, & F_i{}^{\bar\kappa}\frac{\partial z^i}{\partial u^{\bar\beta}} &= 'F_{\bar\beta}{}^{a}\frac{\partial z^{\bar\kappa}}{\partial u^{a}}, \end{aligned} \tag{10.5}$$

where

$$'F_b{}^{a}=\begin{pmatrix} i\delta^{\alpha}_{\beta} & 0 \\ 0 & -i\delta^{\bar\alpha}_{\bar\beta} \end{pmatrix}. \tag{10.6}$$

The equation (10.5) can be written also as

$$F_i{}^{h}\frac{\partial z^i}{\partial u^{b}}='F_b{}^{a}\frac{\partial z^{h}}{\partial u^{a}} \tag{10.7}$$

which says that $F_b{}^a$ is a tensor in the subspace. Since this tensor has the numerical components (10.6) in any complex coordinate system $(u^a)$, the analytic subspace is a complex space.

Now if we represent both the Kähler space $K_{2n}$ and the analytic subspace $K_{2m}$ by the real coordinates $\xi^h$ and $\eta^a$ respectively, we have

$$\xi^h=\xi^h(\eta^a) \tag{10.8}$$

and

$$F_i{}^{h}\frac{\partial \xi^i}{\partial \eta^{b}}='F_b{}^{a}\frac{\partial \xi^{h}}{\partial \eta^{a}}$$

or

$$F_i{}^{h}B_b{}^{i}='F_b{}^{a}B_a{}^{h}, \tag{10.9}$$

where

$$B_b{}^{h}=\partial_b\xi^h \tag{10.10}$$

and

$$'F_c{}^{b}\,'F_b{}^{a}=-A^{a}_{c}. \tag{10.11}$$

The induced fundamental metric is given by

$$'g_{cb} = B_c{}^j B_b{}^i g_{ji}. \tag{10.12}$$

On the other hand, we have

$$'F_c{}^{f\prime} F_b{}^{e\prime} g_{fe} = 'F_c{}^f F_b{}^e (B_f{}^j B_e{}^i g_{ji}) = (F_m{}^j B_c{}^m)(F_l{}^i B_b{}^l) g_{ji} = \\ = B_c{}^m B_b{}^l g_{ml} = 'g_{cb}$$

that is

$$'F_c{}^f \, 'F_b{}^e \, 'g_{fe} = 'g_{cb}, \tag{10.13}$$

hence the metric $'g_{cb}$ is Hermitian.

We denote the Christoffel symbols formed from $'g_{cb}$ by

$$'\left\{ {a \atop c\;b} \right\} = \frac{1}{2} 'g^{ae} (\partial_c 'g_{be} + \partial_b 'g_{ce} - \partial_e 'g_{cb}); \tag{10.14}$$

then we have

$$'\left\{ {a \atop c\;b} \right\} = B^a{}_h \left( B_c{}^j B_b{}^i \left\{ {h \atop j\;i} \right\} + \partial_c B_b{}^h \right). \tag{10.15}$$

Then the van der Waerden–Bortolotti derivative of $B_b{}^h$ given by

$$'\nabla_c B_b{}^h = \partial_c B_b{}^h + B_c{}^j B_b{}^i \left\{ {h \atop j\;i} \right\} - B_a{}^h \, '\left\{ {a \atop c\;b} \right\} \tag{10.16}$$

is, as a vector of $K_{2n}$, orthogonal to $K_{2m}$.

We denote $2n$—$2m$ mutually orthogonal unit normals to $K_{2m}$ by $C_x{}^h$, where

$$x,\ y,\ z = m+1,\ \ldots,\ n,\ \overline{m+1},\ \ldots,\ \bar{n},$$

then we have

$$'\nabla_c B_b{}^h = H_{cbx} C_x{}^h, \tag{10.17}$$

where $H_{cbx}$ are the second fundamental tensors of $K_{2m}$ with respect to normals $C_x{}^h$.

These normals satisfy

$$B_c{}^j C_x{}^i g_{ji} = 0 \qquad C_y{}^j C_x{}^i g_{ji} = \delta_{yx}. \tag{10.18}$$

Moreover since we have

$$B_c{}^j F_i{}^h C_x{}^i g_{jh} = -F_{ji} B_c{}^j C_x{}^i = -'F_c{}^a B_{ai} C_x{}^i = 0,$$

$F_i{}^h C_x^i$ are also normal to $K_{2m}$ and we put

$$F_i{}^h C_x^i = 'F_x{}^y C_y{}^h, \tag{10.19}$$

from which

$$'F_z{}^y F_y{}^x = -\delta_z^x. \tag{10.20}$$

The equations of Weingarten are

$$'\nabla_c C_x{}^h = -H_{cx}{}^a B_a{}^h + L_{cxy} C_y{}^h, \tag{10.21}$$

where

$$L_{cxy} = ('\nabla_c C_x^j) C_y^i g_{ji}. \tag{10.22}$$

Now differentiating (10.9) along the subspace $K_{2m}$, we obtain

$$F_i{}^h H_{cbx} C_x^i = ('\nabla_c F_b{}^a) B_a{}^h + 'F_b{}^a H_{cax} C_x{}^h$$

or

$$H_{cbx}{}' F_x{}^y C_y{}^h = ('\nabla_c{}' F_b{}^a) B_a{}^h + 'F_b{}^a H_{cax} C_x{}^h,$$

from which, $B_a{}^h$ and $C_x{}^h$ being linearly independent,

$$'\nabla_c{}' F_b{}^a = 0 \tag{10.23}$$

and

$$H_{cbx}{}' F_x{}^y = 'F_b{}^a H_{cay}. \tag{10.24}$$

Contracting $'g^{cb}$ to (10.24), we find

$$'g^{cb} H_{cbx}{}' F_x{}^y = 0$$

or

$$'g^{cb} H_{cbx} = 0. \tag{10.25}$$

From (10.12) and what follows we have

THEOREM 10.1. *An analytic subspace in a Kähler space is itself a Kähler space.*

From (10.25) we have

THEOREM 10.2. *An analytic subspace in a Kähler space is minimal.*

(J. A. Schouten and K. Yano [3]).

## § 11. Kähler spaces satisfying the axiom of holomorphic planes

We assume that, when there is given a holomorphic plane element at a point, that is, a plane element determined by a certain vector $u^h$ and its transform $F_i^{\ h}u^i$ by $F_i^{\ h}$ at a point, we can always draw a totally geodesic subspace which is tangent to this holomorphic plane element and passes through this point. If this is the case for any holomorphic plane element and for any point in the space, we say that the space admits the axiom of holomorphic planes. We are going to determine such a Kähler space.

We represent the subspace by

$$\xi^h = \xi^h(\eta^a), \qquad (a,\ b,\ c,\ d = 1,2). \tag{11.1}$$

Then the subspace being totally geodesic, we have

$$\partial_c B_b^{\ h} + B_c^{\ j} B_b^{\ i} \left\{ {h \atop j\ i} \right\} - B_a^{\ h} \left\{ {a \atop c\ b} \right\}' = 0 \tag{11.2}$$

from which, as the integrability conditions of this,

$$B_d^{\ k} B_c^{\ j} B_b^{\ i} K_{kji}^{\ \ \ h} = B_a^{\ h\prime} K_{dcb}^{\ \ \ a}, \tag{11.3}$$

where $'K_{dcb}^{\ \ \ a}$ is the Riemann–Christoffel curvature tensor of the subspace.

If we put

$$B_1^{\ h} = u^h, \qquad B_2^{\ h} = F_i^{\ h} u^i,$$

we get from (11.3)

$$F_s^{\ m} u^s u^j u^i K_{mji}^{\ \ \ h} = \alpha u^h + \beta F_i^{\ h} u^i, \tag{11.4}$$

for $d=2$, $c=1$, $b=1$, where

$$\alpha = {}'K_{211}^{\ \ \ 1}, \qquad \beta = {}'K_{211}^{\ \ \ 2},$$

and

$$F_s^{\ m} u^s u^j F_r^{\ l} u^r K_{mjl}^{\ \ \ h} = \lambda u^h + \mu F_i^{\ h} u^i, \tag{11.5}$$

for $d=2$, $c=1$, $b=2$, where

$$\lambda = {}'K_{212}{}^{1}, \qquad \mu = {}'K_{212}{}^{2}.$$

We can assume that $u^h$ is a unit vector and equations (11.4) and (11.5) should be satisfied for any unit vector $u^h$ and for any arbitrary point.

From (11.4), we have

$$(F_s{}^m K_{mji}{}^h - \alpha g_{sj} A_i^h - \beta g_{sj} F_i{}^h) u^s u^j u^i = 0,$$

where $\alpha$ and $\beta$ are independent of $u$, from which, $F_s{}^m K_{mji}{}^h$ being symmetric with respect to $s$ and $j$,

$$F_s{}^m K_{mji}{}^h + F_j{}^m K_{mis}{}^h + F_i{}^m K_{msj}{}^h$$
$$= \alpha(g_{sj} A_i^h + g_{ji} A_s^h + g_{is} A_j^h) + \beta(g_{sj} F_i{}^h + g_{ji} F_s{}^h + g_{is} F_j{}^h).$$

Transvecting this with $F_k{}^s$, we find

$$-K_{kji}{}^h + F_k{}^s F_j{}^m K_{mis}{}^h + F_k{}^s F_i{}^m K_{mjs}{}^h$$
$$= \alpha(F_{kj} A_i^h + g_{ji} F_k{}^h + F_{ki} A_j^h) + \beta(F_{kj} F_i{}^h - g_{ji} A_k^h + F_{ki} F_j{}^h).$$

Taking the skew-symmetric part of this with respect to $k$ and $j$, and using the relation

$$F_j{}^m F_k{}^s K_{mis}{}^h - F_k{}^m F_j{}^s K_{mis}{}^h = -K_{kji}{}^h$$

we find

$$-4K_{kji}{}^h = \alpha(2F_{kj} A_i^h + g_{ji} F_k{}^h - g_{ki} F_j{}^h + F_{ki} A_j^h - F_{ji} A_k^h)$$
$$+ \beta(2F_{kj} F_i{}^h - g_{ji} A_k^h + g_{ki} A_j^h + F_{ki} F_j{}^h - F_{ji} F_k{}^h).$$

Contracting this equation with respect to $h$ and $i$, we find

$$0 = 4(n+1)\alpha F_{kj} + 0,$$

from which $\alpha = 0$, thus we have

$$K_{kji}{}^h = \frac{\beta}{4}\,[(A_k^h g_{ji} - A_j^h g_{ki}) + (F_k{}^h F_{ji} - F_j{}^h F_{ki}) - 2F_{kj} F_i{}^h].$$

Substituting this into the left hand side of (11.5) we find

$$F_s{}^m u^s u^j F_r{}^l u^r K_{mjl}{}^h = -\beta u^h.$$

Thus (11.5) is satisfied and

$$\lambda = -\beta \quad \text{and} \quad \mu = 0.$$

Thus

THEOREM 11.1. *If a Kähler space admits the axiom of holomorphic planes, then the space is of constant holomorphic sectional curvature.*†

† K. Yano and I. Mogi [1], [2].

CHAPTER V

# ALMOST COMPLEX SPACES†

## § 1. Almost complex spaces

Let us consider a $2n$-dimensional space $M_{2n}$ of differentiability class $C^{r+1}$. If there exists a mixed tensor $F_i{}^h(\xi)$ of class $C^r$ which satisfies

$$F_j{}^i F_i{}^h = -A_j^h, \tag{1.1}$$

we say that the tensor $F_i{}^h$ gives an *almost complex structure* to the space and we call the space an *almost complex space*. (C. Ehresmann [1]).

We consider now the proper value $a$ of the tensor $F_i{}^h$

$$F_i{}^h v^i = a v^h, \tag{1.2}$$

$v^h$ being a proper vector corresponding to $a$. Multiplying (1.2) by $F_h{}^k$ and contracting with respect to $h$, we find

$$-v^k = a F_h{}^k v^h = a^2 v^k,$$

from which $a^2 = -1$. Since the tensor $F_i{}^h$ is real, we see that there are $n$ proper values $+i$ and $n$ proper values $-i$.

We denote by $v_\lambda^h$ $n$ linearly independent proper vectors corresponding to the proper value $+i$ and by $v_{\bar\lambda}^h$ $n$ linearly independent proper vectors corresponding to the proper value

† See E. Calabi [2]; C. Ehresmann [1], [2]; T. Fukami and S. Ishihara [1]; K. Kodaira and D. C. Spencer [1]; G. Legrand [1]; P. Libermann [1]; A. Lichnerowicz [14], [21]; K. Yano [8], [10].

See also W. M. Boothby and H. C. Wang [1]; J. W. Gray [1]; C. J. Hsu [1], [2], [3]; S. Sasaki [2].

$-i$, where $\varkappa$, $\lambda$, $\mu$, ... run over the range 1, 2, 3,..., $n$. We can assume that $v^h_\lambda$ and $v^h_{\bar{\lambda}}$ are complex conjugate to each other.

Now $2n$ vectors $v^h_\lambda$ and $v^h_{\bar{\lambda}}$ are linearly independent. Because, if we have relations of the form

$$p^\lambda v^i_\lambda + p^{\bar{\lambda}} v^i_{\bar{\lambda}} = 0,$$

connecting the $2n$ vectors $v^h_\lambda$ and $v^h_{\bar{\lambda}}$, $p^\lambda$ and $p^{\bar{\lambda}}$ being not all zero, then multiplying the equation by $F_i{}^h$ and contracting, we get

$$ip^\lambda v^h_\lambda - ip^{\bar{\lambda}} v^h_{\bar{\lambda}} = 0$$

on using the assumptions $F_i{}^h v^i_\lambda = +iv^h_\lambda$ and $F_i{}^h v^i_{\bar{\lambda}} = -iv^h_{\bar{\lambda}}$. From these equations, we find

$$p^\lambda v^h_\lambda = 0, \qquad p^{\bar{\lambda}} v^h_{\bar{\lambda}} = 0,$$

which contradict our assumption that $v^h_\lambda$ and $v^h_{\bar{\lambda}}$ are respectively, linearly independent.

Now $n$ vectors $v^h_\lambda$ determine a distribution $\pi_n$ of complex dimension $n$ and the $n$ vectors $v^h_{\bar{\lambda}}$ determine a distribution $\bar{\pi}_n$ of complex dimension $n$ too. The two distributions $\pi_n$ and $\bar{\pi}_n$ have no common direction and they are complementary. The space spanned by $\pi_n$ and $\bar{\pi}_n$ is the complexification of the tangent space $T_{2n}$ to $M_{2n}$.

Since the $2n$ vectors $v^h_\lambda$ and $v^h_{\bar{\lambda}}$ are linearly independent, the determinant of the matrix $(v^h_\lambda, v^h_{\bar{\lambda}})$ is different from zero, and consequently we can define the inverse $(v^\varkappa_i, v^{\bar{\varkappa}}_i)$ of this matrix. Then we have

$$v^i_\lambda v^\varkappa_i = \delta^\varkappa_\lambda, \quad v^i_\lambda v^{\bar{\varkappa}}_i = 0, \quad v^i_{\bar{\lambda}} v^\varkappa_i = 0, \quad v^i_{\bar{\lambda}} v^{\bar{\varkappa}}_i = \delta^{\bar{\varkappa}}_{\bar{\lambda}} \tag{1.3}$$

and

$$v^\lambda_i v^h_\lambda + v^{\bar{\lambda}}_i v^h_{\bar{\lambda}} = A^h_i. \tag{1.4}$$

If we put

$$P_i{}^h = \frac{1}{2}(A^h_i - iF_i{}^h), \qquad Q_i{}^h = \frac{1}{2}(A^h_i + iF_i{}^h), \tag{1.5}$$

we have

$$P_i{}^h v^i = v^h, \qquad Q_i{}^h v^i = 0$$

for a proper vector corresponding to the proper value $+i$, and

$$P_i{}^h v^i = 0, \qquad Q_i{}^h v^i = v^h$$

for a proper vector corresponding to the proper value $-i$, and consequently

$$\left.\begin{aligned} P_i{}^h v^i_\lambda &= v^h_\lambda & Q_i{}^h v^i_\lambda &= 0, \\ P_i{}^h v^i_{\bar\lambda} &= 0, & Q_i{}^h v^i_{\bar\lambda} &= v^h_{\bar\lambda}. \end{aligned}\right\} \tag{1.6}$$

Thus, multiplying $P_i{}^h v^i_{\bar\lambda} = 0$ by $v^{\bar\lambda}_j$ and contracting we get

$$P_i{}^h = v^\lambda_i v^h_\lambda \tag{1.7}$$

by virtue of (1.4) and (1.6). Similarly we get

$$Q_i{}^h = v^{\bar\lambda}_i v^h_{\bar\lambda}. \tag{1.8}$$

From (1.5), (1.7) and (1.8), we get

$$F_i{}^h = i(P_i{}^h - Q_i{}^h) = i(v^\lambda_i v^h_\lambda - v^{\bar\lambda}_i v^h_{\bar\lambda}). \tag{1.9}$$

After all, in an almost complex space, there exists a distribution $\pi_n$ of complex dimension $n$ and another distribution $\bar\pi_n$ of complex dimension $n$ which is conjugate to $\pi_n$. Since $\pi_n$ and $\bar\pi_n$ have no common direction they span a complex linear space of complex dimension $2n$. The projection on $\pi_n$ is given by $P_i{}^h$ and that on $\bar\pi_n$ is given by $Q_i{}^h$, the complex structure $F_i{}^h$ being written as

$$F_i{}^h = i(P_i{}^h - Q_i{}^h).$$

Conversely, suppose that, in a $2n$-dimensional space, there exist a distribution $\pi_n$ of complex dimension $n$ and a distribution $\bar\pi_n$ of complex dimension $n$ which has no common direction with $\pi_n$ and spans with $\pi_n$ a complex linear space of complex dimension $2n$.

We take $n$ linearly independent vectors $v^h_\lambda$ in $\pi_n$, then $n$ vectors $v^h_{\bar\lambda}$ complex conjugate to $v^h_\lambda$ are also linearly independent and span with $v^h_\lambda$ the linear space of complex dimension $2n$. We know that $2n$ vectors $v^h_\lambda$ and $v^h_{\bar\lambda}$ are linearly independent. Thus denoting by $(v^\varkappa_i,\ v^{\bar\varkappa}_i)$ the matrix inverse to $(v^h_\lambda,\ v^h_{\bar\lambda})$, we have (1.3) and (1.4).

Now if we put

$$F_i{}^h = i(v^\lambda_i v^h_\lambda - v^{\bar\lambda}_i v^h_{\bar\lambda}),$$

then $F_i{}^h$ is a real tensor and we have

$$F_j{}^i F_i{}^h = -(v^\lambda_j v^i_\lambda - v^{\bar\lambda}_j v^i_{\bar\lambda})(v^\varkappa_i v^h_\varkappa - v^{\bar\varkappa}_i v^h_{\bar\varkappa}) = -(v^\lambda_j v^h_\lambda + v^{\bar\lambda}_j v^h_{\bar\lambda}),$$

that is,

$$F_j{}^i F_i{}^h = -A^h_j.$$

Thus the space admits an almost complex structure and consequently we have

THEOREM 1.1. *In order that a $2n$-dimensional space* $M_{2n}$ *admits an almost complex structure, it is necessary and sufficient that the space* $M_{2n}$ *contains a distribution* $\pi_n$ *of complex dimension $n$ and a distribution* $\bar\pi_n$, *conjugate to* $\pi_n$, *which has no common direction with* $\pi_n$ *and spans with* $\pi_n$ *a linear space of complex dimension $2n$.*

Now introduce a positive definite Riemannian metric $a_{ji}$ in the almost complex space $M_{2n}$ and put

$$g_{ji} = \frac{1}{2}(a_{ji} + F_j{}^t F_i{}^s a_{ts}), \tag{1.10}$$

then $g_{ji}$ again defines a positive definite Riemannian metric and satisfies

$$F_j{}^t F_i{}^s g_{ts} = g_{ji}. \tag{1.11}$$

Equation (1.11) shows that the operator $F_i{}^h$ applied to contravariant vectors does not change the length of a vector and the angle between two vectors, the length and angle being measured by $g_{ji}$.

If we put

$$F_{ji} = F_j{}^t g_{ti}, \tag{1.12}$$

then we have from (1.1)

$$F_j{}^t F_{ti} = -g_{ji}$$

and from (1.11)

$$F_j{}^t F_{it} = g_{ji}.$$

Comparing these two equations, we find

$$F_{ti} = -F_{it}. \tag{1.13}$$

Equation (1.13) shows that an arbitrary vector $u^h$ and its transform $F_i{}^h u^i$ by $F_i{}^h$ are orthogonal.

Now take a unit vector $u_1^h$ and put

$$u_{n+1}^h = F_i{}^h u_1^i,$$

then $u_{n+1}^h$ is a unit vector orthogonal to $u_1^h$. Take next a unit vector $u_2^h$ orthogonal to $u_1^h$ and $u_{n+1}^h$ and put

$$u_{n+2}^h = F_i{}^h u_2{}^i,$$

then $u_{n+2}^h$ is a unit vector orthogonal not only to $u_2^h$ but also to $u_1^h$ and $u_{n+1}^h$. Because, applying $F_i{}^h$ to three mutually orthogonal vectors $u_1^h$, $u_{n+1}^h$, $u_2^h$, we get $u_{n+1}^h$, $-u_1^h$, $u_{n+2}^h$ which are also mutually orthogonal.

Continuing in this way, we obtain a set of mutually orthogonal unit vectors $u_\lambda^h$, $u_{n+\lambda}^h$ such that

$$u_{n+\lambda}^h = F_i{}^h u_\lambda{}^i .$$

Now if we put

$$g_{cb} = u_c^j u_b^i g_{ji}, \qquad F_{cb} = u_c^j u_b^i F_{ji}, \tag{1.14}$$

then we have

$$g_{cb} = \begin{pmatrix} E_n & 0 \\ 0 & E_n \end{pmatrix}, \qquad F_{cb} = \begin{pmatrix} 0 & E_n \\ -E_n & 0 \end{pmatrix}, \tag{1.15}$$

where $E_n$ denotes the $n \times n$ unit matrix. We call such a set of vectors $u_b^h$ an *adapted frame*.

Take another adapted frame $(\overline{u}_b^h)$ and put

$$\overline{u}_b^h = \gamma_b^a u_a^h,$$

then

$$\bar{g}_{cb}=\gamma_c^f\gamma_b^e g_{fe}\,; \qquad \bar{F}_{cb}=\gamma_c^f\gamma_b^e F_{fe}$$

should have again the form (1.15), from which we conclude that the orthogonal matrix $(\gamma_b^a)$ should be of the form

$$\gamma_b^a=\begin{pmatrix} A_n & B_n \\ -B_n & A_n \end{pmatrix}.$$

This means that $\gamma$ defines a real representation of a unitary transformation. In this case we say that the group of the tangent bundle of the manifold has been reduced to the unitary group U$(n)$.

Conversely suppose that the group of the tangent bundle of a $2n$-dimensional manifold is reduced to the unitary group U$(n)$, then there exists an adapted frame $(u_\lambda^h, u_{n+\lambda}^h)$ subject to a unitary transformation $\bar{u}_b^h=\gamma_b^a u_a^h$. We consider a mixed tensor field $F$ defined by $g_{cb}$ and $F_{cb}$ having components (1.15) with respect to this adapted frame. Since this field has always the same components with respect to any adapted frame, $F$ is globally defined. But

$$F^2=\begin{pmatrix} -E_n & 0 \\ 0 & -E_n \end{pmatrix}=-E_{2n},$$

that is,

$$F_j^{\ i}F_i^{\ h}=-A_j^h.$$

Thus we have

THEOREM 1.2. *A necessary and sufficient condition for a $2n$-dimensional manifold to admit an almost complex structure is that the group of the tangent bundle of the manifold can be reduced to the unitary group* U$(n)$.

## § 2. Integrability of an almost complex structure

Let us consider a $2n$-dimensional manifold $M_{2n}$ with an almost complex structure $F_i^{\ h}(\xi)$. If there exists a system of complex coordinate neighbourhoods $(z^\varkappa)$ with respect to which the tensor $F_i^{\ h}$ has the numerical components

$$F_i^h = \begin{pmatrix} i\delta_\lambda^\varkappa & 0 \\ 0 & -i\delta_{\bar\lambda}^{\bar\varkappa} \end{pmatrix}, \tag{2.1}$$

then we say that the almost complex structure $F_i^{\ h}$ is induced from the complex structure defined by the existence of such a system of complex coordinate neighbourhoods.

Now, suppose that the almost complex structure $F_i^{\ h}$ is induced from a complex structure, then in a complex coordinate system, we have

$$P_i^{\ h} = \frac{1}{2}(A_i^h - iF_i^{\ h}) = \begin{pmatrix} \delta_\lambda^\varkappa & 0 \\ 0 & 0 \end{pmatrix},$$

$$Q_i^{\ h} = \frac{1}{2}(A_i^h + iF_i^{\ h}) = \begin{pmatrix} 0 & 0 \\ 0 & \delta_{\bar\lambda}^{\bar\varkappa} \end{pmatrix}$$

and consequently

$$P_i^{\ h} dz^i = 0 \quad \text{and} \quad Q_i^{\ h} dz^i = 0$$

or

$$dz^\varkappa = 0 \quad \text{and} \quad dz^{\bar\varkappa} = 0$$

are both completely integrable and admit solutions

$$z^\varkappa = \text{const. and } z^{\bar\varkappa} = \text{const.}$$

respectively.

Conversely, suppose that the space is of class $C^\omega$ and the almost complex structure $F_i^{\ h}$ is also of class $C^\omega$, then $F_i^{\ h}(\xi)$ can be regarded as functions of the complex variables $\xi^h$. We assume moreover that

$$P_i^{\ h} d\xi^i = 0 \tag{2.2}$$

are completely integrable. Then the rank of $P_i^{\ h}$ being $n$, the above equations admit $n$ independent solutions $z^\varkappa(\xi)$ such that

$$\frac{\partial z^\varkappa}{\partial \xi^i} d\xi^i = 0$$

for any $d\xi^i$ satisfying (2.2). But we have

$$d\xi^i = P_j^{\ i} d\xi^j + Q_j^{\ i} d\xi^j$$

and consequently

$$\frac{\partial z^\varkappa}{\partial \xi^i} Q_j^{\ i} d\xi^j = 0$$

for arbitrary $d\xi^j$, from which

$$\frac{\partial z^\varkappa}{\partial \xi^i} Q_j^{\ i} = 0,$$

that is,

$$\frac{\partial z^\varkappa}{\partial \xi^i} + i \frac{\partial z^\varkappa}{\partial \xi^h} F_i^{\ h} = 0.$$

Similarly we get, from the complete integrability of $Q_i^{\ h} d\xi^i = 0$,

$$\frac{\partial z^{\bar\varkappa}}{\partial \xi^i} - i \frac{\partial z^{\bar\varkappa}}{\partial \xi^h} F_i^{\ h} = 0.$$

From these equations, we get

$$\frac{\partial z^\varkappa}{\partial \xi^h} \frac{\partial \xi^i}{\partial z^\lambda} F_i^{\ h} = i\delta_\lambda^\varkappa, \qquad \frac{\partial z^{\bar\varkappa}}{\partial \xi^h} \frac{\partial \xi^i}{\partial z^\lambda} F_i^{\ h} = 0,$$

$$\frac{\partial z^\varkappa}{\partial \xi^h} \frac{\partial \xi^i}{\partial z^{\bar\lambda}} F_i^{\ h} = 0, \qquad \frac{\partial z^{\bar\varkappa}}{\partial \xi^h} \frac{\partial \xi^i}{\partial z^{\bar\lambda}} F_i^{\ h} = -i\delta_{\bar\lambda}^{\bar\varkappa},$$

that is, the tensor $F_i^{\ h}$ has numerical components (2.1) in the complex coordinate system $(z^\varkappa, z^{\bar\varkappa})$. Thus the almost complex structure $F_i^{\ h}$ is induced from a complex structure. Thus we have

THEOREM 2.1. *In order that an almost complex structure $F_i^{\ h}$ of class $C^\infty$ be induced from a complex structure, it is necessary and sufficient that*

$$P_i^{\ h} d\xi^i = 0 \quad \textit{and} \quad Q_i^{\ h} d\xi^i = 0$$

*be both completely integrable.*

For this reason, if an almost complex structure is induced by a complex structure, we say that the almost complex structure is *integrable*.

We can express the above fact also in the following way.

Since the space $M_{2n}$ and the almost complex structure $F_i^{\ h}(\xi)$ are both of class $C^\omega$, we can regard the tensor $F_i^{\ h}(\xi)$ as that of a space in $C_n \times \overline{C_n}$.

Now, if an almost complex structure $F_i^{\ h}$ is induced by a complex structure then the equations $P_i^{\ h} d\xi^i = 0$ are completely integrable and admit $n$ independent solutions, $z^\kappa(\xi) = c^\kappa$ and consequently, as follows from $P_i^{\ h} d\xi^i = 0$ or equivalently $F_i^{\ h} d\xi^i = -id\xi^h$, $n$ proper vectors $v^h$ of $F_i^{\ h}$ corresponding to the proper value $-i$ are tangent to the family of subspaces $z^\kappa(\xi) = c^\kappa$ of complex dimension $n$. Similarly we can see that $n$ proper vectors $v_\lambda^h$ of $F_i^{\ h}$ corresponding to the proper value $+i$ are tangent to the other family $z^{\bar\kappa}(\xi) = c^{\bar\kappa}$ of subspaces of complex dimension $n$.

Conversely, if the space $M_{2n}$ admits two families of subspaces of complex dimension $n$ and if $n$ proper vectors $v_{\bar\lambda}^h$ of $F_i^{\ h}$ corresponding to the proper value $-i$ are tangent to the one family and $n$ proper vectors $v_\lambda^h$ of $F_i^{\ h}$ corresponding to the proper value $+i$ are tangent to the other family, then the equations

$$P_i^{\ h} d\xi^i = 0 \quad \text{and} \quad Q_i^{\ h} d\xi^i = 0$$

are completely integrable and consequently the almost complex structure $F_i^{\ h}$ is induced by a complex structure. Hence we have

THEOREM 2.2. *In order that an almost complex structure of class $C^\omega$ be induced by a complex structure, it is necessary and sufficient that $n$ proper vectors $v_{\bar\lambda}^h$ of $F_i^{\ h}$ corresponding to the proper value $-i$ are tangent to a family of subspaces of complex dimension $n$ and $n$ proper vectors $v_\lambda^h$ of $F_i^{\ h}$ corresponding to the value $+i$ are tangent to another family of subspaces of complex dimension $n$.*

Now the almost complex structure $F_i^{\ h}$ being of class $C^\omega$, in order that equations $P_i^{\ h}d\xi^i=0$ be completely integrable, it is necessary and sufficient that

$$d(P_i^{\ h}d\xi^i)=\frac{1}{2}(\partial_j P_i^{\ h}-\partial_i P_j^{\ h})d\xi^j\wedge d\xi^i=0$$

be satisfied by any $d\xi^i$ satisfying $P_i^{\ h}d\xi^i=0$, from which

$$T_{ji}^{\ \ h}\stackrel{def}{=}\frac{1}{2}Q_j^{\ t}Q_i^{\ s}(\partial_t P_s^{\ h}-\partial_s P_t^{\ h})=0. \tag{2.3}$$

Similarly, a necessary and sufficient condition for $Q_i^{\ h}d\xi^i=0$ to be completely integrable is given by

$$\bar{T}_{ji}^{\ \ h}\stackrel{def}{=}\frac{1}{2}P_j^{\ t}P_i^{\ s}(\partial_t Q_s^{\ h}-\partial_s Q_t^{\ h})=0. \tag{2.4}$$

Thus we have

THEOREM 2.3. *In order that an almost complex structure $F_i^{\ h}$ of class $C^\omega$ be integrable, it is necessary and sufficient that*

$$T_{ji}^{\ \ h}=0 \quad \text{and} \quad \bar{T}_{ji}^{\ \ h}=0. \tag{2.5}$$

Now, suppose that we have a differential form

$$w=\frac{1}{p!}w_{i_1i_2\ldots i_p}d\xi^{i_1}\wedge d\xi^{i_2}\wedge\ldots\wedge d\xi^{i_p}.$$

We denote by $Pw$ the form obtained from $w$ by replacing all $d\xi^i$ by $P_j^{\ i}d\xi^j$ and by $Qw$ the form obtained from $w$ by replacing all $d\xi^i$ by $Q_j^{\ i}d\xi^j$. Then for a linear form

$$w=w_id\xi^i$$

we have

$$QdPw=T_{ji}^{\ \ h}w_hd\xi^j\wedge d\xi^i$$

and

$$PdQw=\bar{T}_{ji}^{\ \ h}w_hd\xi^j\wedge d\xi^i,$$

from which we have

THEOREM 2.4. *In order that an almost complex structure $F_i{}^h$ of class $C^\infty$ be integrable, it is necessary and sufficient that*

$$QdPw=0 \quad \text{and} \quad PdQw=0$$

*for any linear form w.*

Now substituting

$$P_i{}^h=\frac{1}{2}(A_i^h-iF_i{}^h), \qquad Q_i{}^h=\frac{1}{2}(A_i^h+iF_i{}^h)$$

into

$$T_{ji}{}^h=\frac{1}{2}Q_j{}^cQ_i{}^b(\partial_cP_b{}^h-\partial_bP_c{}^h),$$

we find

$$T_{ji}{}^h=\frac{1}{16}[F_j{}^t(\partial_tF_i{}^h-\partial_iF_t{}^h)-F_i{}^t(\partial_tF_j{}^h-\partial_jF_t{}^h)]$$

$$-\frac{i}{16}[(\partial_jF_i{}^h-\partial_iF_j{}^h)-F_j{}^tF_i{}^s(\partial_tF_s{}^h-\partial_sF_t{}^h)].$$

Thus, if we put

$$N_{ji}{}^h=F_j{}^t(\partial_tF_i{}^h-\partial_iF_t{}^h)-F_i{}^t(\partial_tF_j{}^h-\partial_jF_t{}^h), \tag{2.6}$$

we have

$$F_j{}^tN_{ti}{}^h=-(\partial_jF_i{}^h-\partial_iF_j{}^h)+F_j{}^tF_i{}^s(\partial_tF_s{}^h-\partial_sF_t{}^h), \tag{2.7}$$

and consequently

$$T_{ji}{}^h=\frac{1}{16}(N_{ji}{}^h+iF_j{}^tN_{ti}{}^h). \tag{2.8}$$

Similarly we have

$$\overline{T}_{ji}{}^h=\frac{1}{16}(N_{ji}{}^h-iF_j{}^tN_{ti}{}^h). \tag{2.9}$$

Thus, from Theorem 2.3, (2.8) and (2.9), we get

THEOREM 2.5. *In order that an almost complex structure $F_i{}^h$ of class $C^\infty$ be integrable, it is necessary and sufficient* *that*

$$N_{ji}{}^h = 0. \qquad (2.10)$$

(E. Calabi and D. C. Spencer [1]; B. Eckmann and A. Frölicher [1]; W. V. D. Hodge [1]; K. Yano [8]; J. A. Schouten and K. Yano [2]).

A. Newlander and L. Nirenberg [1] proved that Theorem 2.5 holds also for the case where the almost complex structure is of class $C^r$.†

The tensor $N_{ji}{}^h$ defined in (2.6) is called *Nijenhuis' tensor*. (A. Nijenhuis [1], J. A. Schouten [2]).

Now when a differential form

$$w = \frac{1}{p!} w_{i_1 i_2 \dots i_p} d\xi^{i_1} \wedge d\xi^{i_2} \wedge \dots \wedge d\xi^{i_p}$$

is given, we denote by $Fw$ the form obtained from $w$ by replacing all $d\xi^i$ by $F_j{}^i d\xi^j$. Then for a scalar field $f$, we have

$$(dFdF)f = \frac{1}{2}(\partial_j F_i{}^h \partial_h f + F_i{}^h \partial_j \partial_h f - \partial_i F_j{}^h \partial_h f - F_j{}^h \partial_i \partial_h f) d\xi^j \wedge d\xi^i$$

and

$$(FdFd)f = \frac{1}{2} F_j{}^t F_i{}^s (\partial_t F_s{}^h \partial_h f + F_s{}^h \partial_t \partial_h f - \partial_s F_t{}^h \partial_h f - F_t{}^h \partial_s \partial_h f) d\xi^j \wedge d\xi^i,$$

and consequently

$$(dFdF - FdFd) \cdot f = -\frac{1}{2} F_j{}^t N_{ti}{}^h \partial_h f d\xi^j \wedge d\xi^i.$$

Hence we have

THEOREM 2.6. *In order that an almost complex structure $F_i^h$ of class $C^\infty$ be integrable, it is necessary and sufficient that*

$$(dFdF - FdFd)f = 0$$

*holds good for any scalar $f$.*

W. V. D. Hodge [3] proved that Theorem 2.6 holds good also for any differential form $f$.

---

† Nijenhuis, A. and W. B. Woolf [1]

## § 3. Another form of the integrability condition of an almost complex structure

If an almost complex structure is induced by a complex structure, then there exists a complex coordinate system $(z^\varkappa)$ in which the almost complex structure tensor $F_i^{\ h}$ has the components

$$F_i^{\ h} = \begin{pmatrix} i\delta_\lambda^\varkappa & 0 \\ 0 & -i\delta_{\bar\lambda}^{\bar\varkappa} \end{pmatrix}, \tag{3.1}$$

and consequently, as proper vectors $v_\lambda^h$ and $v_{\bar\lambda}^h$ corresponding to the proper values $+i$ and $-i$ respectively, we can take

$$v_\lambda^h = \delta_\lambda^h, \qquad v_{\bar\lambda}^h = \delta_{\bar\lambda}^h.$$

Then the partial differential equations

$$X_\lambda f \overset{def}{=} v_\lambda^h \partial_h f = 0 \quad \text{and} \quad X_{\bar\lambda} f \overset{def}{=} v_{\bar\lambda}^h \partial_h f = 0$$

are completely integrable and admit the solutions $z^{\bar\varkappa}$ and $z^\varkappa$ respectively.

Conversely, suppose that, in an almost complex space with $F_i^{\ h}$ of class $C^\omega$, the systems of partial differential equations

$$X_\lambda f = v_\lambda^h \partial_h f = 0 \quad \text{and} \quad X_{\bar\lambda} f = v_{\bar\lambda}^h \partial_h f = 0$$

are both completely integrable and admit solutions $z^{\bar\varkappa}$ and $z^\varkappa$ respectively. Then we have

$$v_\lambda^h \frac{\partial z^{\bar\varkappa}}{\partial \xi^h} = 0 \tag{3.2}$$

and

$$v_{\bar\lambda}^h \frac{\partial z^\varkappa}{\partial \xi^h} = 0. \tag{3.3}$$

Multiplying (3.2) by $v_i^\lambda$ and taking account of $v_i^\lambda v_\lambda^h = P_i^{\ h} =$

$=\frac{1}{2}(A_i^h - iF_i^h)$, we find

$$\frac{\partial z^{\bar{\varkappa}}}{\partial \xi^i} - i\frac{\partial z^{\bar{\varkappa}}}{\partial \xi^h} F_i{}^h = 0.$$

We get similarly from (3.3)

$$\frac{\partial z^{\varkappa}}{\partial \xi^i} + i\frac{\partial z^{\varkappa}}{\partial \xi^h} F_i{}^h = 0.$$

The equations show that, in the complex coordinate system $(z^{\varkappa}, z^{\bar{\varkappa}})$, the tensor $F_i{}^h$ has numerical components (3.1). Hence we have

THEOREM 3.1. *In order that an almost complex structure* $F_i{}^h$ *of class* $C^\omega$ *be induced by a complex structure, it is necessary and sufficient that the systems of partial differential equations*

$$X_\lambda f = 0 \quad \textit{and} \quad X_{\bar{\lambda}} f = 0$$

*be completely integrable.*

Consider now $n$ linearly independent proper vectors $v_\lambda^h$ of $F_i{}^h$ corresponding to the proper values $+i$ and $v_{\bar{\lambda}}^h$ corresponding to the proper values $-i$. We denote these $2n$ linearly independent vectors by

$$v_a^h = (v_\lambda^h,\ v_{\bar{\lambda}}^h) \tag{3.4}$$

where the indices $a, b, c, \ldots$ run over the range $1, 2, \ldots n;\ \bar{1}, \bar{2}, \ldots, \bar{n}$.

If we put

$$X_a f = v_a^h \partial_h f, \tag{3.5}$$

we have

$$(X_c X_b - X_b X_c) f = \Omega_{cb}{}^a X_a f, \tag{3.6}$$

where

$$\Omega_{cb}{}^a = (v_c^t \partial_t v_b^r - v_b^t \partial_t v_c^r) v_r^a, \tag{3.7}$$

$v_r^a$ being elements of the matrix inverse to $v_a^h$.

The quantity $\Omega_{cb}{}^{a}$ defined by (3.7) is called the *non-holonomy object* of the set of $2n$ linearly independent vectors $v_a^h$.

From (3.6), we find

$$(X_{\mu}X_{\lambda}-X_{\lambda}X_{\mu})f=\Omega_{\mu\lambda}{}^{\kappa}X_{\kappa}f+\Omega_{\mu\lambda}{}^{\bar\kappa}X_{\bar\kappa}f$$

and

$$(X_{\bar\mu}X_{\bar\lambda}-X_{\bar\lambda}X_{\bar\mu})f=\Omega_{\bar\mu\bar\lambda}{}^{\kappa}X_{\kappa}f+\Omega_{\bar\mu\bar\lambda}{}^{\bar\kappa}X_{\bar\kappa}f.$$

Thus a necessary and sufficient condition that $X_{\lambda}f=0$ and $X_{\bar\lambda}f=0$ are completely integrable is given by

$$\Omega_{\mu\lambda}{}^{\bar\kappa}=0 \quad \text{and} \quad \Omega_{\bar\mu\bar\lambda}{}^{\kappa}=0 \tag{3.8}$$

respectively, and hence we have

THEOREM 3.2. *In order that an almost complex structure $F_i{}^h$ of class $C^\omega$ be integrable, it is necessary and sufficient that $\Omega_{\mu\lambda}{}^{\bar\kappa}=0$ and $\Omega_{\bar\mu\bar\lambda}{}^{\kappa}=0$.* (K. Yano [10]).

This theorem may also be obtained in the following way.

Following Theorem 2.1, in order that an almost complex structure $F_i{}^h$ of class $C^\omega$ be induced by a complex structure, it is necessary and sufficient that

$$P_i{}^h d\xi^i=0 \quad \text{and} \quad Q_i{}^h d\xi^i=0$$

are completely integrable.

But these equations can be written respectively as

$$v_i^{\lambda}v_{\lambda}^h d\xi^i=0 \quad \text{and} \quad v_i^{\bar\lambda}v_{\bar\lambda}^h d\xi^i=0,$$

from which, $v_{\lambda}^h$ and $v_{\bar\lambda}^h$ being linearly independent, we have

$$w^{\kappa}=v_i^{\kappa}d\xi^i=0 \quad \text{and} \quad w^{\bar\kappa}=v_i^{\bar\kappa}d\xi^i=0.$$

Applying exterior differentiation, we find

$$dw^{\kappa}=\frac{1}{2}(\partial_j v_i^{\kappa}-\partial_i v_j^{\kappa})d\xi^j\wedge d\xi^i \tag{3.9}$$

and

$$dw^{\bar{\varkappa}} = \frac{1}{2}(\partial_j v_i^{\bar{\varkappa}} - \partial_i v_j^{\bar{\varkappa}}) d\xi^j \wedge d\xi^i \tag{3.10}$$

respectively. On the other hand, $d\xi^i$ may be written as

$$d\xi^i = A_j^i d\xi^j = (P_j{}^i + Q_j{}^i) d\xi^j = (v_j^\lambda v_\lambda^i + v_j^{\bar{\lambda}} v_{\bar{\lambda}}^i) d\xi^j = v_\lambda^i w^\lambda + v_{\bar{\lambda}}^i w^{\bar{\lambda}},$$

and consequently, from (3.9) and (3.10), we get

$$dw^\varkappa = \frac{1}{2}(\Omega_{\mu\lambda}{}^\varkappa w^\mu \wedge w^\lambda + \Omega_{\mu\bar{\lambda}}{}^\varkappa w^\mu \wedge w^{\bar{\lambda}} + \Omega_{\bar{\mu}\lambda}{}^\varkappa w^{\bar{\mu}} \wedge w^\lambda + \Omega_{\bar{\mu}\bar{\lambda}}{}^\varkappa w^{\bar{\mu}} \wedge w^{\bar{\lambda}}) \tag{3.11}$$

and

$$dw^{\bar{\varkappa}} = \frac{1}{2}(\Omega_{\mu\lambda}{}^{\bar{\varkappa}} w^\mu \wedge w^\lambda + \Omega_{\mu\bar{\lambda}}{}^{\bar{\varkappa}} w^\mu \wedge w^{\bar{\lambda}} + \Omega_{\bar{\mu}\lambda}{}^{\bar{\varkappa}} w^{\bar{\mu}} \wedge w^\lambda + \Omega_{\bar{\mu}\bar{\lambda}}{}^{\bar{\varkappa}} w^{\bar{\mu}} \wedge w^{\bar{\lambda}}) \tag{3.12}$$

respectively. Thus, following a theorem of Frobenius, a necessary and sufficient condition that $w^\varkappa = 0$ and $w^{\bar{\varkappa}} = 0$ be completely integrable is given by

$$\Omega_{\bar{\mu}\bar{\lambda}}{}^\varkappa = 0 \quad \text{and} \quad \Omega_{\mu\lambda}{}^{\bar{\varkappa}} = 0$$

respectively. This proves Theorem 3.2.

Now, substituting

$$P_i{}^h = v_i^\lambda v_\lambda^h \quad \text{and} \quad Q_i{}^h = v_i^{\bar{\lambda}} v_{\bar{\lambda}}^h$$

into (2.3) and (2.4), we find

$$T_{ji}{}^h = -\frac{1}{2} v_j^{\bar{\mu}} v_i^{\bar{\lambda}} v_\varkappa^h \Omega_{\bar{\mu}\bar{\lambda}}{}^\varkappa \tag{3.13}$$

and

$$\bar{T}_{ji}{}^h = -\frac{1}{2} v_j^\mu v_i^\lambda v_{\bar{\varkappa}}^h \Omega_{\mu\lambda}{}^{\bar{\varkappa}} \tag{3.14}$$

respectively, from which

$$\Omega_{\bar{\mu}\bar{\lambda}}{}^\varkappa = -2 v_{\bar{\mu}}^j v_{\bar{\lambda}}^i v_h^\varkappa T_{ji}{}^h \tag{3.15}$$

and

$$\Omega_{\mu\lambda}{}^{\bar{\varkappa}} = -2v_{\mu}^{j}v_{\lambda}^{i}v_{h}^{\bar{\varkappa}}\bar{T}_{ji}{}^{h}. \tag{3.16}$$

These equations show the equivalence of

$$T_{ji}{}^{h}=0, \qquad \bar{T}_{ji}{}^{h}=0$$

and

$$\Omega_{\mu\lambda}{}^{\bar{\varkappa}}=0, \qquad \Omega_{\bar{\mu}\bar{\lambda}}{}^{\varkappa}=0.$$

Of course, these are also equivalent to

$$N_{ji}{}^{h}=0.$$

## § 4. Almost Hermite and almost Kähler spaces

Let us consider a $2n$-dimensional space of class $C^r$ with the almost complex structure $F_i^h$

$$F_j^i F_i^h = -A_j^h, \tag{4.1}$$

and assume that there is given, in $M_{2n}$, a positive definite Riemannian metric

$$ds^2 = g_{ji}(\xi)d\xi^j d\xi^i. \tag{4.2}$$

If the Riemannian metric satisfies

$$F_j^t F_i^s g_{ts} = g_{ji}, \tag{4.3}$$

then we call this Riemannian metric a *Hermite metric* in an almost complex space $M_{2n}$ and call an *almost Hermite space* an almost complex space with a Hermite metric.

If we put

$$F_{ji} = F_j^t g_{ti}, \tag{4.4}$$

we get, from (4.1) and (4.3),

$$F_j^i F_{ih} = -g_{jh} \quad \text{and} \quad F_j^i F_{hi} = g_{jh},$$

from which

$$F_{ji} = -F_{ij}. \tag{4.5}$$

Conversely, if a skew-symmetric tensor $F_{ji}$ satisfies

$$F_j^{\ t}F_i^{\ s}g_{ts}=g_{ji} \quad (F_j^{\ t}=F_{ji}g^{it})$$

hen we have

$$F_j^{\ t}F_t^{\ i}=-A_j^i$$

and hence $F_j^{\ t}$ defines an almost complex structure.

Now, if a Hermite metric $g_{ji}$ in an almost complex space satisfies

$$\nabla_j F_i^{\ h}=0 \quad \text{or} \quad \nabla_j F_{ih}=0, \tag{4.6}$$

then we call this Hermite metric a *Kähler metric* in $M_{2n}$, where $\nabla_k$ denotes the covariant differentiation with respect to the Christoffel symbols $\left\{ {h \atop j\ i} \right\}$ formed with $g_{ji}$.

On the other hand, the Nijenhuis tensor $N_{ji}^{\ \ h}$ can be written as

$$N_{ji}^{\ \ h}=F_j^{\ t}(\nabla_t F_i^{\ h}-\nabla_i F_t^{\ h})-F_i^{\ t}(\nabla_t F_j^{\ h}-\nabla_j F_t^{\ h}), \tag{4.7}$$

and consequently, if a Hermite metric in an almost complex space is Kähler then $N_{ji}^{\ \ h}=0$ and the almost complex space is a complex space and is a Kähler space.

Consider now a $2n$-dimensional irreducible Riemannian space with positive definite metric $ds^2=g_{ji}d\xi^j d\xi^i$ and suppose that there exists a skew-symmetric tensor field $F_{ji}$ of rank $2n$ such that

$$\nabla_k F_{ji}=0. \tag{4.8}$$

Then defining $F_j^{\ i}=F_{jt}g^{ti}$, we have

$$\nabla_k(F_j^{\ t}F_i^{\ s}g_{ts})=0,$$

and consequently, $F_j^{\ t}F_i^{\ s}g_{ts}$ being also positive definite,

$$F_j^{\ t}F_i^{\ s}g_{ts}=c^2g_{ji}$$

by virtue of the irreducibility of the space, where $c^2$ is a positive constant. Thus writing $F_{ji}$ and $F_j^{\ i}$ instead of

$\frac{1}{c} F_{ji}$ and $\frac{1}{c} F_j{}^i$ respectively, we get

$$F_j{}^t F_i{}^s g_{ts} = g_{ji}$$

and moreover

$$F_{ji} = -F_{ij}, \qquad \nabla_k F_{ji} = 0.$$

Thus the space is Kähler, and we have

THEOREM 4.1. *If a 2n-dimensional irreducible Riemannian space with positive definite metric contains a skew-symmetric tensor field of rank 2n whose covariant derivative vanishes, the space is Kählerian.*

Now coming back to an almost general Hermite space $M_{2n}$, consider a vector $u^i$ and its transform $F_i{}^h u^i$ by $F_i{}^h$. Since

$$g_{ji}(F_t{}^j u^t)(F_s{}^i u^s) = g_{ts} u^t u^s,$$

the vectors $u^h$ and $F_i{}^h u^i$ have the same length, and since

$$g_{ji} u^j (F_t{}^i u^t) = F_{tj} u^t u^j = 0,$$

the vectors $u^h$ and $F_i{}^h u^i$ are orthogonal to each other.

Take a unit vector $u_1^h$, then

$$u_{n+1}^h = F_i{}^h u_1^i$$

is also a unit vector and is orthogonal to $u_1^h$. Then take a unit vector $u_2^h$ which is orthogonal to $u_1^h$ and $u_{n+1}^h$; then

$$u_{n+2}^h = F_i{}^h u_2^i$$

is also a unit vector and is orthogonal to $u_1^h$, $u_2^h$ and $u_{n+1}^h$, by virtue of

$$g_{ts} u_{n+2}^t u_1^s = g_{ts} F_i{}^t u_2^i u_1^s = -g_{ts} u_2^t F_i{}^s u_1^i = -g_{ts} u_2^t u_{n+1}^s = 0$$

and

$$g_{ts} u_{n+2}^t u_{n+1}^s = g_{ts} F_j{}^t u_2^j F_i{}^s u_1^i = g_{ji} u_2^j u_1^i = 0.$$

Continuing in this way, we can find $2n$ mutually orthogonal vectors $u_\lambda^h$ and $u_{\bar\lambda}^h$ which satisfy

$$F_i{}^h u_\lambda^i = +u_{\bar\lambda}^h \quad \text{and} \quad F_i{}^h u_{\bar\lambda}^i = -u_\lambda^h. \tag{4.9}$$

Now, putting

$$v^h_\lambda = \frac{1}{\sqrt{2}}(u^h_\lambda - iu^h_{\bar\lambda}), \qquad v^h_{\bar\lambda} = \frac{1}{\sqrt{2}}(u^h_\lambda + iu^h_{\bar\lambda}), \tag{4.10}$$

we find

$$F_i^{\ h} v^i_\lambda = +iv^h_\lambda, \qquad F_i^{\ h} v^i_{\bar\lambda} = -iv^h_{\bar\lambda}, \tag{4.11}$$

and

$$g_{ji} v^j_\mu v^i_\lambda = 0, \qquad g_{ji} v^j_\mu v^i_{\bar\lambda} = \delta_{\mu\lambda}, \qquad g_{ji} v^j_{\bar\mu} v^i_{\bar\lambda} = 0. \tag{4.12}$$

The last equations show that the plane $\pi_n$ spanned by $v^h_\lambda$ and the plane $\bar\pi_n$ spanned by $v^h_{\bar\lambda}$ are both null planes.

Denoting by $(v^\lambda_i, v^{\bar\lambda}_i)$ the matrix inverse to $(v^h_\lambda, v^h_{\bar\lambda})$, we obtain

$$g_{ji} = v^\lambda_j v^{\bar\lambda}_i + v^{\bar\lambda}_j v^\lambda_i \tag{4.13}$$

and

$$F_i^{\ h} = i(v^\lambda_i v^h_\lambda - v^{\bar\lambda}_i v^h_{\bar\lambda}), \tag{4.14}$$

and consequently

$$F_{ji} = i(v^\lambda_i v^{\bar\lambda}_j - v^{\bar\lambda}_j v^\lambda_i). \tag{4.15}$$

The equation (4.13) shows that

$$ds^2 = g_{ji} d\xi^j d\xi^i = 2w^\lambda w^{\bar\lambda} \tag{4.16}$$

and the equation (4.15) that

$$F_{ji} d\xi^j \wedge d\xi^i = 2iw^\lambda \wedge w^{\bar\lambda}. \tag{4.17}$$

Now, we consider the so-called non-holonomic coordinate system defined by $v^h_a = (v^h_\lambda, v^h_{\bar\lambda})$, and put

$$g_{cb} = g_{ji} v^j_c v^i_b,$$
$$F_{cb} = F_{ji} v^j_c v^i_b;$$

then we obtain

$$g_{cb} = \begin{pmatrix} 0 & \delta_{\mu\lambda} \\ \delta_{\mu\lambda} & 0 \end{pmatrix} \tag{4.18}$$

and

$$F_{cb}=\begin{pmatrix} 0 & i\delta_{\mu\lambda} \\ -i\delta_{\mu\lambda} & 0 \end{pmatrix}. \tag{4.19}$$

Now the components $\Gamma^a_{cb}$ of the Riemannian connexion $\left\{ {h \atop j\ i} \right\}$ with respect to this non-holonomic coordinate system are given by

$$\Gamma^a_{cb}=\left( v^j_c v^i_b \left\{ {h \atop j\ i} \right\} + v^r_c \partial_r v^h_b \right) v^a_h, \tag{4.20}$$

from which

$$\Omega_{cb}{}^a = \Gamma^a_{cb} - \Gamma^a_{bc}. \tag{4.21}$$

Now the fact that $\nabla_k g_{ji}=0$ can be expressed as

$$v^r_d \partial_r g_{cb} - \Gamma^a_{dc} g_{ab} - \Gamma^a_{db} g_{ca} = 0$$

or

$$\Gamma^a_{cb} + \Gamma^{\bar{b}}_{c\bar{a}} = 0, \tag{4.22}$$

by virtue of (4.18).

Consequently, from

$$\Gamma^a_{cb} - \Gamma^a_{bc} = \Omega_{cb}{}^a, \quad -\Gamma^{\bar{b}}_{c\bar{a}} + \Gamma^{\bar{b}}_{\bar{a}c} = -\Omega_{c\bar{a}}{}^{\bar{b}}, \quad \Gamma^{\bar{c}}_{b\bar{a}} - \Gamma^{\bar{c}}_{\bar{a}b} = \Omega_{b\bar{a}}{}^{\bar{c}}$$

and (4.22), we find

$$\Gamma^a_{cb} = \frac{1}{2} (\Omega_{cb}{}^a + \Omega_{\bar{a}c}{}^{\bar{b}} + \Omega_{\bar{a}b}{}^{\bar{c}}), \tag{4.23}$$

from which

$$\Gamma^{\bar{\varkappa}}_{\mu\lambda} = \frac{1}{2} (\Omega_{\mu\lambda}{}^{\bar{\varkappa}} + \Omega_{\varkappa\mu}{}^{\bar{\lambda}} + \Omega_{\varkappa\lambda}{}^{\bar{\mu}}). \tag{4.24}$$

Thus, if an almost Hermite space is integrable, then we have $\Omega_{\mu\lambda}{}^{\bar{\varkappa}}=0$ and consequently $\Gamma^{\bar{\varkappa}}_{\mu\lambda}=0$. Conversely, if $\Gamma^{\bar{\varkappa}}_{\mu\lambda}=0$, then from (4.21) we see that $\Omega_{\mu\lambda}{}^{\bar{\varkappa}}=0$ and consequently the almost Hermite structure is integrable. Hence

THEOREM 4.2. *In order that an almost complex structure with a Hermite metric be integrable, it is necessary and sufficient that $\Gamma^{\bar{\varkappa}}_{\mu\lambda}=0$.*

The fact that $\nabla_k F_{ji}=0$ can be expressed as

$$v^r_c \partial_r F_{ba} - \Gamma^d_{cb} F_{da} - \Gamma^d_{ca} F_{bd} = 0$$

or by

$$\Gamma^{\varkappa}_{c\lambda} + \Gamma^{\bar{\lambda}}_{c\bar{\varkappa}} = 0, \qquad \Gamma^{\bar{\varkappa}}_{c\lambda} - \Gamma^{\bar{\lambda}}_{c\varkappa} = 0. \tag{4.25}$$

If the conditions $\nabla_k g_{ji}=0$ and $\nabla_k F_{ji}=0$ are both satisfied, then we get, from (4.22) and (4.25),

$$\Gamma^{\varkappa}_{c\bar{\lambda}} = 0, \qquad \Gamma^{\bar{\varkappa}}_{c\lambda} = 0 \tag{4.26}$$

and consequently

$$\Omega_{\bar{\mu}\bar{\lambda}}{}^{\varkappa} = \Gamma^{\varkappa}_{\bar{\mu}\bar{\lambda}} - \Gamma^{\varkappa}_{\bar{\lambda}\bar{\mu}} = 0 \tag{4.27}$$

and then the almost complex structure is integrable.

Equations (4.26) show that the two planes $\pi_n$ and $\bar{\pi}_n$ are both parallel.

Conversely, if the planes $\pi_n$ and $\bar{\pi}_n$ are both parallel, then we have (4.26) and consequently $\nabla_k F_{ji}=0$, which shows that the metric is Kählerian. Thus we have

THEOREM 4.3. *In order that an almost Hermite space be Kählerian, it is necessary and sufficient that $\Gamma^{\varkappa}_{c\bar{\lambda}}=0$, $\Gamma^{\bar{\varkappa}}_{c\lambda}=0$.*

THEOREM 4.4. *In an almost Hermite space, if the planes $\pi_n$ and $\bar{\pi}_n$ are both parallel, the space is Kählerian.* (E. M. Patterson [1]).

CHAPTER VI

# LINEAR CONNEXIONS IN ALMOST COMPLEX SPACES

## § 1. A lemma†

Suppose that, in a $2n$-dimensional space $M_{2n}$ with an almost complex structure $F_i^{\ h}$, there is given an affine connexion $\mathring{\Gamma}_{j\ i}^{\ h}$. For example, since we can always introduce a Riemannian metric in the space $M_{2n}$, the affine connexion $\mathring{\Gamma}_{j\ i}^{\ h}$ may be that given by Levi-Civita's parallelism defined by this Riemannian metric. Then an arbitrary affine connexion in $M_{2n}$ may be written as

$$\Gamma_{j\ i}^{\ h} = \mathring{\Gamma}_{j\ i}^{\ h} + T_{ji}^{\ \ h}, \tag{1.1}$$

where $T_{ji}^{\ \ h}$ is a tensor of the type indicated by the indices. Denoting by $\nabla_j$ and $\mathring{\nabla}_j$ the covariant differentiations with respect to $\Gamma_{j\ i}^{\ h}$ and $\mathring{\Gamma}_{j\ i}^{\ h}$ respectively, we have

$$\nabla_j F_i^{\ h} = \mathring{\nabla}_j F_i^{\ h} + T_{jr}^{\ \ h} F_i^{\ r} - T_{ji}^{\ \ r} F_r^{\ h}. \tag{1.2}$$

We shall discuss the possibility of finding an affine connexion $\Gamma_{j\ i}^{\ h}$ such that

$$\nabla_j F_i^{\ h} = 0. \tag{1.3}$$

From (1.2) and (1.3), we find

$$*O_{ir}^{sh} T_{js}^{\ \ r} = -\frac{1}{2} (\mathring{\nabla}_j F_i^{\ t}) F_t^{\ h}. \tag{1.4}$$

† See T. Fukami [3]; K. Yano [14].

Thus we see that the existence of an affine connexion $\Gamma_{j\,i}^{\,h}$ which satisfies the equation (1.3) reduces to the existence of a tensor $T_{ji}{}^{h}$ which satisfies (1.4).

To discuss the existence of solutions of the algebraic equation (1.4), we need a lemma.

We defined operators $O_{ir}^{sh}$ and $*O_{ir}^{sh}$ by

$$O_{ir}^{sh}=\frac{1}{2}(A_i^s A_r^h - F_i{}^s F_r{}^h), \qquad *O_{ir}^{sh}=\frac{1}{2}(A_i^s A_r^h + F_i{}^s F_r{}^h). \quad (1.5)$$

These operators operate on a general tensor $T$ in the following way:

$$\begin{aligned} OT: &\qquad O_{ir}^{sh} T_{\cdots s \cdots}^{\cdots r \cdots} \\ *OT: &\qquad *O_{ir}^{sh} T_{\cdots s \cdots}^{\cdots r \cdots}. \end{aligned} \quad (1.6)$$

It is easy to verify that these operators satisfy

$$\begin{aligned} O+*O=A, \qquad O\cdot O=O, \qquad O\cdot *O=0, \\ *O\cdot O=0, \qquad *O*O=*O, \end{aligned} \quad (1.7)$$

$A$ being the identity operator.

LEMMA. *A tensor $V$ being given, in order that the linear equations*

$$OT=V \quad \Big| \quad *OT=V \quad (1.8)$$

*with unknown tensor $T$ admit a solution, it is necessary and sufficient that $V$ satisfies*

$$*OV=0 \quad \Big| \quad OV=0 \quad (1.9)$$

*and if $V$ satisfies this condition, then the general solution of* (1.8) *is given by*

$$T=V+*OW \quad \Big| \quad T=V+OW, \quad (1.10)$$

*$W$ being an arbitrary tensor of the same type as $T$.*

*Proof.* From (1.8), applying

$$*O \quad \Big| \quad O$$

from the left, we find

$$*OV=0 \qquad \Big| \qquad OV=0$$

by virtue of (1.7). Thus the condition (1.9) is necessary.

To prove the sufficiency, we assume that $V$ satisfies (1.9). Then we have

$$0=*OV=(A-O)V \qquad \Big| \qquad 0=OV=(A-*O)V$$

from which

$$V=OV \qquad \Big| \qquad V=*OV.$$

Thus, equations (1.8) become

$$OT=OV, \qquad \Big| \qquad *OT=*OV,$$
$$O(T-V)=0. \qquad \Big| \qquad *O(T-V)=0.$$

Thus, if we put

$$T-V=W$$

then $W$ satisfies

$$OW=0 \qquad \Big| \qquad *OW=0$$

or

$$W=*OW \qquad \Big| \qquad W=OW$$

and consequently we have

$$T=V+*OW \qquad \Big| \qquad T=V+OW. \tag{1.11}$$

Then $T$ given by (1.11) satisfies the original equation (1.8), $W$ being an arbitrary tensor of the same type as $T$ and $V$.

The latter part of the lemma is now obvious.

As in the case of a complex manifold, if a tensor $T^{\cdots r \cdots}_{\cdots s \cdots}$ satisfies

$$O^{sh}_{ir}T^{\cdots r \cdots}_{\cdots s \cdots}=0$$

we say that $T_{\cdots s \cdots}^{\cdots r \cdots}$ is *hybrid* in $r$ and $s$ and if it satisfies

$$*O_{ir}^{sh} T_{\cdots s \cdots}^{\cdots r \cdots} = 0,$$

we say that $T_{\cdots s \cdots}^{\cdots r \cdots}$ is *pure* in $r$ and $s$.

Also if a tensor $T_{\ldots t \ldots s \ldots}$ satisfies

$$O_{ji}^{ts} T_{\ldots t \ldots s \ldots} = 0,$$

we say that $T_{\ldots t \ldots s \ldots}$ is *hybrid* in $t$ and $s$ and if it satisfies

$$*O_{ji}^{ts} T_{\ldots t \ldots s \ldots} = 0,$$

we say that it is *pure* in $t$ and $s$.

## § 2. Affine connexions in almost complex spaces

In the previous section, we have seen that the problem of finding an affine connexion $\Gamma_{j\ i}^{h}$ such that $\nabla_j F_i^{\ h} = 0$ reduces to the problem of finding a tensor $T_{ji}^{\ \ h}$ such that

$$*O_{ir}^{sh} T_{js}^{\ \ r} = -\frac{1}{2} (\mathring{\nabla}_j F_i^{\ t}) F_t^{\ h} \tag{2.1}$$

where $\mathring{\nabla}_j$ denotes covariant differentiation with respect to a given affine connexion $\mathring{\Gamma}_{j\ i}^{h}$, and that the required connexion is given by

$$\Gamma_{j\ i}^{h} = \mathring{\Gamma}_{j\ i}^{h} + T_{ji}^{\ \ h}. \tag{2.2}$$

Now the equation (2.1) has the form $*OT = V$, and consequently, to apply the Lemma, we compute the $OV$:

$$O_{ir}^{sh} \left[ -\frac{1}{2} (\mathring{\nabla}_j F_s^{\ t}) F_t^{\ r} \right] = -\frac{1}{4} (A_i^s A_r^h - F_i^{\ s} F_r^{\ h}) (\mathring{\nabla}_j F_s^{\ t}) F_t^{\ r}$$

$$= -\frac{1}{4} [(\mathring{\nabla}_j F_i^{\ t}) F_t^{\ h} + F_i^{\ s} (\mathring{\nabla}_j F_s^{\ h})] = 0.$$

Thus by the Lemma, the equation (2.1) admits a solution, and the general solution is given by

$$T_{ji}{}^{h} = -\frac{1}{2}(\mathring{\nabla}_j F_i{}^{t})F_t^{h} + O_{ir}^{sh} W_{js}{}^{r}, \tag{2.3}$$

$W_{js}{}^{r}$ being an arbitrary tensor. Thus, if we take as $W$

$$W_{ji}{}^{h} = -\frac{1}{4}(\mathring{\nabla}_i F_j{}^{t})F_t{}^{h} - \frac{1}{4}(\mathring{\nabla}_t F_j{}^{h})F_i{}^{t}, \tag{2.4}$$

then we have

$$*O_{ir}^{sh} W_{js}{}^{r} = \frac{1}{2}(A_i^s A_r^h + F_i{}^{s} F_r{}^{h})W_{js}{}^{r}$$

$$= -\frac{1}{8}[(\mathring{\nabla}_i F_j^{t})F_t{}^{h} + (\mathring{\nabla}_t F_j{}^{h})F_i^{t} - F_i{}^{s}(\mathring{\nabla}_s F_j{}^{h}) - F_r{}^{h}(\mathring{\nabla}_i F_j{}^{r})] = 0,$$

from which

$$O_{ir}^{sh} W_{js}{}^{r} = W_{ji}{}^{h}.$$

Thus (2.3) gives

$$T_{ji}{}^{h} = -\frac{1}{4}(\mathring{\nabla}_j F_i + \mathring{\nabla}_i F_j^{t})F_t{}^{h} + \frac{1}{4}(\mathring{\nabla}_j F_t{}^{h} - \mathring{\nabla}_t F_j{}^{h})F_i{}^{t}, \tag{2.5}$$

from which

$$T_{ji}^{h} - T_{ij}^{h} = \frac{1}{4}N_{ji}^{h}. \tag{2.6}$$

Thus we see that the affine connexion

$$\Gamma \quad -{}_{j\,i} - \frac{1}{4}(\mathring{\nabla}_j F_i + \mathring{\nabla}_i F_j^{t})F_t{}^{h} + \frac{1}{4}(\mathring{\nabla}_j F_t{}^{h} - \mathring{\nabla}_t F_j{}^{h})F_i^{t} \tag{2.7}$$

satisfies

$$\nabla_j F_i^{h} = 0 \qquad S_{ji}^{h} = \frac{1}{8}N_{ji}^{h}, \tag{2.8}$$

where

$$S_{ji}^{h} = \frac{1}{2}(\Gamma_j^h{}_i - \Gamma_i^h{}_j) \tag{2.9}$$

is the torsion tensor of the affine connexion $\Gamma_j^h{}_i$.

Now suppose that the almost complex structure is integrable, then we have $N_{ji}{}^{h}=0$ and consequently, the affine connexion introduced above satisfies $\nabla_j F_i{}^h=0$ and has no torsion.

Conversely, if it is possible to introduce, in an almost complex space, an affine connexion such that

$$\nabla_j F_i^h=0 \quad \text{and} \quad \Gamma_i{}^h{}_j=\Gamma_j{}^h{}_i,$$

then

$$N_{ji}^h=F_j^t(\nabla_t F_i^h-\nabla_i F_t^h)-F_i^t(\nabla_t F_j^h-\nabla_j F_t^h)=0,$$

and consequently the almost complex structure is integrable. Thus we have

THEOREM 2.1. *In order that an almost complex space be integrable, it is necessary and sufficient that it is possible to introduce a symmetric affine connexion with respect to which the structure tensor is covariantly constant.* (B. Eckmann [1], [2]; A. G. Walker [1]; K. Yano [14])

## § 3. Walker differentiation†

Following Walker, we shall introduce a covariant differentiation of the type

$$\nabla_{kj}U_i{}^h=N_{kj}{}^t\partial_t U_i{}^h+\Gamma_{kjt}{}^hU_i{}^t-\Gamma_{kji}{}^tU_t{}^h \tag{3.1}$$

for a general tensor $U$ provided that the Nijenhuis tensor $N_{kj}{}^h$ is not identically zero.

We shall discuss the possibility of finding coefficients $\Gamma_{kji}{}^h$ such that

$$\nabla_{kj}F_i{}^h=N_{kj}{}^t\partial_t F_i{}^h+\Gamma_{kjt}{}^hF_i{}^t-\Gamma_{kji}{}^tF_t{}^h=0. \tag{3.2}$$

Introducing an arbitrary affine connexion $\Gamma_j{}^h{}_i$ in the space, we can write (3.2) in the form

$$\nabla_{kj}F_i{}^h=N_{kj}{}^t\nabla_t F_i{}^h+T_{kjt}{}^hF_i{}^t-T_{kji}{}^tF_t{}^h=0, \tag{3.3}$$

† See A. G. Walker [2]; T. J. Willmore [2]; K. Yano [14].

where

$$T_{kji}{}^h = \Gamma_{kji}{}^h - N_{kj}{}^t \Gamma_t{}^h{}_i. \tag{3.4}$$

From (3.3), we have

$$O^{*sh}_{ir} T_{kjs}{}^r = -\frac{1}{2} N_{kj}{}^u (\nabla_u F_i{}^t) F_t{}^h. \tag{3.5}$$

This equation has the form $*OT = V$ and consequently, to apply the Lemma, we compute the $OV$:

$$O^{sh}_{ir}\left[-\frac{1}{2} N_{kj}{}^u (\nabla_u F_s{}^t) F_t{}^r\right] = -\frac{1}{4}(A^s_i A^h_r - F_i{}^s F_r{}^h) N_{kj}{}^u (\nabla_u F_s{}^t) F_t^r$$

$$= -\frac{1}{4}[N_{kj}{}^u (\nabla_u F_i{}^t) F_t{}^h + N_{ki}{}^u F_i{}^s (\nabla_u F^h)] = 0.$$

Thus, by the Lemma, the equation (3.5) admits a solution and the general solution is given by

$$T_{kji}{}^h = -\frac{1}{2} N_{kj}{}^u (\nabla_u F_i{}^t) F_t{}^h + O^{sh}_{ir} W_{kjsr}, \tag{3.6}$$

$W_{kjs}{}^r$ being an arbitrary tensor. Thus, if we take as $W$

$$W_{kji}{}^h = -O^{uh}_{it} \nabla_u N_{kj}{}^t + \frac{1}{2}(N_{ku}{}^t \nabla_j F_i{}^u - N_{ju}{}^t \nabla_k F_i{}^u) F_t{}^h, \tag{3.7}$$

then we have

$$*O^{sh}_{ir} W_{kjs}{}^r = \frac{1}{2}(A^s_i A^h_r + F_i{}^s F_r{}^h) W_{kjs}{}^r$$

$$= \frac{1}{4}[(N_{ku}{}^t \nabla_j F_i{}^u - N_{ju}{}^t \nabla_k F_i{}^u) F_t^h - F_i{}^s (N_{ku}{}^h \nabla_j F_s{}^u - N_{ju}{}^h \nabla_k F_s{}^u)] = 0$$

by virtue of the relation

$$N_{ku}{}^t F_t{}^h + N_{kt}{}^h F_u{}^t = 0.$$

Thus we have

$$O^{sh}_{ir} W_{kjs}{}^r = W_{kji}{}^h,$$

and hence (3.6) gives

$$T_{kji}{}^{h} = -\frac{1}{2} N_{kj}{}^{u}(\nabla_u F_i{}^t) F_t{}^h - O_{it}^{uh} \nabla_u N_{kj}$$

$$+\frac{1}{2}(N_{ku}{}^t \nabla_j F_i{}^u - N_{ju}{}^t \nabla_k F_i{}^u) F_t{}^h. \qquad (3.8)$$

Thus the coefficients $\Gamma_{kji}{}^h$ of Walker differentiation are given by

$$\Gamma_{kji}{}^h = N_{kj}{}^t \Gamma_t{}^h{}_i - \frac{1}{2} N_{kj}{}^u (\nabla_u F_i{}^t) F_t{}^h - O_{it}^{uh} \nabla_u N_{kj}{}^t$$

$$+\frac{1}{2}(N_{ku}{}^t \nabla_j F_i{}^u - N_{ju}{}^t \nabla_k F_i{}^u) F_t{}^h. \qquad (3.9)$$

These of course satisfy (3.2). If we take as $\Gamma_j{}^h{}_i$ a symmetric affine connexion $\mathring{\Gamma}_j{}^h{}_i$ and compute $\mathring{\Gamma}_{kji}{}^h$ above, then we find

$$\Gamma_{kji}{}^h = -O_{ir}^{sh} \partial_s N_{kj}{}^r - \frac{1}{2}(N_{kj}{}^s \partial_s F_i{}^r - N_{ks}{}^r \partial_j F_i{}^s + N_{js}{}^r \partial_k F_i{}^s) F_r{}^h \qquad (3.10)$$

which contains only the structure tensor $F_i{}^h$ and the Nijenhuis tensor $N_{ji}{}^h$ constructed from it.

Now using the $\Gamma_j{}^h{}_i$ given by (2.7) and the relation

$$S_{ji}{}^h = \frac{1}{2}(\Gamma_j{}^h{}_i - \Gamma_i{}^h{}_j) = \frac{1}{8} N_{ji}{}^h,$$

we transcribe (3.10) in the following form:

$$\Gamma_{kji}{}^h = N_{kj}{}^t \Gamma_t{}^h{}_i$$

$$-O_{ir}^{sh}\left[\nabla_s N_{kj}{}^r + \frac{1}{4} N_{kj}{}^t N_{ts}{}^r + \frac{1}{4} N_{sk}{}^t N_{tj}{}^r - \frac{1}{4} N_{sj}{}^t N_{tk}{}^r\right]. \qquad (3.11)$$

Then the Walker differentiation of the Nijenhuis tensor $N_{ji}{}^h$ takes the form

$$\nabla_{lk} N_{ji}{}^h = N_{lk}{}^t \nabla_t N_{ji}{}^h + T_{lkt}{}^h N_{ji}{}^t - T_{lkj}{}^t N_{ti}{}^h - T_{lki}{}^t N_{jt}{}^h$$

$$= N_{lk}{}^t \nabla_t N_{ji}{}^h - (\nabla_s N_{lk}{}^r + \frac{1}{4} N_{lk}{}^t N_{ts}{}^r + \frac{1}{4} N_{sl}{}^t N_{tk}{}^r$$

$$-\frac{1}{4} N_{sk}{}^t N_{tl}{}^r)(O_{ur}^{sh} N_{ji}{}^u - O_{jr}^{su} N_{ui}{}^h - O_{ir}^{su} N_{ju}{}^h). \qquad (3.12)$$

Since we know by (3.10) that $\nabla_{lk}N_{ji}{}^{h}$ does not depend on any particular choice of affine connexion, equation (3.12) gives a new differential invariant formed uniquely from $F_i{}^h$.

## § 4. Almost Kähler and almost Tachibana spaces†

Let us consider an almost Hermite space, that is, a $2n$-dimensional space with almost complex structure $F_i{}^h$:

$$F_j{}^iF_i{}^h = -A_j^h, \tag{4.1}$$

and with positive definite metric $ds^2 = g_{ji}d\xi^j d\xi^i$ satisfying

$$F_j{}^tF_i{}^s g_{ts} = g_{ji}. \tag{4.2}$$

If we put

$$F_j{}^t g_{ti} = F_{ji}, \tag{4.3}$$

then $F_{ji}$ is skew-symmetric.

The tensor

$$F_{jih} = \partial_j F_{ih} + \partial_i F_{hj} + \partial_h F_{ji} \tag{4.4}$$

is also skew-symmetric in its three covariant indices. Denoting by $\mathring{\nabla}_j$ the operation of covariant differentiation with respect to the Christoffel symbols

$$\left\{ {h \atop j\ i} \right\} = \frac{1}{2} g^{ht}(\partial_j g_{it} + \partial_i g_{jt} - \partial_t g_{ji}), \tag{4.5}$$

we can write (4.4) also in the form

$$F_{jih} = \mathring{\nabla}_j F_{ih} + \mathring{\nabla}_i F_{hj} + \mathring{\nabla}_h F_{ji}. \tag{4.6}$$

Transvecting (4.6) with $F^{ih}$, we find

$$F_{jih}F^{ih} = (\mathring{\nabla}_i F_{hj})F^{ih} + (\mathring{\nabla}_h F_{ji})F^{ih} = -F_j{}^h \mathring{\nabla}_i F^i{}_h - F_j{}^i \mathring{\nabla}_h F_i{}^h$$

† See S. S. Chern [6]; T. Fukami and S. Ishihara [1]; S. Tachibana [1], [2], [3], [4], [5], [8], [9], [10], [11]; K. Yano and M. Ako [1].

or

$$F_j{}^i F_i = +\frac{1}{2} F_{jih} F^{ih}. \tag{4.7}$$

where

$$F_i = -\mathring{\nabla}_h F_i{}^h. \tag{4.8}$$

If, in an almost Hermite space, the tensor $F_{jih}$ vanishes identically, the space is called an *almost Kähler space*. From (4.7), we have $F_i = 0$ in an almost Kähler space. Thus we have

THEOREM 4.1. *In an almost Kähler space, the tensor* $F_{ji}$ *is harmonic.*

Now the covariant components

$$N_{jih} = N_{ji}{}^t g_{th} = F_j{}^t(\mathring{\nabla}_t F_{ih} - \mathring{\nabla}_i F_{th}) - F_i{}^t(\mathring{\nabla}_t F_{jh} - \mathring{\nabla}_j F_{th}) \tag{4.9}$$

of the Nijenhuis tensor are also written in the form

$$N_{jih} = 2F_{jt}\mathring{\nabla}_h F_i{}^t + F_j{}^t F_{tih} - F_i{}^t F_{tjh}, \tag{4.10}$$

and hence

THEOREM 4.2. *The Nijenhuis tensor* $N_{jih}$ *in an almost Kähler space has the form*

$$N_{jih} = 2F_{jt}\mathring{\nabla}_h F_i{}^t. \tag{4.11}$$

Thus if the almost complex structure is integrable, then the almost Kählerian space reduces to a Kähler space.

Now we put

$$G_{ji}{}^h = \mathring{\nabla}_j F_i{}^h + \mathring{\nabla}_i F_j{}^h, \tag{4.12}$$

from which

$$G_{jt}{}^t = -F_j. \tag{4.13}$$

If in an almost Hermite space, the tensor $G_{ji}{}^h$ vanishes identically, the space is called an *almost Tachibana space*. In an almost Tachibana space, the tensor $F_i$ vanishes too.

By a straightforward calculation, we find

$$3\mathring{\nabla}_j F_{ih} - F_{jih} = G_{jih} - G_{jhi}, \tag{4.14}$$

from which

THEOREM 4.3. *In an almost Tachibana space, we have*

$$3\mathring{\nabla}_j F_{ih} = F_{jih}$$

*and consequently* $\mathring{\nabla}_j F_{ih}$ *is skew symmetric in all its indices.*

Now the Nijenhuis tensor $N_{i}{}^{h}$ may be written in the form

$$N_{ji}{}^{h} = -4(\mathring{\nabla}_j F_i^t) F_t^h + 2G_{ji}{}^t F_t^h + F_j^t G_{ti}{}^h - F_i^t G_{tj}{}^h, \qquad (4.15)$$

from which we get

THEOREM 4.4. *In an almost Tachibana space, the Nijenhuis tensor has the form*

$$N_{ji}^h = -4(\mathring{\nabla}_j F_i{}^t) F_t{}^h \qquad (4.16)$$

$$N_{jih} = -4(\mathring{\nabla}_j F_i{}^t) F_{th} = +4(\mathring{\nabla}_j F_h{}^t) F_{ti} \qquad (4.17)$$

*and consequently,* $N_{jih}$ *is skew-symmetric in all its indices.*

If the almost complex structure of an almost Tachibana space is integrable, we get, from (4.16), $\nabla_j F_i{}^t = 0$ and the space reduces to a Kähler space.

## § 5. Affine connexions in almost Hermite spaces†

Let us consider an almost Hermite space, that is, a $2n$-dimensional space with almost complex structure $F_i{}^h$:

$$F_j{}^i F_i{}^h = -A_j^h \qquad (5.1)$$

and with positive definite Riemannian metric $g_{ji}$ satisfying

$$F_j{}^t F_i{}^s g_{ts} = g_{ji}. \qquad (5.2)$$

We denote by

$$\left\{ {h \atop j\ i} \right\} = \frac{1}{2} g^{ht}(\partial_j g_{it} + \partial_i g_{jt} - \partial_t g_{ji}) \qquad (5.3)$$

† See S. Bochner [3], [10]; S. Bochner and K. Yano [1]; A. Fröhlicher [1]; S. I. Goldberg [1], [2]; A. Lichnerowicz [14], [21]; M. Obata [1] [2], [3]; S. Sawaki and S. Kotō [2]; J. A. Schouten and K. Yano [1]; H. Wakakuwa [1]; K. Yano [9].

the Christoffel symbols formed with $g_{ji}$. If we introduce an affine connexion $\Gamma_{j\ i}^{\ h}$ in the almost Hermite space and put

$$\Gamma_{j\ i}^{\ h} = \left\{ {h \atop j\ i} \right\} + T_{ji}^{\ \ h}, \tag{5.4}$$

then $T_{ji}^{\ \ h}$ is a tensor of type indicated by the indices and the torsion tensor of this connexion is given by

$$S_{ji}^{\ \ h} = \frac{1}{2}(\Gamma_{j\ i}^{\ h} - \Gamma_{i\ j}^{\ h}) = \frac{1}{2}(T_{ji}^{\ \ h} - T_{ij}^{\ \ h}). \tag{5.5}$$

We denote by $\nabla_j$ the operation of covariant differentiation with respect to $\Gamma_{j\ i}^{\ h}$ and by $\mathring{\nabla}_j$ that with respect to $\left\{ {h \atop j\ i} \right\}$. Then for the covariant derivative of $g_{ji}$, we have

$$\nabla_j g_{ih} = \mathring{\nabla}_j g_{ih} - T_{ji}^{\ \ t} g_{th} - T_{jh}^{\ \ t} g_{it},$$

and consequently, in order to have

$$\nabla_j g_{ih} = 0,$$

it is necessary and sufficient that we have

$$T_{jih} + T_{jhi} = 0, \tag{5.6}$$

where we have put

$$T_{jih} = T_{ji}^{\ \ t} g_{th}. \tag{5.7}$$

Now putting

$$S_{jih} = S_{ji}^{\ \ t} g_{th}, \tag{5.8}$$

we have from (5.5)

$$S_{jih} = \frac{1}{2}(T_{jih} - T_{ijh}),$$

from which

$$S_{hji} = \frac{1}{2}(T_{hji} - T_{jhi}),$$

$$S_{hij} = \frac{1}{2}(T_{hij} - T_{ihj}).$$

Taking account of (5.6), we have, from these three equations

$$T_{jih} = S_{jih} + S_{hji} + S_{hij}. \tag{5.9}$$

Thus, an affine connexion $\Gamma_{j\ i}^{h}$ satisfying $\nabla_j g_{ih} = 0$ is given by (5.4) where $T_{ji}{}^h$ satisfies (5.6), or by

$$\Gamma_{j\ i}^{h} = \left\{ {h \atop j\ i} \right\} + S_{ji}{}^h - S^h{}_{ji} - S^h{}_{ij}, \tag{5.10}$$

$S_{ji}{}^h$ being the torsion tensor. (H. A. Hayden [1]).

On the other hand, in order to have $\nabla_j F_{ih} = 0$, it is necessary and sufficient that we have

$$\mathring{\nabla}_j F_{ih} - T_{ji}{}^t F_{th} - T_{jh}{}^t F_{it} = 0, \tag{5.11}$$

which is equivalent to

$$\mathring{\nabla}_j F_{ih} + T_{jit} F_h{}^t - T_{jht} F_i{}^t = 0. \tag{5.12}$$

The $\Gamma_{j\ i}^{h}$ are not completely determined by (5.6) and (5.12). So we can introduce some other condition on $T_{jih}$.

We formulate this condition with respect to the tensor

$$'T_{jih} = T_{jit} F_h{}^t, \tag{5.13}$$

and we try to solve the equation (5.12) with respect to $T_{jih}$. From now on, we assume only $\nabla_j F_{ih} = 0$ and make no use of $\nabla_j g_{ih} = 0$. This latter equation will be satisfied in special cases as a consequence of the condition introduced.

## § 6. The first connexion

If we define $'T_{jih}$ by (5.13), then equation (5.12) can be written as

$$\mathring{\nabla}_j F_{ih} + {}'T_{jih} - {}'T_{jhi} = 0. \tag{6.1}$$

We first assume that

$$'T_{jih} + {}'T_{jhi} = 0, \tag{6.2}$$

then from these two equations, we have

$$'T_{jih} = -\frac{1}{2}\mathring{\nabla}_j F_{ih}, \tag{6.3}$$

from which

$$T_{ji}{}^h = -\frac{1}{2}(\mathring{\nabla}_j F_i{}^t) F_t{}^h. \tag{6.4}$$

Thus the torsion tensor of the connexion

$$\Gamma_j{}^h{}_i = \left\{ {h \atop j\ i} \right\} + T_{ji}{}^h \tag{6.5}$$

is given by

$$S_{ji}{}^h = -\frac{1}{4}(\mathring{\nabla}_j F_i{}^t - \mathring{\nabla}_i F_j{}^t) F_t{}^h . \tag{6.6}$$

Since we have, from (6.4),

$$T_{jih} = -\frac{1}{2}(\mathring{\nabla}_j F_i{}^t) F_{th} = +\frac{1}{2}(\mathring{\nabla}_j F_h{}^t) F_{ti}$$

or

$$T_{jih} = -T_{jhi}$$

the connexion (6.5) is metric and we have

$$\nabla_j g_{ih} = 0, \qquad \nabla_j F_{ih} = 0, \tag{6.7}$$

for this connexion.

From (4.10) and (6.4), we find

$$T_{jih} = \frac{1}{4}(N_{hij} + F_h{}^t F_{tji} - F_i{}^t F_{thj}). \tag{6.8}$$

Thus, in the case of an almost Kähler space, we have

$$T_{jih} = \frac{1}{4} N_{hij} \tag{6.9}$$

and, in the case of a Hermite space,

$$T_{jih}=\frac{1}{4}(F_h{}^tF_{tji}-F_i{}^tF_{tjh}). \tag{6.10}$$

In the case of a Kähler space, we have

$$T_{jih}=0 \tag{6.11}$$

and our connexion coincides with the Riemannian connexion.

On the other hand, from (4.15) and (6.4), we find

$$T_{ji}{}^h=\frac{1}{8}N_{ji}{}^h-\frac{1}{8}(2G_{ji}{}^tF_t{}^h+F_j{}^tG_{ti}{}^h-F_i{}^tG_{tj}{}^h). \tag{6.12}$$

Thus, in the case of an almost Tachibana space, we have

$$T_{ji}{}^h=\frac{1}{8}N_{ji}{}^h \tag{6.13}$$

and, in the case of a Hermite space,

$$T_{ji}{}^h=-\frac{1}{8}(2G_{ji}{}^tF_t{}^h+F_j{}^tG_{ti}{}^h-F_i{}^tG_{tj}{}^h)=-\frac{1}{2}(\mathring{\nabla}_jF_i{}^t)F_t{}^h \tag{6.14}$$

by virtue of (4.15).

In a Hermite case, we take the complex coordinate system; then we have

$$\left.\begin{aligned}\begin{Bmatrix}\varkappa\\ \mu\ \lambda\end{Bmatrix}&=\frac{1}{2}g^{\varkappa\bar{\varrho}}(\partial_\mu g_{\lambda\bar{\varrho}}+\partial_\lambda g_{\mu\bar{\varrho}}),\ \text{conj.}\\ \begin{Bmatrix}\varkappa\\ \mu\ \bar{\lambda}\end{Bmatrix}&=\frac{1}{2}g^{\varkappa\bar{\varrho}}(\partial_{\bar{\lambda}} g_{\mu\bar{\varrho}}-\partial_{\bar{\varrho}} g_{\mu\bar{\lambda}}),\ \text{conj.}\end{aligned}\right\} \tag{6.15}$$

the other $\begin{Bmatrix}h\\ j\ i\end{Bmatrix}$ being zero. Thus, for the components of $\mathring{\nabla}_jF_i{}^t$, we have

$$\mathring{\nabla}_\mu F_{\bar{\lambda}}{}^{\varkappa}=-2i\begin{Bmatrix}\varkappa\\ \mu\ \bar{\lambda}\end{Bmatrix},\ \text{conj.} \tag{6.16}$$

the other components being zero. Hence

$$T_{\mu\bar{\lambda}}{}^{\varkappa} = -\left\{\begin{matrix} \varkappa \\ \mu\ \bar{\lambda} \end{matrix}\right\}, \text{ conj.} \tag{6.17}$$

the other components being zero, and consequently

$$\Gamma_{\mu}{}^{\varkappa}{}_{\lambda} = \left\{\begin{matrix} \varkappa \\ \mu\ \lambda \end{matrix}\right\}, \qquad \Gamma_{\bar{\mu}}{}^{\varkappa}{}_{\lambda} = \left\{\begin{matrix} \varkappa \\ \bar{\mu}\ \lambda \end{matrix}\right\}, \text{ conj.} \tag{6.18}$$

the other components being all zero.

This is the affine connexion introduced by Lichnerowicz ([8], [19]).

## § 7. The second connexion

To solve the equation

$$\mathring{\nabla}_j F_{ih} + 'T_{jih} - 'T_{jhi} = 0, \tag{7.1}$$

we assume next

$$'T_{jih} + 'T_{hij} = 0. \tag{7.2}$$

From (7.1), we have

$$\mathring{\nabla}_j F_{ih} + 'T_{jih} - 'T_{jhi} = 0,$$
$$\mathring{\nabla}_i F_{jh} + 'T_{ijh} - 'T_{ihj} = 0,$$
$$\mathring{\nabla}_h F_{ji} + 'T_{hji} - 'T_{hij} = 0,$$

and consequently, taking account of (7.2), we find

$$'T_{jih} = -\frac{1}{2}(\mathring{\nabla}_j F_{ih} + \mathring{\nabla}_i F_{jh} + \mathring{\nabla}_h F_{ji}), \tag{7.3}$$

from which

$$T_{jih} = \frac{1}{2}(\mathring{\nabla}_j F_{it} + \mathring{\nabla}_i F_{jt} + \mathring{\nabla}_t F_{ji}) F_h{}^t, \tag{7.4}$$

and consequently

$$T_{ji}{}^{h} = -\frac{1}{2}(\mathring{\nabla}_j F_i{}^t + \mathring{\nabla}_i F_j{}^t + \mathring{\nabla}^t F_{ji}) F_t{}^h. \tag{7.5}$$

Thus the torsion tensor of the connexion

$$\Gamma_j{}^h{}_i = \left\{ {h \atop j\ i} \right\} + T_{ji}{}^h \tag{7.6}$$

is given by

$$S_{ji}{}^h = -\frac{1}{2}(\mathring{\nabla}_t F_{ji}) F^{th}. \tag{7.7}$$

From (7.4), we have

$$T_{jih} = \left( \mathring{\nabla}_i F_{jt} + \frac{1}{2} F_{jit} \right) F_h{}^t. \tag{7.8}$$

Thus combining (4.10) and (7.8), we get

$$T_{jih} = \frac{1}{2} N_{hji} + \frac{1}{2} F_j{}^t F_{hit}. \tag{7.9}$$

Thus, in an almost Kähler space, we have

$$T_{jih} = \frac{1}{2} N_{hji}, \tag{7.10}$$

and in a Hermite space, we have

$$T_{jih} = \frac{1}{2} F_j{}^t F_{hit}. \tag{7.11}$$

In a Hermite space, the $T_{jhi}$ satisfies

$$T_{jhi} + T_{jih} = 0$$

and consequently the connexion is metric, that is,

$$\nabla_j g_{ih} = 0, \qquad \nabla_j F_{ih} = 0. \tag{7.12}$$

In a Kähler space, we have

$$T_{jih} = 0 \tag{7.13}$$

and the connexion reduces to Riemannian connexion.

On the other hand, from (4.15) and (7.8), we find

$$T_{jih} = -\frac{1}{4} N_{jih} + \frac{1}{2} F_{jit} F_h{}^t - \frac{1}{4} (2G_{ji}{}^t F_{th} - F_j{}^t G_{tih} + F_i{}^t G_{tjh}). \quad (7.14)$$

Thus, in an almost Tachibana space, we have

$$T_{ji}{}^h = -\frac{1}{4} N_{ji}{}^h - \frac{1}{2} F_{ji}{}^t F_t{}^h, \quad (7.15)$$

and in a Hermite space

$$T_{ji}{}^h = -\frac{1}{2} F_{ji}{}^t F_t{}^h - \frac{1}{4} (2G_{ji}{}^t F_t{}^h - F_j{}^t G_{ti}{}^h + F_i{}^t G_{tj}{}^h). \quad (7.16)$$

For a Hermite case, taking a complex coordinate system, we have (6.15) and (6.16), and consequently, from (7.5),

$$\left.\begin{aligned} T_{\mu\lambda}{}^{\varkappa} &= \frac{1}{2} g^{\varkappa\bar{\varrho}}(\partial_\mu g_{\lambda\bar{\varrho}} - \partial_\lambda g_{\mu\bar{\varrho}}), \quad \text{conj.} \\ T_{\mu\lambda}{}^{\bar{\varkappa}} &= -\frac{1}{2} g^{\varkappa\bar{\varrho}}(\partial_{\bar{\lambda}} g_{\bar{\varrho}\mu} - \partial_{\bar{\varrho}} g_{\bar{\lambda}\mu}), \quad \text{conj.} \\ T_{\mu\bar{\lambda}}{}^{\bar{\varkappa}} &= -\frac{1}{2} g^{\bar{\varkappa}\varrho}(\partial_\mu g_{\varrho\bar{\lambda}} - \partial_\varrho g_{\mu\bar{\lambda}}), \quad \text{conj.} \end{aligned}\right\} \quad (7.17)$$

the other $T_{ji}{}^h$ being zero, and hence

$$\Gamma_\mu{}^\varkappa{}_\lambda = g^{\varkappa\bar{\varrho}} \partial_\mu g_{\lambda\bar{\varrho}}, \quad \text{conj.} \quad (7.18)$$

the other $\Gamma_j{}^h{}_i$ being zero.

This is the affine connexion introduced by J. A. Schouten and D. van Dantzig [1] and used recently by S. S. Chern [1] and P. Libermann [1].

## § 8. The third connexion

To solve the equation

$$\mathring{\nabla}_j F_{ih} + 'T_{jih} - 'T_{jhi} = 0, \tag{8.1}$$

we assume now

$$'T_{jih} + 'T_{ijh} = 0. \tag{8.2}$$

From (8.1), we have

$$\mathring{\nabla}_j F_{ih} + 'T_{jih} - 'T_{jhi} = 0,$$
$$\mathring{\nabla}_i F_{hj} + 'T_{ihj} - 'T_{ijh} = 0,$$
$$\mathring{\nabla}_h F_{ji} - 'T_{hji} + 'T_{hij} = 0,$$

and consequently, taking account of (8.2), we find

$$'T_{jih} = -\frac{1}{2}(\mathring{\nabla}_j F_{ih} + \mathring{\nabla}_i F_{hj} - \mathring{\nabla}_h F_{ji}), \tag{8.3}$$

from which

$$T_{jih} = \frac{1}{2}(\mathring{\nabla}_j F_{it} - \mathring{\nabla}_i F_{jt} - \mathring{\nabla}_t F_{ji}) F_h{}^t \tag{8.4}$$

and consequently

$$T_{ji}{}^h = -\frac{1}{2}(\mathring{\nabla}_j F_i{}^t - \mathring{\nabla}_i F_j{}^t - \mathring{\nabla}^t F_{ji}) F_t{}^h. \tag{8.5}$$

Thus the torsion tensor of the connexion

$$\Gamma_j{}^h{}_i = \left\{ {h \atop j\ i} \right\} + T_{ji}{}^h \tag{8.6}$$

is given by

$$S_{ji}{}^h = T_{ji}{}^h. \tag{8.7}$$

Now, the Nijenhuis tensor is written in the form

$$N_{jih} = (\mathring{\nabla}_j F_{it} - \mathring{\nabla}_i F_{jt}) F_h{}^t + F_j{}^t \mathring{\nabla}_t F_{ih} - F_i{}^t \mathring{\nabla}_t F_{jh}. \tag{8.8}$$

Thus combining (8.4) and (8.8), we obtain

$$T_{jih} = \frac{1}{2} N_{jih} - \frac{1}{2}(F_j{}^t \mathring{\nabla}_t F_{ih} + F_i{}^t \mathring{\nabla}_t F_{hj} + F_h{}^t \mathring{\nabla}_t F_{ji}). \tag{8.9}$$

On the other hand, we have

$$F_j^{\ t}F_i^{\ s}F_h^{\ r}F_{tsr} = -F_j^{\ t}\mathring{\nabla}_t F_{ih} - F_i^{\ t}\mathring{\nabla}_t F_{hj} - F_h^{\ t}\mathring{\nabla}_t F_{ji}. \tag{8.10}$$

Substituting this into (8.9), we find

$$T_{jih} = \frac{1}{2}N_{jih} + \frac{1}{2}F_j^{\ t}F_i^{\ s}F_h^{\ r}F_{tsr}. \tag{8.11}$$

Thus, in an almost Kähler space, we have

$$T_{jih} = \frac{1}{2}N_{jih}, \tag{8.12}$$

and in a Hermite space, we have

$$T_{jih} = \frac{1}{2}F_j^{\ t}F_i^{\ s}F_h^{\ r}F_{tsr}. \tag{8.13}$$

In a Hermite space, the tensor $T_{jih}$ satisfies

$$T_{jih} + T_{jhi} = 0$$

and consequently the connexion is metric, that is,

$$\nabla_j g_{ih} = 0, \qquad \nabla_j F_{ih} = 0. \tag{8.14}$$

In a Kähler space, we have

$$T_{jih} = 0 \tag{8.15}$$

and the connexion reduces to Riemannian connexion.

For a Hermite case, taking a complex coordinate system, we have (6.15) and (6.16), consequently, from (8.5),

$$\left.\begin{aligned}
T_{\mu\lambda}^{\ \ \varkappa} &= -\frac{1}{2}g^{\varkappa\bar{\varrho}}(\partial_\mu g_{\lambda\bar{\varrho}} - \partial_\lambda g_{\mu\bar{\varrho}}), \quad \text{conj.}\\
T_{\mu\bar{\lambda}}^{\ \ \varkappa} &= -\frac{1}{2}g^{\varkappa\bar{\varrho}}(\partial_{\bar{\lambda}} g_{\bar{\varrho}\mu} - \partial_{\bar{\varrho}} g_{\bar{\lambda}\mu}), \quad \text{conj.}\\
T_{\mu\bar{\lambda}}^{\ \ \bar{\varkappa}} &= +\frac{1}{2}g^{\bar{\varkappa}\varrho}(\partial_\mu g_{\varrho\bar{\lambda}} - \partial_\varrho g_{\mu\bar{\lambda}}), \quad \text{conj.}
\end{aligned}\right\} \tag{8.16}$$

the other $T_{ji}^{\ \ h}$ being zero, and hence

$$\left.\begin{aligned}
\Gamma_{\mu\ \lambda}^{\ \varkappa} &= g^{\varkappa\bar{\varrho}}\partial_\lambda g_{\mu\bar{\varrho}}, \quad && \text{conj.}\\
\Gamma_{\bar{\mu}\ \lambda}^{\ \varkappa} &= g^{\varkappa\bar{\varrho}}(\partial_{\bar{\mu}} g_{\bar{\varrho}\lambda} - \partial_{\bar{\varrho}} g_{\bar{\mu}\lambda}), \quad && \text{conj.}
\end{aligned}\right\} \tag{8.17}$$

CHAPTER VII

# ALMOST KÄHLER SPACES

## § 1. Almost Kähler spaces†

An almost Kähler space is first of all an almost complex space, that is, a $2n$-dimensional space with an almost complex structure $F_i^{\ h}$:

$$F_j^{\ i}F_i^{\ h} = -A_j^h, \tag{1.1}$$

and it has a Hermite metric $g_{ji}$, that is, a positive definite Riemannian metric $g_{ji}$ satisfying

$$F_j^{\ t}F_i^{\ s}g_{ts} = g_{ji}, \tag{1.2}$$

from which

$$F_{ji} = -F_{ij}, \tag{1.3}$$

where

$$F_{ji} = F_j^{\ t}g_{ti}, \tag{1.4}$$

and finally has the property that the differential form

$$F_{ji}d\xi^j \wedge d\xi^i$$

is closed, that is,

$$F_{jih} \stackrel{def}{=} \nabla_j F_{ih} + \nabla_i F_{hj} + \nabla_h F_{ji} = 0. \tag{1.5}$$

We have already proved the identity

$$F_j^{\ i}F_i = \frac{1}{2}F_{jih}F^{ih}, \tag{1.6}$$

† H. Guggenheimer [1]; E. H. Lee [1]; A. Lichnerowicz [4], [5], [6], [7], [8].

where

$$F_i = -\nabla_j F_i^{\ j}. \tag{1.7}$$

Equation (1.6) shows that, in an almost Kähler space, $F_i = 0$, that is, *the tensor* $F_{ji}$ *is harmonic.*

Since we have

$$*O_{ir}^{sh} F_s^{\ r} = 0 \tag{1.8}$$

the tensor $F_i^{\ h}$ is pure in $h$ and $i$. Equation (1.2) may be written as

$$O_{ji}^{ts} g_{ts} = 0, \tag{1.9}$$

which shows that $g_{ji}$ is hybrid in $j$ and $i$.

From (1.8), we get

$$O_{ji}^{ts} F_{ts} = 0, \tag{1.10}$$

which shows that $F_{ji}$ is hybrid in $j$ and $i$.

Now the covariant components of the Nijenhuis tensor $N_{jih}$ may be written in the form

$$N_{jih} = F_j^{\ t} F_{tih} - F_i^{\ t} F_{tjh} + 2F_j^{\ t}(\nabla_h F_{it}) \tag{1.11}$$

and consequently, in the case of an almost Kähler space, we have

$$N_{jih} = 2F_j^{\ t}(\nabla_h F_{it}), \tag{1.12}$$

from which

$$N_{jih} = 2F_j^{\ t}(\nabla_t F_{ih} - \nabla_i F_{th}). \tag{1.13}$$

On the other hand, we have

$$*O_{ih}^{sr} N_{jsr} = 0, \tag{1.14}$$

that is, $N_{jih}$ is pure in $i$ and $h$. Thus, equation (1.12) gives

THEOREM 1.1. *The tensor* $\nabla_j F_i^{\ h}$ *in an almost Kähler space is pure in* $j$ *and* $i$.

In general the Nijenhuis tensor $N_{jih}$ satisfies

$$N_{jih} + N_{jhi} = -F_{jit} F_h^{\ t} - F_{jht} F_i^{\ t} + 2F_j^{\ t} G_{iht}. \tag{1.15}$$

Thus, if, in an almost Kähler space, we have

$$N_{jih}+N_{jhi}=0,$$

then we deduce from it $G_{jih}=0$ and the space is an almost Tachibana space. Thus we have

$$3\nabla_j F_{ih}=F_{jih}=0$$

and consequently the space is a Kähler space. It is evident that in a Kähler space we have $N_{jih}+N_{jhi}=0$. Thus we have

THEOREM 1.2. *An almost Kähler space is a Kähler space if and only if*

$$N_{jih}+N_{jhi}=0. \tag{1.16}$$

Also the tensor $G_{ji}{}^{h}$ satisfies

$$2{}^*O_{ji}^{ts}G_{tsh}=-(F_j{}^tF_{tsi}+F_i{}^tF_{tjs})F_h{}^s, \tag{1.17}$$

from which

$${}^*O_{ji}^{ts}G_{tsh}=0. \tag{1.18}$$

in an almost Kähler space, which shows that $G_{ji}{}^{h}$ is pure in $j$ and $i$.

## § 2. Curvature tensors in an almost Kähler space

We denote the curvature tensor of an almost Kähler space by

$$K_{kji}{}^{h}=\partial_k\left\{{h \atop j\ i}\right\}-\partial_j\left\{{h \atop k\ i}\right\}+\left\{{h \atop k\ t}\right\}\left\{{t \atop j\ i}\right\}-\left\{{h \atop j\ t}\right\}\left\{{t \atop k\ i}\right\}, \tag{2.1}$$

its covariant components by

$$K_{kjih}=K_{kji}{}^{t}g_{th}, \tag{2.2}$$

and the Ricci tensor and the curvature scalar by

$$K_{ji}=K_{tji}{}^{t} \quad \text{and} \quad K=g^{ji}K_{ji} \tag{2.3}$$

respectively. Moreover we put

$$H_{kj}=\frac{1}{2}K_{kjih}F^{ih} \tag{2.4}$$

and

$$K^*_{ji}=-H_{jt}F_i^{\ t} \tag{2.5}$$

or

$$H_{ji}=K^*_{jt}F_i^{\ t}. \tag{2.6}$$

Now applying the Ricci identity to the tensor $F_i^{\ h}$, we find

$$\nabla_k\nabla_j F_i^{\ h}-\nabla_j\nabla_k F_i^{\ h}=K_{kjt}{}^{h}F_i^{\ t}-K_{kji}{}^{t}F_t^{\ h}. \tag{2.7}$$

Contraction with respect to $k$ and $h$ in this equation gives

$$\begin{aligned}\nabla_t\nabla_j F_i^{\ t}=K_{jt}F_i^{\ t}-K_{sjit}F^{ts}&=K_{jt}F_i^{\ t}-\frac{1}{2}(K_{sjit}-K_{tjis})F^{ts}\\&=K_{jt}F_i^{\ t}-\frac{1}{2}K_{tsji}F^{ts},\end{aligned}$$

from which

$$\nabla_t\nabla_j F_i^{\ t}=K_{jt}F_i^{\ t}-H_{ji} \tag{2.8}$$

or

$$\nabla_t\nabla_j F_i^{\ t}=(K_{jt}-K^*_{jt})F_i^{\ t} \tag{2.9}$$

by virtue of (2.4) and (2.6). Hence,

THEOREM 2.1. *In order that we have*

$$K_{ji}=K^*_{ji}$$

*in an almost Kähler space, it is necessary and sufficient that*

$$\nabla_t\nabla_j F_i^{\ t}=0. \tag{2.10}$$

Now, the tensor $F_{ji}$ in an almost Kähler space is harmonic and consequently it satisfies

$$g^{sr}\nabla_s\nabla_r F_{ji}-K_j^{\ r}F_{ri}-K_i^{\ r}F_{jr}-K_{ji}{}^{sr}F_{sr}=0 \tag{2.11}$$

or

$$g^{sr}\nabla_s\nabla_r F_{ji} + (K_{jr} - K^*_{jr})F_i{}^r - (K_{ir} - K^*_{ir})F_j{}^r = 0 \qquad (2.12)$$

by virtue of (2.4) and (2.6).

In the following we shall often meet the expression

$$F^{kj}\nabla_k\nabla_j w_i,$$

which may be written as

$$F^{kj}\nabla_k\nabla_j w_i = \frac{1}{2} F^{kj}(\nabla_k\nabla_j w_i - \nabla_j\nabla_k w_i)$$

$$= -\frac{1}{2} F^{kj}K_{kji}{}^t w_t = -H_{it}w^t,$$

from which

$$F^{kj}\nabla_k\nabla_j w_i = -K^*_{it}F_s{}^t w^s. \qquad (2.13)$$

## § 3. Contravariant almost analytic vectors

Let us consider a self-conjugate contravariant vector field $v^h = (v^\varkappa, v^{\bar\varkappa})$ in a complex manifold. The condition for $v^h$ to be a contravariant analytic vector field is given by

$$\partial_{\bar\lambda} v^\varkappa = 0, \qquad \partial_\lambda v^{\bar\varkappa} = 0 \qquad (3.1)$$

or by

$$^*O_{ir}^{sh}\partial_s v^r = 0; \qquad (3.2)$$

that is, by the fact that $\partial_i v^h$ is pure. Equation (3.2) is equivalent to

$$-F_i{}^t\partial_t v^h + F_t{}^h\partial_i v^t = 0.$$

If we introduce an affine connexion $\Gamma^h_{ji}$ without torsion in the manifold and denote by $\nabla_j$ the covariant derivative with respect to $\Gamma^h_{ji}$, the above equation can be written as

$$-F_i{}^t\nabla_t v^h + F_t{}^h\nabla_i v^t + F_i{}^t\Gamma^h_{ts}v^s - F_t{}^h\Gamma^t_{is}v^s = 0.$$

On the other hand, we have

$$\nabla_s F_i^{\ h} = F_i^{\ t}\Gamma_{st}^h - F_t^{\ h}\Gamma_{si}^t$$

by virtue of the fact that $F_i^{\ h}$ has numerical components.

Thus, from the above two equations, we get

$$v^s\nabla_s F_i^{\ h} - F_i^{\ t}\nabla_t v^h + F_t^{\ h}\nabla_i v^t = 0$$

or

$$\underset{v}{\mathfrak{L}} F_i^{\ h} = v^t\partial_t F_i^{\ h} - F_i^{\ t}\partial_t v^h + F_t^{\ h}\partial_i v^t = 0, \tag{3.3}$$

which is a tensor equation, where $\underset{v}{\mathfrak{L}}$ denotes the Lie derivative with respect to the vector field $v^h$.

Thus, in an almost complex space, we define a *contravariant almost analytic vector field* as a vector which satisfies (3.3), that is, as a vector $v^h$, the infinitesimal transformation with respect to which does not change the almost complex structure $F_i^{\ h}$.

In an almost Hermite space, the equation (3.3) may be written as

$$\underset{v}{\mathfrak{L}} F_i^{\ h} = v^t\nabla_t F_i^{\ h} - F_i^{\ t}\nabla_t v^h + F_t^{\ h}\nabla_i v^t = 0, \tag{3.4}$$

from which

$$v^t\nabla_t F_{ih} - F_i^{\ t}\nabla_t v_h - F_h^{\ t}\nabla_i v_t = 0, \tag{3.5}$$

where the covariant derivative is with respect to $\left\{ {h \atop j\ i} \right\}$.

Taking the symmetric part of (3.5) with respect to $i$ and $h$, we find

$$O_{ji}^{ts}(\nabla_t v_s + \nabla_s v_t) = 0 \quad \text{or} \quad O_{ji}^{ts}(\underset{v}{\mathfrak{L}}\, g_{ts}) = 0, \tag{3.6}$$

and also

$$O_{ts}^{ji}(\nabla^t v^s + \nabla^s v^t) = 0 \quad \text{or} \quad O_{ts}^{ji}(\underset{v}{\mathfrak{L}}\, g^{ts}) = 0, \tag{3.7}$$

from which

THEOREM 3.1. *For a contravariant almost analytic vector field $v^h$ in an almost Hermite space, the Lie derivatives $\underset{v}{\mathfrak{L}}\, g_{ji}$ and $\underset{v}{\mathfrak{L}}\, g^{ji}$ are both hybrid tensors.*

We shall here prove a

LEMMA. *If $S_{ji}$ is pure in $j$ and $i$ and $T^{ji}$ is hybrid in $j$ and $i$, then we have*

$$S_{ji}T^{ji}=0. \tag{3.8}$$

In fact, if $S_{ji}$ is pure in $j$ and $i$, we have

$$S_{ji}=O_{ji}^{ts}S_{ts},$$

and if $T^{ji}$ is hybrid, we have

$$O_{ts}^{ji}T^{ts}=0,$$

and consequently

$$S_{ji}T^{ji}=O_{ji}^{ts}S_{ts}T^{ji}=S_{ts}(O_{ji}^{ts}T^{ji})=0,$$

which proves the Lemma.

Now we know that $G_{ji}{}^{h}$ is pure in $j$ and $i$ and $\underset{v}{\mathfrak{L}}g^{ji}$ is hybrid for a contravariant almost analytic vector field $v^h$ and consequently by the lemma above, we have

$$G_{ji}{}^{h}(\underset{v}{\mathfrak{L}}g^{ji})=0 \tag{3.9}$$

or

$$G_{ji}{}^{h}(\nabla^{j}v^{i})=0, \tag{3.10}$$

for a contravariant almost analytic vector $v^h$.

From (3.5), we have

$$v^t\nabla_tF_{ji}-F_j{}^t\nabla_tv_i-F_i{}^t\nabla_jv_t=0.$$

Substituting

$$\nabla_tF_{ji}=-\nabla_jF_{it}-\nabla_iF_{tj}$$

in this equation, we get

$$\nabla_j\tilde{v}_i-\nabla_i\tilde{v}_j=F\,{}^t(\nabla_tv_i+\nabla_iv_t), \tag{3.11}$$

which is valid for a contravariant almost analytic vector field $v^h$, where we have put

$$\tilde{v}_i=-F_i{}^tv_t. \tag{3.12}$$

From (3.11), we have

THEOREM 3.2. *A necessary and sufficient condition that a contravariant almost analytic vector field $v^h$ be a Killing vector is that the vector $\tilde{v}_i = -F_i{}^t v_t$ be closed.* (S. Tachibana [3]).

Now we apply the operator $\nabla^i$ to (3.4), then we get

$$F_t{}^h[\nabla^i\nabla_i v^t + K_i{}^t v^i - G_{ji}{}^s F_s{}^t(\nabla^j v^i)] = 0$$

or

$$\nabla^i\nabla_i v^h + K_i{}^h v^i - G_{ji}{}^s F_s{}^h(\nabla^j v^i) = 0$$

or by virtue of (3.10),

$$\nabla^i\nabla_i v^h + K_i{}^h v^i = 0, \tag{3.13}$$

which is a necessary condition for a vector $v^h$ in an almost Kähler space to be contravariant almost analytic.

The same result may be obtained also in the following way. In general, we have

$$\underset{v}{\mathfrak{L}}\nabla_j F_i{}^h - \nabla_j \underset{v}{\mathfrak{L}} F_i{}^h = \left(\underset{v}{\mathfrak{L}}\left\{ {h \atop j\ t} \right\}\right) F_i{}^t - \left(\underset{v}{\mathfrak{L}}\left\{ {t \atop j\ i} \right\}\right) F_t{}^h,$$

but, for a contravariant almost analytic vector $v^h$ we have $\underset{v}{\mathfrak{L}} F_i{}^h = 0$ and consequently

$$\underset{v}{\mathfrak{L}}\nabla_j F_i{}^h = \left(\underset{v}{\mathfrak{L}}\left\{ {h \atop j\ t} \right\}\right) F_i{}^t - \left(\underset{v}{\mathfrak{L}}\left\{ {t \atop j\ i} \right\}\right) F_t{}^h. \tag{3.14}$$

On the other hand, the Nijenhuis tensor has, following (1.13), the form

$$N_{ji}{}^h = 2F_j{}^t(\nabla_t F_i{}^h - \nabla_i F_t{}^h) \tag{3.15}$$

and consequently

$$\underset{v}{\mathfrak{L}} N_{ji}{}^h = 2F_j{}^t(\underset{v}{\mathfrak{L}}\nabla_t F_i{}^h - \underset{v}{\mathfrak{L}}\nabla_i F_t{}^h).$$

Substituting (3.14) in this equation, we find

$$\underset{v}{\mathfrak{L}} N_{ji}{}^h = 2\left[\underset{v}{\mathfrak{L}}\left\{ {h \atop j\ i} \right\} + F_j{}^t F_i{}^s \underset{v}{\mathfrak{L}}\left\{ {h \atop t\ s} \right\}\right] \tag{3.16}$$

or

$$\underset{v}{\mathfrak{L}} N_{ji}{}^{h} = 4{}^{*}O_{ji}^{ts} \underset{v}{\mathfrak{L}} \left\{ {h \atop t\ s} \right\}. \tag{3.17}$$

But the left hand side of this equation is skew-symmetric in $j$ and $i$ and the right hand side is symmetric in $j$ and $i$, and consequently we have

$$\underset{v}{\mathfrak{L}} N_{ji}{}^{h} = 0 \tag{3.18}$$

and

$${}^{*}O_{ji}^{ts} \underset{v}{\mathfrak{L}} \left\{ {h \atop t\ s} \right\} = 0. \tag{3.19}$$

Transvecting (3.19) with $g^{ji}$ and remembering that $g^{ji}{}^{*}O_{ji}^{ts} = g^{ts}$, we have

$$g^{ji} \underset{v}{\mathfrak{L}} \left\{ {h \atop j\ i} \right\} = g^{ji}\nabla_j\nabla_i v^h + K_i{}^h v^i = 0 \tag{3.20}$$

for the contravariant almost analytic vector $v^h$.

From (3.18), we have

THEOREM 3.3. *The Lie derivative of the Nijenhuis tensor with respect to a contravariant almost analytic vector field in an almost Kähler space vanishes.*

## § 4. A necessary and sufficient condition for a vector to be contravariant almost analytic

If we put

$$S^{ji} = g^{jt}(\underset{v}{\mathfrak{L}} F_t^{\,i}), \tag{4.1}$$

we find by a straightforward computation

$$\frac{1}{2} S^{ji}S_{ji} = \frac{1}{2} v^t v^s (\nabla_t F^{ji})(\nabla_s F_{ji}) - v^t(\nabla_t F^{ji}) F_j{}^s (\nabla_s v_i)$$

$$+ v^t(\nabla_t F^{ji}) F_{si}(\nabla_j v^s) + (\nabla^j v^i)(\nabla_j v_i) - F^{jt}F_{si}(\nabla_t v^i)(\nabla_j v^s) \tag{4.2}$$

and

$$\nabla^j\{(\underset{v}{\mathfrak{L}} F_j{}^t)F_t{}^i v_i\} + (\nabla^i\nabla_i v^h + K_i{}^h v^i)v_h + \frac{1}{2} S^{ji}S_{ji} = 0. \quad (4.3)$$

Assuming that the almost Kähler space is compact and applying Green's theorem to (4.3), we find

$$\int_M \left[\{\nabla^i\nabla_i v^h + K_i{}^h v^i\} v_h + \frac{1}{2} S^{ji}S_{ji}\right] d\sigma = 0, \quad (4.4)$$

where $d\sigma$ denotes the volume element of the space.

Thus, if

$$\nabla^i\nabla_i v^h + K_i{}^h v^i = 0,$$

then we have

$$S^{ji} = g^{jt}(\underset{v}{\mathfrak{L}} F_t{}^i) = 0,$$

from which

$$\underset{v}{\mathfrak{L}} F_t{}^h = 0,$$

which shows that vector $v^h$ is contravariant analytic. Thus we have

THEOREM 4.1. *A necessary and sufficient condition for a vector field $v^h$ in a compact almost Kähler space to be contravariant almost analytic is*

$$\nabla^i\nabla_i v^h + K_i{}^h v^i = 0.$$

(S. Tachibana [3], K. Yano and M Ako [1]).

Since a necessary and sufficient condition for a vector field $v^h$ in a compact orientable Riemannian space to be a Killing vector is

$$\nabla^i\nabla_i v^h + K_i{}^h v^i = 0, \qquad \nabla_i v^i = 0,$$

we have the following two corollaries to the above theorem.

COROLLARY 1. *A Killing vector field in a compact almost Kähler space is contravariant almost analytic.*

COROLLARY 2. *An infinitesimal almost analytic transformation in an almost Kähler space which does not change the volume of the space is a motion.*

Also since we have

$$\frac{1}{2}\nabla^i\nabla_i(v^h v_h) = (\nabla^i\nabla_i v^h)v_h + (\nabla^j v^i)(\nabla_j v_i) = -K_{ji}v^j v^i + (\nabla^j v^i)(\nabla_j v_i)$$

for a contravariant almost analytic vector in a compact almost Kähler space, integration of the above equation gives

$$\int K_{ji}v^j v^i d\sigma = \int (\nabla^j v^i)(\nabla_j v_i)d\sigma,$$

and hence we have

THEOREM 4.2. *In a compact almost Kähler space, we have*

$$\int K_{ji}v^j v^i d\sigma \geqq 0$$

*for a contravariant almost analytic vector field and if the equality sign occurs, then $v^h$ is parallel.*

As a corollary to this theorem we have

COROLLARY 1. *In a compact almost Kähler space with negative definite Ricci curvature, there exists no contravariant almost analytic vector field other than the zero vector.*

*In a compact almost Kähler space with negative semi-definite Ricci curvature, a contravariant almost analytic vector field is parallel.*

## § 5. Projective and conformal Killing vectors

Let $v^h$ be a projective Killing vector, then we have

$$\underset{v}{\mathfrak{L}}\left\{ {h \atop j\ i} \right\} = \nabla_j\nabla_i v^h + K_{kji}{}^h v^k = p_j A_i^h + p_i A_j^h, \qquad (5.1)$$

$p_j$ being a certain covariant vector. From (5.1), we get, by contraction with respect to $h$ and $i$,

$$p_j=\frac{1}{2n+1}\nabla_j\nabla_r v^r. \tag{5.2}$$

Consequently transvecting (5.1) by $g^{ji}$ we find

$$\nabla^i\nabla_i v^h+K_i{}^h v^i=\frac{2}{2n+1}\nabla^h\nabla_r v^r. \tag{5.3}$$

Thus, if a contravariant almost analytic vector in an almost Kähler space is a projective Killing vector, we have

$$\nabla^h\nabla_r v^r=0,$$

from which

$$\nabla_r v^r=\text{const.}$$

But, when the space is compact and orientable, Green's theorem

$$\int(\nabla_r v^r)d\sigma=0$$

gives us

$$\nabla_r v^r=0,$$

and consequently the vector $v^h$ is a Killing vector and we have

THEOREM 5.1. *If a contravariant almost analytic vector in a compact almost Kähler space is a projective Killing vector, it is a Killing vector*. (S. Tachibana [3]).

Next let $v^h$ be a conformal Killing vector, then we have

$$\underset{v}{\mathfrak{L}}\, g_{ji}=2\Phi g_{ji}, \tag{5.4}$$

from which

$$\underset{v}{\mathfrak{L}}\left\{{h \atop j\ i}\right\}=\nabla_j\nabla_i v^h+K_{kji}{}^h v^k=\Phi_j A_i^h+\Phi_i A_j^h-g_{ji}\Phi^h, \tag{5.5}$$

where

$$\Phi_j=\nabla_j\Phi \quad\text{and}\quad \Phi^h=\Phi_i g^{ih}.$$

From (5.5), we get, by contraction with respect to $h$ and $i$,

$$\Phi_j = \frac{1}{2n} \nabla_j \nabla_r v^r. \tag{5.6}$$

Consequently, transvecting (5.5) with $g^{ji}$, we find

$$\nabla^i \nabla_i v^h + K_i{}^h v^i = -\frac{n-1}{n} \nabla^h \nabla_r v^r. \tag{5.7}$$

Substituting this into (4.4), we get

$$\int \left[ -\frac{n-1}{n} v_h \nabla^h \nabla_r v^r + \frac{1}{2} S^{ji} S_{ji} \right] d\sigma = 0$$

or

$$\int \left[ \frac{n-1}{n} (\nabla_r v^r)^2 + \frac{1}{2} S^{ji} S_{ji} \right] d\sigma = 0, \tag{5.8}$$

from which, for $2n > 2$,

$$\nabla_r v^r = 0 \quad \text{and} \quad S_{ji} = 0,$$

and, for $2n = 2$,

$$S_{ji} = 0.$$

Thus we have

THEOREM 5.2. *In a compact almost Kähler space, a conformal Killing vector is a Killing vector for $2n > 2$ and is contravariant almost analytic* for $2n \geqq 2$.

(S. I. Goldberg [4], [5]; A. Lichnerowicz [18], [21]; K. Yano [17]; S. Tachibana [3]).

## § 6. Covariant almost analytic vectors

Let us consider a self-conjugate covariant vector field $w_i = (w_\lambda, w_{\bar\lambda})$ in a complex manifold. The condition for $w_i$ to be a covariant analytic vector field is given by

$$\partial_{\bar\mu} w_\lambda = 0, \qquad \partial_\mu w_{\bar\lambda} = 0 \tag{6.1}$$

or by

$$*O_{ji}^{ts}\partial_t w_s = 0; \tag{6.2}$$

that is, by the fact that $\partial_j w_i$ is pure. Equation (6.2) is equivalent to

$$-F_j{}^t\partial_t w_i + F_i{}^t\partial_j w_t = 0.$$

If we introduce an affine connexion $\Gamma_j{}^h{}_i$ without torsion in the manifold and denote by $\nabla_j$ the covariant derivative with respect to $\Gamma_j{}^h{}_i$, the above equation can be written as

$$-F_j{}^t\nabla_t w_i + F_i{}^t\nabla_j w_t - F_j{}^t\Gamma_t{}^r{}_i w_r + F_i{}^t\Gamma_j{}^r{}_t w_r = 0. \tag{6.3}$$

On the other hand, we have

$$\nabla_j F_i{}^r = F_i{}^t\Gamma_j{}^r{}_t - F_t{}^r\Gamma_j{}^t{}_i$$

by virtue of the fact that $F_i{}^h$ has numerical components. From this equation, we get

$$\nabla_j F_i{}^r - \nabla_i F_j{}^r = F_i{}^t\Gamma_j{}^r{}_t - F_j{}^t\Gamma_i{}^r{}_t$$

and consequently from (6.3) and this equation, we get

$$(\nabla_j F_i{}^r - \nabla_i F_j{}^r)w_r - F_j{}^t\nabla_t w_i + F_i{}^t\nabla_j w_t = 0$$

or

$$(\partial_j F_i{}^r - \partial_i F_j{}^r)w_r - F_j{}^t\partial_t w_i + F_i{}^t\partial_j w_t = 0, \tag{6.4}$$

which is a tensor equation.

Thus, in an almost complex space, we define a *covariant almost analytic vector field* as a vector which satisfies (6.4).

In an almost Hermite space, the equation (6.4) may be written as

$$(\nabla_j F_i{}^h - \nabla_i F_j{}^h)w_h - F_j{}^t\nabla_t w_i + F_i{}^t\nabla_j w_t = 0, \tag{6.5}$$

where the covariant derivative is with respect to the Christoffel symbols $\left\{ {h \atop j\ i} \right\}$.

The equation (6.5) can also be written as

$$\nabla_j(F_i{}^h w_h) - \nabla_i(F_j{}^h w_h) - F_j{}^t(\nabla_t w_i - \Delta_i w_t) = 0. \tag{6.6}$$

But in an almost Kähler space, we have

$$\nabla_j F_i^{\ h} - \nabla_i F_j^{\ h} = -\nabla^h F_{ji} \tag{6.7}$$

and consequently

$$\nabla_j (F_i^{\ h} w_h) - \nabla_i (F_j^{\ h} w_h) = -\underset{w}{\mathfrak{L}} F_{ji}. \tag{6.8}$$

Thus (6.5) may be written as

$$w^t \nabla_t F_{ji} + F_j^{\ t} \nabla_t w_i - F_i^{\ t} \nabla_j w_t = 0 \tag{6.9}$$

and (6.6) as

$$\underset{w}{\mathfrak{L}} F_{ji} + F_j^{\ t} (\nabla_t w_i - \nabla_i w_t) = 0. \tag{6.10}$$

Equations (6.4), (6.5), (6.6), (6.9) and (6.10) are all equivalent in an almost Kähler space.

Taking the symmetric part of (6.10) with respect to $j$ and $i$, we find

$$*O_{ji}^{ts} (\nabla_t w_s - \nabla_s w_t) = 0 \tag{6.11}$$

which shows that $\nabla_j w_i - \nabla_i w_j$ is pure for a covariant almost analytic vector, and consequently, $F^{ji}$ being hybrid in $j$ and $i$, we have

$$F^{ji} (\nabla_j w_i - \nabla_i w_j) = 0. \tag{6.12}$$

Now if we put

$$F_i^{\ h} w_h = -\tilde{w}_i, \tag{6.13}$$

then (6.6) becomes

$$\nabla_j \tilde{w}_i - \nabla_i \tilde{w}_j + F_j^{\ t} (\nabla_t w_i - \nabla_i w_t) = 0. \tag{6.14}$$

This is of course satisfied when $w_i$ and $\tilde{w}_i$ are both closed Thus we have

THEOREM 6.1. *If $w_i$ and $\tilde{w}_i$ are both closed in an almost Kähler space, then $w_i$ is covariant almost analytic.*

The equation (6.14) is equivalent to (S. Tachibana [3])

$$\nabla_j (-w_i) - \nabla_i (-w_j) + F_j^{\ t} (\nabla_t \tilde{w}_i - \nabla_i \tilde{w}_t) = 0,$$

from which we have

THEOREM 6.2. *If $w_i$ is covariant almost analytic in an almost Kähler space, then $\tilde{w}_i$ is also covariant almost analytic.* (S. Tachibana [3]).

From (6.10) we have

THEOREM 6.3. *If a covariant almost analytic vector field $w_i$ in an almost Kähler space is closed, then we have*

$$\underset{w}{\mathfrak{L}} F_{ji}=0, \tag{6.15}$$

*that is, the infinitesimal transformation defined by $w^h$ does not change the exterior differential form $F_{ji}\, d\xi^j \wedge d\xi^i$.*

Now for the divergence of $\tilde{w}_i$, we have

$$\nabla^j\tilde{w}_j = -\nabla^j(F_j{}^i w_i) = -F^{ji}\nabla_j w_i = -\frac{1}{2}F^{ji}(\nabla_j w_i - \nabla_i w_j) = 0$$

by virtue of (6.12). Thus from (6.14) and this equation, we have

THEOREM 6.4. *If a covariant almost analytic vector field $w_i$ in an almost Kähler space is closed, the vector $\tilde{w}_i$ is harmonic.* (S. Tachibana [3]).

But we know that under the assumptions of Theorem 6.4, the vector $\tilde{w}_i$ is also covariant almost analytic. The vector $\tilde{w}_i$, being harmonic, is closed. Thus applying again Theorem 6.4 we see that $w_i$ is harmonic and hence we have

THEOREM 6.5. *If a covariant almost analytic vector field in an almost Kähler space is closed, it is harmonic.*

Transvecting (6.9) with $\nabla_k F^{ji}$, we find

$$(\nabla_k F^{ji})(\nabla_t F_{ji}) w^t = 0 \tag{6.16}$$

from which

$$(\nabla_k F^{ji}) w^k (\nabla_t F_{ji}) w^t = 0. \tag{6.17}$$

The metric being positive definite, we have from (6.17)

$$w^t \nabla_t F_{ji} = 0. \tag{6.18}$$

Thus from (6.9), we get

$$F_j{}^t\nabla_t w_i - F_i{}^t \nabla_j w_t = 0 \tag{6.19}$$

Conversely if (6.18) and (6.19) are satisfied, then (6.9) is satisfied and $w_i$ is covariant almost analytic and hence

THEOREM 6.6. *A necessary and sufficient condition for a vector field $w_i$ in an almost Kähler space to be covariant almost analytic is that*

$$w^t \nabla_t F_{ji} = 0, \qquad F_j{}^t \nabla_t w_i - F_i{}^t \nabla_j w_t = 0.$$

(S. Sawaki [2]; K. Yano and M. Ako [1]).

From (1.12) and Theorem 6.6, we have

THEOREM 6.7. *For a covariant almost analytic vector $w_i$ in an almost Kähler space, we have*

$$N_{ji}{}^h w_h = 0. \tag{6.20}$$

Suppose that a vector $w^h$ is contravariant and at the same time covariant almost analytic in an almost Kähler space, then we have

$$w^t \nabla_t F_i{}^h \cdot \ F_i{}^t \nabla_t w^h + F_t{}^h \nabla_i w^t = 0$$

or

$$w^t \nabla_t F_{ih} - F_i{}^t \nabla_t w_h - F_h{}^t \nabla_i w_t = 0$$

and

$$w^t \nabla_t F_{ih} + F_i{}^t \nabla_t w_h - F_h{}^t \nabla_i w_t = 0,$$

from which

$$F_i{}^t \nabla_t w_h = 0,$$

and hence we have

THEOREM 6.8. *If a vector is contravariant and at the same time covariant almost analytic in an almost Kähler space, it is covariantly constant.* (S. Tachibana [3]).

Now the Nijenhuis tensor satisfies

$$N_{ji}{}^h = -2O_{ji}^{ts}(\nabla_t F_s{}^r - \nabla_s F_t{}^r) F_r{}^h. \tag{6.21}$$

Thus transvecting (6.21) with $\nabla^j w^i$ and remembering that

$$O_{ji}^{ts}(\nabla^j w^i - \nabla^i w^j) = \nabla^t w^s - \nabla^s w^t,$$

for a covariant almost analytic vector, we find

$$N_{ji}{}^h (\nabla^j w^i) = -2(\nabla_j F_i{}^r - \nabla_i F_j{}^r)(\nabla^j w^i) F_r{}^h, \tag{6.22}$$

from which

$$(\nabla_j F_i^{\ r} - \nabla_i F_j^{\ r})(\nabla^j w^i) F_r^{\ h} w_h = 0 \tag{6.23}$$

by virtue of (6.20).

Now applying the operator $F_h^{\ i}\nabla_j$ to (6.9), we get

$$\nabla^j \nabla_j w_i - (2K^*_{ji} - K_{ji}) w^j + (\nabla^t w^s) G_{tsr} F_i^{\ r} = 0, \tag{6.24}$$

which is a necessary condition for a vector $w_i$ in an almost Kähler space to be covariant almost analytic.

## § 7. A necessary and sufficient condition for a vector to be covariant almost analytic

We defined $\tilde{w}_i$ by

$$\tilde{w}_i = -F_i^{\ t} w_t, \tag{7.1}$$

and consequently

$$\nabla_j \tilde{w}_i = -(\nabla_j F_i^{\ t}) w_t - F_i^{\ t} \nabla_j w_t$$

and

$$\nabla^j \nabla_j \tilde{w}_i = -(\nabla^j \nabla_j F_i^{\ t}) w_t - 2(\nabla_j F_{it}) \nabla^j w^t - F_i^{\ t} \nabla^j \nabla_j w_t \, .$$

Substituting (2.12) into this equation, we find

$$\begin{aligned} \nabla^j \nabla_j \tilde{w}_i = {} & [(K_{ir} - K^*_{ir}) F_t^{\ r} - (K_{tr} - K^*_{tr}) F_i^{\ r}] w^t \\ & - 2(\nabla_j F_{it}) \nabla^j w^t - F_i^{\ t} \nabla^j \nabla_j w_t \end{aligned}$$

from which

$$\begin{aligned} \nabla^j \nabla_j \tilde{w}_i - K_i^{\ r} \tilde{w}_r = {} & -F_i^{\ t} [\nabla^j \nabla_j w_t + K_t^{\ r} w_r] - (K^*_{ir} F_t^{\ r} - K^*_{tr} F_i^{\ r}) w^t \\ & - 2(\nabla_j F_{it}) \nabla^j w^t , \end{aligned}$$

and consequently

$$\begin{aligned} & (\nabla^j \nabla_j \tilde{w}_i - K_i^{\ r} \tilde{w}_r) \tilde{w}^i \\ & = [\nabla^j \nabla_j w_i - (2K^*_{ji} - K_{ji}) w^j] w^i + 2(\nabla_j F_{it})(\nabla^j w^i) F_s^{\ t} w^s , \end{aligned} \tag{7.2}$$

by virtue of the relation $K^*_{ts} F_j^{\ t} F_i^{\ s} w^j w^i = K^*_{ji} w^j w^i$ .

Using the relation

$$2\nabla_j F_{it} = \nabla_j F_{it} + \nabla_i F_{jt} - \nabla_t F_{ji},$$

we can write (7.2) in the form

$$(\nabla^j \nabla_j \tilde{w}_i - K_i{}^r \tilde{w}_r)\tilde{w}^i = [\nabla^j \nabla_j w_i - \\ -(2K^*_{ji} - K_{ji})w^j + G_{tsr}(\nabla^t w^s)F_i{}^r]w^i - (\nabla_t F_{ji})(\nabla^j w^i)F_s{}^t w^s. \quad (7.3)$$

On the other hand, for the tensor $T_{ji}$ defined by

$$T_{ji} = w^t \nabla_t F_{ji} + F_j{}^t \nabla_t w_i - F_i{}^t \nabla_j w_t, \quad (7.4)$$

we have the identity

$$\nabla^j(T_{ji}\tilde{w}^i) + [\nabla^j \nabla_j w_i - (2K^*_{ji} - K_{ji})w^j + G_{tsr}(\nabla^t w^s)F_i{}^r]w^i \\ + (\nabla_t F_{ji})(\nabla^j w^i)F_s{}^t w^s + \frac{1}{2}T^{ji}T_{ji} = 0. \quad (7.5)$$

Forming $\frac{1}{2}[(7.3)+(7.5)]$, we obtain

$$\frac{1}{2}(\nabla^j \nabla_j \tilde{w}_i - K_i{}^r \tilde{w}_r)\tilde{w}^i + \frac{1}{2}\nabla^j(T_{ji}\tilde{w}^i)$$

$$= [\nabla^j \nabla_j w_i - (2K^*_{ji} - K_{ji})w^j + G_{tsr}(\nabla^t w^s)F_i{}^r]w^i + \frac{1}{4}T^{ji}T_{ji} = 0. \quad (7.6)$$

Supposing that the almost Kähler space is compact, and integrating (7.6), we find

$$\frac{1}{2}\int (\nabla^j \nabla_j \tilde{w}_i - K_i{}^r \tilde{w}_r)\tilde{w}^i d\sigma$$

$$= \int \left[ \{\nabla^j \nabla_j w_i - (2K^*_{ji} - K_{ji})w^j + G_{tsr}(\nabla^t w^s)F_i{}^r\}w^i + \frac{1}{4}T^{ji}T_{ji} \right] d\sigma.$$

On the other hand, we have

$$\int \Big[ (\nabla^j \nabla_j \tilde{w}_i - K_i{}^r \tilde{w}_r)\tilde{w}^i \\ + \frac{1}{2}(\nabla^j \tilde{w}^i - \nabla^i \tilde{w}^j)(\nabla_j \tilde{w}_i - \nabla_i \tilde{w}_j) + (\nabla_i \tilde{w}^i)^2 \Big] d\sigma = 0.$$

Thus, from these two integral formulae, we get

$$\int\Big[\{\nabla^j\nabla_j w_i-(2K^*_{ji}-K_{ji})w^j+G_{tsr}(\nabla^t w^s)F_i{}^r\}w^i+\frac{1}{4}T^{ji}T_{ji}$$
$$+\frac{1}{4}(\nabla^j\tilde{w}^i-\nabla^i\tilde{w}^j)(\nabla_j\tilde{w}_i-\nabla_i\tilde{w}_j)+\frac{1}{2}(\nabla_i\tilde{w}^i)^2\Big]d\sigma=0. \quad (7.7)$$

Thus, if the condition (6.24) is satisfied, we conclude from (7.7) that

$$T_{ji}=0, \quad (7.8)$$

$$\nabla_j\tilde{w}_i-\nabla_i\tilde{w}_j=0 \quad (7.9)$$

and

$$\nabla_i\tilde{w}^i=0. \quad (7.10)$$

Thus we have

THEOREM 7.1. *In order that a vector field $w_i$ in a compact almost Kähler space be covariant almost analytic, it is necessary and sufficient that*

$$\nabla^j\nabla_j w_i-(2K^*_{ji}-K_{ji})w^j+G_{tsr}(\nabla^t w^s)F_i{}^r=0. \quad (7.11)$$

(S. Tachibana [3]).

On account of (2.9) and the first equation in Theorem 6.6, we have

$$\begin{aligned}-2(K^*_{ji}-K_{ji})w^j&+G_{tsr}(\nabla^t w^s)F_i{}^r\\ &=-2F_i{}^l(\nabla_a\nabla_j F_l{}^a)w^j+F_i{}^l(\nabla^a w^j)G_{ajl}\\ &=2F_i{}^l(\nabla_j F_{la})(\nabla^a w^j)+F_i{}^l(\nabla^a w^j)(\nabla_a F_{jl}+\nabla_j F_{al})\\ &=F_i{}^l(\nabla^a w^j)(\nabla_a F_{jl}-\nabla_j F_{al})\end{aligned}$$

and consequently, from (7.11),

$$(\nabla^a\nabla_a w_i-K_{ji}w^j)w^i=-F_i{}^l(\nabla^a w^j)(\nabla_a F_{jl}-\nabla_j F_{al})w^i=0$$

by virtue of (6.23). Thus, from the integral formula

$$\int_M\Big[(\nabla^a\nabla_a w_i-K_{ji}w^j)w^i+\frac{1}{2}(\nabla^j w^i-\nabla^i w^j)(\nabla_j w_i-\nabla_i w_j)$$
$$+(\nabla_i w^i)^2\Big]d\sigma=0,$$

we have

$$\nabla_j w_i - \nabla_i w_j = 0, \qquad \nabla_i w^i = 0,$$

and hence

THEOREM 7.2. *A covariant almost analytic vector in a compact almost Kähler space is harmonic.*

Take a covariant almost analytic vector $w_i$ in a compact almost Kähler space, then, following Theorem 6.2 $\tilde{w}_i$ is also covariant almost analytic. Thus, following Theorem 7.2, $w_i$ and $\tilde{w}_i$ are both harmonic.

Conversely, if $w_i$ and $\tilde{w}_i$ are both harmonic, they are both closed and consequently, following Theorem 6.1 $w_i$ is covariant almost analytic. Hence we have

THEOREM 7.3. *In order that a vector $w_i$ in a compact almost Kähler space be covariant almost analytic, it is necessary and sufficient that $w_i$ and $\tilde{w}_i$ are both harmonic.* (S. Tachibana [3])

Thus applying a theorem of Bochner, we have

THEOREM 7.4. *In a compact almost Kähler space with positive definite Ricci curvature, there exists no covariant almost analytic vector field other than the zero vector.*

*In a compact almost Kähler space with positive semi-definite Ricci curvature, a covariant almost analytic vector is parallel.*

Suppose that $w^h$ is contravariant almost analytic, then from (3.10) and (3.13) and (7.7), we find

$$\int K^*_{ji} w^j w^i d\sigma$$

$$= \frac{1}{8}\int [T^{ji}T_{ji} + (\nabla^j \tilde{w}^i - \nabla^i \tilde{w}^j)(\nabla_j \tilde{w}_i - \nabla_i \tilde{w}_j) + 2(\nabla_i \tilde{w}^i)^2]\, d\sigma \quad (7.12)$$

from which we have

THEOREM 7.5. *In a compact almost Kähler space we have*

$$\int K^*_{ji} w^j w^i d\sigma \geqq 0$$

*for a contravariant almost analytic vector field $w^h$.*

*If the equality sign occurs, then $w^h$ is covariant almost analytic and consequently, being contravariant and at the same time covariant almost analytic, $w^h$ is parallel.* (S. Tachibana [3])

## § 8. Lie algebra of analytic vectors

Let us consider a compact almost Kähler–Einstein space with positive curvature scalar, that is, a compact almost Kähler space satisfying

$$K_{ji} = cg_{ji}, \tag{8.1}$$

where

$$c = \frac{K}{2n} \tag{8.2}$$

is supposed to be positive.

Then a necessary and sufficient condition that $v^h$ is a contravariant almost analytic vector is given by

$$\nabla^j \nabla_j v^h + cv^h = 0, \tag{8.3}$$

from which we have

$$\nabla^j \nabla_j (\nabla_i v^i) + 2c \nabla_i v^i = 0; \tag{8.4}$$

that is, the function

$$f = \nabla_i v^i \tag{8.5}$$

satisfies

$$\nabla^j \nabla_j f + 2cf = 0. \tag{8.6}$$

From this equation, we get

$$\nabla^j \nabla_j \nabla_i f + c \nabla_i f = 0 \tag{8.7}$$

and hence $\nabla_i f$ is also a contravariant almost analytic vector.

If we put

$$p_i = v_i + \frac{1}{2c} \nabla_i f, \tag{8.8}$$

then $p_i$ is contravariant almost analytic and

$$\nabla^i p_i = \nabla^i v_i + \frac{1}{2c} \nabla^i \nabla_i f = \nabla^i v_i - f = 0, \tag{8.9}$$

and consequently $p_i$ is a Killing vector.

Thus from (8.8), we have

$$v_i = p_i - \frac{1}{2c} \nabla_i f, \tag{8.10}$$

where $p_i$ is a Killing vector and $\nabla_i f$ is a contravariant almost analytic vector.

Conversely, if $p_i$ is a Killing vector and $f$ is a solution of (8.6), then $v_i$ given by (8.10) is a contravariant almost analytic vector.

We denote by $L$ and $L_1$ the Lie algebra of contravariant almost analytic vectors and that of Killing vectors respectively and by $L_2$ the vector space of the gradient of solutions of (8.6). Then we have

$$L = L_1 + L_2, \tag{8.11}$$

where the plus sign denotes the direct sum.

Take a Killing vector $p^h$ and a solution $f$ of (8.6) and put

$$f_i = \nabla_i f. \tag{8.12}$$

Then we have

$$[p, f]_i = p^j \nabla_j f_i - f^j \nabla_j p_i = p^j \nabla_i f_j + f^j \nabla_i p_j = \nabla_i (p^j f_j). \tag{8.13}$$

On the other hand, we have

$$\nabla^k \nabla_k (p^j f_j) = (\nabla^k \nabla_k p^j) f_j + 2(\nabla^k p^j) \nabla_k f_j + p^j (\nabla^k \nabla_k f_j).$$

Substituting

$$\nabla^k \nabla_k p^j = -c p^j \quad \text{and} \quad \nabla^k \nabla_k f_j = -c f_j$$

in this equation and taking account of

$$\nabla^k p^j = -\nabla^k p^j \quad \text{and} \quad \nabla_k f_j = \nabla_j f_k,$$

we find

$$\nabla^k \nabla_k (p^j f_j) + 2c(p^j f_j) = 0. \tag{8.14}$$

Equations (8.13) and (8.14) show that

$$[L_1, L_2] \subset L_2. \tag{8.15}$$

Take next a solution $g$ of (8.6) and put

$$g_i = \nabla_i g. \tag{8.16}$$

Since $f^h$ and $g^h$ are both contravariant almost analytic, $[f, g]^h$ is also contravariant almost analytic. On the other hand, we have

$$\nabla_i [f, g]^i = \nabla_i (f^j \nabla_j g^i - g^j \nabla_j f^i) = f^j \nabla_i \nabla_j g^i - g^j \nabla_i \nabla_j f^i.$$

But,

$$\nabla_i \nabla_j g^i - \nabla_j \nabla_i g^i = K_{ji} g^i = c g_j,$$

from which

$$\nabla_i \nabla_j g^i = -c g_j$$

by virtue of

$$\nabla_i g^i = \nabla_i \nabla^i g = -2cg.$$

Similarly

$$\nabla_i \nabla_j f^i = -c f_j.$$

Thus we have

$$\nabla_i [f, g]^i = 0,$$

which shows that $[f, g]^h$ is a Killing vector and consequently

$$[L_2, L_2] \subset L_1. \tag{8.17}$$

Thus we have proved

THEOREM 8.1. *In a compact almost Kähler–Einstein space with positive curvature scalar, we have*

$$L = L_1 + L_2$$

*where* $L$ and $L_1$ *are the Lie algebras of contravariant almost analytic vectors and of Killing vectors respectively and* $L_2$ *is the vector space of gradients of solutions of*

$$\nabla^i \nabla_i f + 2cf = 0, \qquad c = \frac{k}{2n},$$

*and also*

$$[L_1, L_2] \subset L_2, \qquad [L_2, L_2] \subset L_1.$$

(S. Tachibana [3]).

CHAPTER VIII

# ALMOST TACHIBANA SPACES

## § 1. Almost Tachibana spaces

An almost Tachibana space is first of all an almost Hermite space, that is, a $2n$-dimensional space with an almost complex structure $F$:

$$F_j^{\ i}F_i^{\ h}=-A_j^h, \tag{1.1}$$

and with a Riemannian metric $g_{ji}$ satisfying

$$F_j^{\ t}F_i^{\ s}g_{ts}=g_{ji}, \tag{1.2}$$

from which

$$F_{ji}=-F_{ij}, \tag{1.3}$$

where

$$F_{ji}=F_j^{\ t}g_{ti} \tag{1.4}$$

and finally has the property that the skew-symmetric tensor $F_{ih}$ is a Killing tensor:

$$\nabla_j F_{ih}+\nabla_i F_{jh}=0, \tag{1.5}$$

from which

$$\nabla_j F_i^{\ h}+\nabla_i F_j^{\ h}=0 \tag{1.6}$$

and

$$F_i=-\nabla_j F_i^{\ j}=0. \tag{1.7}$$

Now the Nijenhuis tensor $N_{ji}^{\ \ h}$ is written in the form

$$N_{ji}^{\ \ h}=-4(\nabla_j F_i^{\ t})F_t^{\ h}+2G_{ji}^{\ \ t}F_t^{\ h}+F_j^{\ t}G_{ti}^{\ \ h}-F_i^{\ t}G_{tj}^{\ \ h}, \tag{1.8}$$

and consequently, we have

THEOREM 1.1. *In an almost Tachibana space, we have*

$$N_{ji}{}^{h} = -4(\nabla_j F_i{}^t) F_t{}^h \tag{1.9}$$

*and consequently* $\nabla_j F_i{}^h$ *is pure in j and i.*

When the Nijenhuis tensor vanishes, the almost Tachibana space is called a Tachibana space. In this case, we have from (1.9)

$$\nabla_j F_i{}^h = 0;$$

hence we have

THEOREM 1.2. *A Tachibana space is a Kähler space.*

By a straightforward calculation, we find

$$3\nabla_j F_{ih} - F_{jih} = G_{jih} - G_{jhi}, \tag{1.10}$$

from which

THEOREM 1.3. *In an almost Tachibana space, we have*

$$3\nabla_j F_{ih} = F_{jih} \tag{1.11}$$

*and consequently* $\nabla_j F_{ih}$ *is skew-symmetric in all its indices.*

In general the Nijenhuis tensor $N_{jih}$ satisfies

$$N_{jih} + N_{jhi} = -F_{jit} F_h{}^t - F_{jht} F_i{}^t + 2F_j{}^t G_{iht} \tag{1.12}$$

and the tensor $G_{jih}$ satisfies

$$2{}^*O_{ji}^{ts} G_{tsh} = -(F_j{}^t F_{tsi} + F_i{}^t F_{tsj}) F_h{}^s. \tag{1.13}$$

From these two equations we get

$$N_{jih} + N_{jhi} = 2F_j{}^t O_{ih}^{sr} G_{srt}, \tag{1.14}$$

and hence

THEOREM 1.4. *In an almost Tachibana space, the Nijenhuis tensor* $N_{jih}$ *is skew-symmetric in all its indices.*

Now, from (1.9) and (1.11), we get

$$N_{jih} = \frac{4}{3} F_{jit} F_h{}^t \tag{1.15}$$

or equivalently

$$F_{jih} = -\frac{3}{4} N_{jit} F_h{}^t \tag{1.16}$$

and hence

THEOREM 1.5. *In an almost Tachibana space, the tensor* $N_{jit} F_h{}^t$ *is skew-symmetric in all its indices.*

## § 2. Curvature tensors in an almost Tachibana space

Applying the Ricci indentity to the tensor $F_i^{\ h}$, we find

$$\nabla_k\nabla_j F_i^{\ h} - \nabla_j\nabla_k F_i^{\ h} = K_{kjt}^{\ \ \ h} F_i^{\ t} - K_{kji}^{\ \ \ t} F_t^{\ h}. \tag{2.1}$$

Contraction with respect to $k$ and $h$ in this equation gives

$$\nabla_t\nabla_j F_i^{\ t} = K_{jt} F_i^{\ t} - H_{ji}, \tag{2.2}$$

from which

$$\nabla_t G_{ji}^{\ \ t} = K_{jt} F_i^{\ t} + K_{it} F_j^{\ t}; \tag{2.3}$$

hence we have

THEOREM 2.1. *In an almost Tachibana space, the tensor $K_{jt}F_i^{\ t}$ is skew-symmetric and consequently $K_{ji}$ is hybrid.*

The equation (2.2) can be written as

$$\nabla_t\nabla_j F_i^{\ t} = (K_{jt} - K^*_{jt}) F_i^{\ t}, \tag{2.4}$$

from which

THEOREM 2.2. *In order that $K_{ji} = K^*_{ji}$ in an almost Tachibana space, it is necessary and sufficient that*

$$\nabla^t F_{tih} = 0 \quad or \quad \nabla^t\nabla_t F_{ih} = 0. \tag{2.5}$$

Now $F_{ji}$ in an almost Tachibana space is a Killing tensor and consequently it satisfies

$$g^{sr}\nabla_s\nabla_r F_{ji} + \frac{1}{2}(K_j^{\ t} F_{ti} + K_i^{\ t} F_{jt}) + \frac{1}{2} K_{ji}^{\ \ ts} F_{ts} = 0 \tag{2.6}$$

or

$$\nabla^t\nabla_t F_{ji} + K_j^{\ t} F_{ti} + H_{ji} = 0. \tag{2.7}$$

This is written as

$$\nabla^t\nabla_t F_{ji} = (K_{jt} - K^*_{jt}) F_i^{\ t} \tag{2.8}$$

and is equivalent to (2.4) by virtue of the relation

$$\nabla_j F_{it} = \nabla_t F_{ji}\,. \tag{2.9}$$

From (2.7), we get, taking the symmetric part,

$$K_j^{\ t} F_{ti} + K_i^{\ t} F_{tj} = 0$$

or

$$2O_{ji}^{ts}K_{ts}=K_{ji}-F_j{}^tF_i{}^sK_{ts}=0, \tag{2.10}$$

and hence, again

THEOREM 2.3. *The Ricci tensor* $K_{ji}$ *in an almost Tachibana space is hybrid.*

Transvecting (2.9) with $F^{ji}$, we find

$$(\nabla_j F_{it})F^{ji}=0,$$

from which, by operating $\nabla_k$,

$$(\nabla_k\nabla_j F_{it})F^{ji}+(\nabla_j F_{it})(\nabla_k F^{ji})=0$$

or

$$(\nabla_k\nabla_j F_{ih})F^{ji}+(\nabla_k F_{ji})(\nabla_h F^{ji})=0 \tag{2.11}$$

by virtue of the skew-symmetry of $\nabla_k F_{ji}$.

On the other hand, we have, from the Ricci identity

$$\nabla_k\nabla_j F_{ih}=\nabla_j\nabla_k F_{ih}-K_{kji}{}^tF_{th}-K_{kjh}{}^tF_{it}$$

or

$$\nabla_k\nabla_j F_{ih}=\nabla_j\nabla_i F_{hk}+K_{kjit}F_h{}^t-K_{kjht}F_i{}^t.$$

Substituting this into (2.11), we obtain

$$(\nabla_j\nabla_i F_{hk})F^{ji}+K_{kjit}F_h{}^tF^{ji}-K_{kh}+(\nabla_k F_{ji})(\nabla_h F^{ji})=0.$$

Substituting

$$\begin{aligned}(\nabla_j\nabla_i F_{hk})F^{ji}&=-\frac{1}{2}(K_{jih}{}^tF_{tk}+K_{jik}{}^tF_{ht})F^{ji}\\&=-(H_h{}^tF_{tk}+H_k{}^tF_{ht})=-K^*_{hk}+K^*_{kh}\end{aligned}$$

and

$$K_{kjit}F^{ji}=\frac{1}{2}(K_{kjit}-K_{kijt})F^{ji}=-H_{kt}=-K^*_{ks}F_t{}^s$$

into the above equation, we find

$$(\nabla_k F_{ji})(\nabla_h F^{ji})=K_{kh}-2K^*_{kh}+K^*_{hk}. \tag{2.12}$$

Since the left hand side of this equation is symmetric with respect to $k$ and $h$, we get

$$K^*_{kh} = K^*_{hk}, \tag{2.13}$$

and (2.12) becomes

$$(\nabla_k F_{ji})(\nabla_h F^{ji}) = K_{kh} - K^*_{kh}. \tag{2.14}$$

Since

$$(\nabla_k F_{ji})(\nabla_h F^{ji}) v^k v^h \geqq 0,$$

we have

THEOREM 2.4. *In an almost Tachibana space, we have*

$$K_{ji} v^j v^i \geqq K^*_{ji} v^j v^i. \tag{2.15}$$

Now suppose that an almost Tachibana space reduces to a Kähler space, then from (2.14), we have

$$K_{kh} = K^*_{kh}.$$

Conversely, if this equation is satisfied in an almost Tachibana space, then we have from (2.14)

$$(\nabla_k F_{ji})(\nabla_h F^{ji}) = 0,$$

from which

$$\nabla_k F_{ji} = 0;$$

that is, the space is a Kähler space, and hence

THEOREM 2.5. *A necessary and sufficient condition for an almost Tachibana space to reduce to a Kähler space is that*

$$K_{kh} = K^*_{kh}.$$

(S. Kotō [1]).

Now from

$$K^*_{jt} F_i{}^t = H_{ji},$$

we get

$$K^*_{jt} F_i{}^t + K^*_{it} F_j{}^t = 0$$

or

$$K^*_{ji} - F_j{}^t F_i{}^s K^*_{st} = 0.$$

But $K^*_{ji}$ is symmetric and consequently

$$2O^{ts}_{ji}K^*_{ts}=K^*_{ji}-F_j{}^tF_i{}^sK^*_{ts}=0, \tag{2.16}$$

and hence

THEOREM 2.6. *The tensor $K^*_{ji}$ in an almost Tachibana space is hybrid.*

Now suppose that our almost Tachibana space is conformally flat. Then the curvature has the form

$$K_{kjih}=-[g_{kh}C_{ji}-g_{jh}C_{ki}+C_{kh}g_{ji}-C_{jh}g_{ki}],$$

where

$$C_{ji}=-\frac{K_{ji}}{2n-2}+\frac{Kg_{ji}}{2(2n-1)(2n-2)}.$$

Transvecting this equation with $\frac{1}{2}F^{kj}$, we find

$$H_{ji}=F_j{}^tC_{ti}-F_i{}^tC_{tj},$$

from which

$$K^*_{ji}=-H_{jt}F_i{}^t=-2{}^*O^{ts}_{ji}C_{ts}. \tag{2.17}$$

But, $K_{ji}$ and $g_{ji}$ being both hybrid, we have

$${}^*O^{ts}_{ji}C_{ts}=C_{ji},$$

and hence

$$K^*_{ji}=-2C_{ji}, \tag{2.18}$$

from which

$$K_{ji}-K^*_{ji}=\frac{2n-4}{2n-2}K_{ji}+\frac{1}{(2n-1)(2n-2)}Kg_{ji}. \tag{2.19}$$

Thus, if $K_{ji}v^jv^i$ is negative definite, then

$$(K_{ji}-K^*_{ji})v^jv^i$$

is negative definite, which contradicts Theorem 2.4, and consequently we have

THEOREM 2.7. *If an almost Tachibana space is conformally flat, then the Ricci quadratic form cannot be negative definite.*

## § 3. Contravariant almost analytic vectors

Let us consider a contravariant almost analytic vector field $v^h$. By definition it satisfies

$$\underset{v}{\mathfrak{L}} F_i{}^h = v^t\partial_t F_i{}^h - F_i{}^t\partial_t v^h + F_t{}^h\partial_i v^t = 0, \tag{3.1}$$

which can be written as

$$\underset{v}{\mathfrak{L}} F_i{}^h = v^t\nabla_t F_i{}^h - F_i{}^t\nabla_t v^h + F_t{}^h\nabla_i v^t = 0 \tag{3.2}$$

or

$$v^t\nabla_t F_{ih} - F_i{}^t\nabla_t v_h - F_h{}^t\nabla_i v_t = 0, \tag{3.3}$$

in our almost Tachibana space.

We know already that equation (3.3) gives

$$O_{ji}^{ts}(\underset{v}{\mathfrak{L}} g_{ts}) = 0 \tag{3.4}$$

and

$$O_{ts}^{ji}(\underset{v}{\mathfrak{L}} g^{ts}) = 0 \tag{3.5}$$

proving that both $\underset{v}{\mathfrak{L}} g_{ji}$ and $\underset{v}{\mathfrak{L}} g^{ji}$ are hybrid.

Now, by a straightforward computation, we can prove that the tensor

$$F_j{}^tF_{ti}{}^h + F_i{}^tF_{tj}{}^h$$

is pure in $j$ and $i$ and consequently we have

$$(F_j{}^tF_{ti}{}^h + F^{\,t}F\;{}_j{}^h)(\underset{v}{\mathfrak{L}} g^{ji}) = 0,$$

from which

$$F_{ji}{}^h(\underset{v}{\mathfrak{L}} F^{ji}) = 0 \tag{3.6}$$

for a contravariant almost analytic vector $v^h$.

Equation (3.6) can be written also in the form

$$(\nabla_k F_{ji})(\underset{v}{\mathfrak{L}} F^{ji}) = 0, \tag{3.7}$$

or

$$(\nabla_k F_{ji})(v^h \nabla_h F^{ji} - F^{ti} \nabla_t v^j - F^{jt} \nabla_t v^i) = 0. \tag{3.8}$$

On the other hand, we have from (2.14)

$$(\nabla_k F_{ji})(\nabla_h F^{ji}) v^h = (K_{kh} - K^*_{kh}) v^h.$$

Substituting this into (3.9), we obtain

$$(K_{kh} - K^*_{kh}) v^h = -\frac{1}{2} N_{kji} (\nabla^j v^i) \tag{3.9}$$

by virtue of (1.9).

Equations (3.7) or (3.9) is a necessary condition for a vector field $v^h$ to be contravariant almost analytic.

Now applying $\nabla^i$ to (3.2), we have

$$F_t^{\ h} [\nabla^i \nabla_i v^t + K_i^{\ t} v^i] = 0$$

or

$$\nabla^i \nabla_i v^h + K_i^{\ h} v^i = 0. \tag{3.10}$$

This is another necessary condition for a vector $v^h$ to be contravariant almost analytic in an almost Tachibana space. Thus we get

THEOREM 3.1. *In order that $v^h$ be contravariant almost analytic in an almost Tachibana space, it is necessary that*

$$F_{ij}^{\ \ h} (\underset{v}{\mathfrak{L}} F^{ji}) = 0 \quad \textit{or} \quad (K_{kh} - K^*_{kh}) v^h + \frac{1}{2} N_{kji} (\nabla^j v^i) = 0$$

*and*

$$\nabla^i \nabla_i v^h + K_i^{\ h} v^i = 0.$$

## § 4. A necessary and sufficient condition for a vector to be contravariant almost analytic

If we put

$$S^{ji} = g^{jt} (\underset{v}{\mathfrak{L}} F_t^{\ i}), \tag{4.1}$$

we have

$$\frac{1}{2} S^{ji} S_{ji} = \frac{1}{2} v^t v^s (\nabla_t F^{ji})(\nabla_s F_{ji}) - v^t (\nabla_t F^{ji}) F_j^{\ s} (\nabla_s v_i)$$
$$+ v^t (\nabla_t F^{ji}) F_{si} (\nabla_j v^s) + (\nabla^j v^i)(\nabla_j v_i) - F^{jt} F_{si} (\nabla_t v^i)(\nabla_j v^s) \tag{4.2}$$

and

$$\nabla^j\{(\underset{v}{\mathfrak{L}} F_j{}^t)F_t{}^i v_i\} + \left[\nabla^i\nabla_i v^h + K_i{}^h v^i - \frac{1}{2}F_{ji}{}^h(\underset{v}{\mathfrak{L}} F^{ji})\right] v_h$$

$$+\frac{1}{2}S^{ji}S_{ji}=0. \tag{4.3}$$

Thus assuming that our almost Tachibana space is compact, we have, from (4.3)

$$\iint\left[\left\{\nabla^i\nabla_i v^h + K_i{}^h v^i - \frac{1}{2}F_{ji}{}^h(\underset{v}{\mathfrak{L}} F^{ji})\right\}v_h + \frac{1}{2}S^{ji}S_{ji}\right]d\sigma = 0. \tag{4.4}$$

Thus we have

THEOREM 4.1. *A necessary condition for a vector field $v^h$ in an almost Tachibana space to be contravariant almost analytic is that*

$$F_{ji}{}^h(\underset{v}{\mathfrak{L}} F^{ji}) = 0, \qquad \nabla^i\nabla_i v^h + K_i{}^h v^i = 0;$$

*and a sufficient condition for $v^h$ in a compact almost Tachibana space to be contravariant almost analytic is that*

$$\nabla^i\nabla_i v^h + K_i{}^h v^i - \frac{1}{2}F_{ji}{}^h(\underset{v}{\mathfrak{L}} F^{ji}) = 0. \tag{4.5}$$

In exactly the same way as in the case of almost Kähler space, we have

COROLLARY 1. *A Killing vector in a compact almost Tachibana space satisfying $F_{ji}{}^h(\underset{v}{\mathfrak{L}} F^{ji}) = 0$ is contravariant almost analytic.*

COROLLARY 2. *An infinitesimal almost analytic transformation in a compact almost Tachibana space which does not change the volume of the space is a motion.*

THEOREM 4.2. *In a compact almost Tachibana space, we have*

$$\int K_{ji}v^j v^i d\sigma \geqq 0$$

*for a contravariant almost analytic vector field* $v^h$. *If the equality sign occurs, then* $v^h$ *is parallel.*

COROLLARY 1. *In a compact almost Tachibana space with negative definite Ricci curvature, there exists no contravariant almost analytic vector field other than the zero vector.*

*In a compact almost Tachibana space with negative semi-definite Ricci curvature a contravariant almost analytic vector field is parallel.*

## § 5. Projective and conformal Killing vectors

An infinitesimal projective transformation $v^h$ satisfies

$$\underset{v}{\mathfrak{L}}\left\{ {h \atop j\ i} \right\} = \nabla_j\nabla_i v^h + K_{kji}{}^h v^k = A_j^h p_i + A_i^h p_j, \tag{5.1}$$

from which

$$g^{ji}\nabla_j\nabla_i v^h + K_i{}^h v^i - \frac{2}{2n+1}\nabla^h\nabla_t v^t = 0. \tag{5.2}$$

Thus, if the vector $v^h$ defining a projective transformation is at the same time contravariant almost analytic, we have

$$\nabla^h\nabla_t v^t = 0, \qquad \nabla_t v^t = \text{const.}$$

Supposing that our space is compact, we have from this

$$\nabla_t v^t = 0. \tag{5.3}$$

Thus from (5.2) we have

$$g^{ji}\nabla_j\nabla_i v^h + K_i{}^h v^i = 0 \tag{5.4}$$

Equations (5.3) and (5.4) show that $v^h$ is a Killing vector. Thus we have

THEOREM 5.1. *If, in a compact almost Tachibana space, a vector* $v^h$ *defining a projective transformation is at the same time a contravariant almost analytic vector field, then it is a Killing vector.*

An infinitesimal conformal transformation $v^h$ satisfies

$$\underset{v}{\mathfrak{L}}\, g_{ji} = \nabla_j v_i + \nabla_i v_j = 2\Phi g_{ji}, \tag{5.5}$$

from which

$$\underset{v}{\mathfrak{L}} \left\{ {h \atop j\ i} \right\} = \nabla_j \nabla_i v^h + K_{kji}{}^h v^k = A_j^h \Phi_i + A_i^h \Phi_j - \Phi^h g_{ji}, \tag{5.6}$$

and

$$g^{ji} \nabla_j \nabla_i v^h + K_i{}^h v^i + \frac{n-1}{n} \nabla^h \nabla_t v^t = 0. \tag{5.7}$$

Thus if the vector $v^h$ defining a conformal transformation is at the same time contravariant almost analytic, we have

$$\nabla_t v^t = \text{const.}$$

Supposing that our space is compact, we have

$$\nabla_t v^t = 0.$$

Thus from (5.7) we have

$$g^{ji} \nabla_j \nabla_i v^h + K_i{}^h v^i = 0,$$

and hence

THEOREM 5.2.† *If, in a compact almost Tachibana space, a vector $v^h$ defining a conformal transformation is at the same time a contravariant almost analytic vector field, then it is a Killing vector.*

## § 6. Covariant almost analytic vectors

In an almost complex space, a covariant almost analytic vector field is defined as a vector field satisfying

$$(\partial_j F_i{}^r - \partial_i F_j{}^r) w_r - F_j{}^t \partial_t w_i + F_i{}^t \partial_j w_t = 0. \tag{6.1}$$

which is a tensor equation.

† See S. Sawaki [7].

In an almost Hermite space, (6.1) can be written as

$$(\nabla_j F_i{}^h - \nabla_i F_j{}^h) w_h - F_j{}^t \nabla_t w_i + F_i{}^t \nabla_j w_t = 0. \tag{6.2}$$

Taking the symmetric part of (6.2) with respect to $j$ and $i$, we find

$$*O_{ji}^{ts}(\nabla_t w_s - \nabla_s w_t) = 0, \tag{6.3}$$

which shows that $\nabla_j w_i - \nabla_i w_j$ is pure for a covariant almost analytic vector, and consequently, $F^{ji}$ being hybrid in $j$ and $i$, we have

$$F^{ji}(\nabla_j w_i - \nabla_i w_j) = 0. \tag{6.4}$$

Equation (6.2) can be written as

$$\nabla_j(F_i{}^h w_h) - \nabla_i(F_j{}^h w_h) = F_j{}^t(\nabla_t w_i - \nabla_i w_t), \tag{6.5}$$

or

$$\nabla_j \tilde{w}_i - \nabla_i \tilde{w}_j + F_j{}^t(\nabla_t w_i - \nabla_i w_t) = 0 \tag{6.6}$$

where

$$\tilde{w}_i = -F_i{}^t w_t. \tag{6.7}$$

Equations (6.1) or (6.2) and (6.6) are equivalent.
From (6.6) and

$$\nabla_j(-w_i) - \nabla_i(-w_j) + F_j^t(\nabla_t \tilde{w}_i - \nabla_i \tilde{w}_t) = 0 \tag{6.8}$$

derived from (6.6), we have

THEOREM 6.1. *If $w_i$ and $\tilde{w}_i$ are both closed in an almost Tachibana space, then $w_i$ is covariant almost analytic.*

THEOREM 6.2. *If $w_i$ is covariant almost analytic in an almost Tachibana space, then $\tilde{w}_i$ is also covariant almost analytic.*

Now for the divergence of $\tilde{w}_i$, we have

$$\nabla^j \tilde{w}_j = -\nabla^j(F_j{}^i w_i) = -F^{ji}\nabla_j w_i = -\frac{1}{2} F^{ji}(\nabla_j w_i - \nabla_i w_j) = 0$$

by virtue of (6.4). Thus from this and (6.6) we get

THEOREM 6.3. *If a covariant almost analytic vector field $w_i$ in an almost Tachibana space is closed, then the vector $\tilde{w}_i$ is harmonic.*

In just the same way as in the case of almost Kähler space, we get from this

THEOREM 6.4. *If a covariant almost analytic vector in an almost Tachibana space is closed, then it is harmonic.*

We can also prove the following theorems.

THEOREM 6.5. *A necessary and sufficient condition for a vector field $w_i$ in an almost Tachibana space to be covariant almost analytic is that*

$$w^t\nabla_t F_{ji}=0, \qquad F_j{}^t\nabla_t w^i - F_i{}^t\nabla_j w_t=0. \tag{6.9}$$

(S. Sawaki [2], [3]).

THEOREM 6.6. *For a covariant almost analytic vector $w_i$ in an almost Tachibana space, we have*

$$N_{ji}{}^h w_h=0. \tag{6.10}$$

THEOREM 6.7. *If a vector is contravariant and at the same time covariant almost analytic in an almost Tachibana space, it is covariantly constant.*

Now the Nijenhuis tensor satisfies

$$N_{ji}{}^h = -2O_{ji}^{ts}(\nabla_t F_s{}^r - \nabla_s F_t{}^r)F_r{}^h. \tag{6.11}$$

Thus transvecting (6.11) with $\nabla^j w^i$ and remembering that

$$O_{ji}^{ts}(\nabla^j w^i - \nabla^i w^j) = \nabla^t w^s - \nabla^s w^t$$

for a covariant almost analytic vector, we find

$$N_{ji}{}^h(\nabla^j w^i) = -2(\nabla_j F_i{}^r - \nabla_i F_j{}^r)(\nabla^j w^i)F_r{}^h,$$

from which

$$(\nabla_j F_i{}^r)(\nabla^j w^i)F_r{}^h w_h=0 \tag{6.12}$$

by virtue of (6.10).

Now applying the operator $F_h{}^i\nabla_j$ to (6.2), we get

$$\nabla^t\nabla_t w_i - (2K^*_{ji} - K_{ji})w^j + F_i{}^t\nabla^s(F_{tsr}w^r)=0. \tag{6.13}$$

On the other hand, from (2.14) and (6.9), we have

$$(K_{ji}-K^*_{ji})w^j=0, \tag{6.14}$$

and from (6.9)

$$F_{tsr}w^r=3(\nabla_r F_{ts})w^r=0. \tag{6.15}$$

Thus (6.13) becomes

$$\nabla^t\nabla_t w_i-K_{ji}w^j=0. \tag{6.16}$$

Equations (6.12) and (6.16) are necessary conditions for $w_i$ in an almost Tachibana space to be covariant almost analytic.

## § 7. A necessary and sufficient condition for a vector to be covariant almost analytic

For the tensor $S_{ji}$ defined by

$$S_{ji}=(\nabla_j F_i{}^t-\nabla_i F_j{}^t)w_t-F_j{}^t\nabla_t w_i+F_i{}^t\nabla_j w_t, \tag{7.1}$$

we have the identity

$$\nabla^j(T_{ji}F_t{}^i w^t)+[\nabla^t\nabla_t w_i-(2K^*_{ji}-K_{ji})w^j$$
$$+F_i{}^t\nabla^s(F_{tsr}w^r)-2\nabla_t F_{sr}(\nabla^t w^s)F_i{}^r]w^i+\frac{1}{2}S^{ji}S_{ji}=0. \tag{7.2}$$

Thus, in a compact almost Tachibana space, we have

$$\int\Big[\{\nabla^t\nabla_t w_i-(2K^*_{ji}-K_{ji})w^j+F_i{}^t\nabla^s(F_{tsr}w^r) \tag{7.3}$$
$$-2\nabla_t F_{sr}(\nabla^t w^s)F_i{}^r\}w^i+\frac{1}{2}S^{ji}S_{ji}\Big]d\sigma=0.$$

Thus, from the result obtained at the end of the last section and this integral formula, we get

THEOREM 7.1. *A necessary condition for a vector field in an almost Tachibana space to be covariant almost analytic is that*

$$(\nabla_j F_{ir})(\nabla^j w^i) F_h{}^r w^h = 0$$

*and*

$$\nabla^t \nabla_t w_i - K_{ji} w^j = 0$$

*and a sufficient condition for $w_i$ in a compact almost Tachibana space to be covariant almost analytic is that*

$$\nabla^t \nabla_t w_i - (2K^*_{ji} - K_{ji}) w^j + F_i{}^t \nabla^s (F_{tsr} w^r) - 2 \nabla_t F_{sr} (\nabla^t w^s) F_i{}^r = 0. \quad (7.4)$$

(S. Sawaki [2]; K. Yano and M. Ako [1]).

Thus in a compact almost Tachibana space, a covariant almost analytic vector is harmonic and consequently closed.

Consequently, following Theorem 6.3, the vector $\tilde{w}_i$ is harmonic.

Conversely, if $w_i$ and $\tilde{w}_i$ are both harmonic, then following Theorem 6.1, $w_i$ is covariant almost analytic. Thus we have

THEOREM 7.2. *A necessary and sufficient condition for $w_i$ in a compact almost Tachibana space to be covariant almost analytic is that $w_i$ and $\tilde{w}_i$ are both harmonic.* (K. Yano and M. Ako [1]).

S. Sawaki [2] proved the following theorems.

THEOREM 7.3. *In a compact almost Tachibana space, a necessary and sufficient condition that a covariant vector $u_i$ be almost analytic is that $u_i$ satisfy*

$$\nabla^t \nabla_t u_k - u^t K_{kt} = 0, \qquad u^t (K_{tk} - K^*_{tk}) = 0;$$

*In an almost Tachibana space, if the rank of the matrix $K_{tk} - K^*_{tk}$ is $2n$, then there exists no covariant almost analytic vector other than the zero vector.*

THEOREM 7.4. *In a compact almost Tachibana space, a harmonic vector $u_i$ is almost analytic if and only if $u^t(K_{tk} - K^*_{tk}) = 0$.*

CHAPTER IX

# ALMOST HERMITE SPACES

## § 1. Almost Hermite spaces

An almost Hermite space is first of all a space with an almost complex structure $F_i{}^h$:

$$F_j{}^i F_i{}^h = -A_j^h. \tag{1.1}$$

A necessary and sufficient condition that the almost complex structure $F_i^h$ is induced by a complex structure is given by the vanishing of the Nijenhuis tensor

$$N_{ji}{}^h = F_j{}^t(\partial_t F_i{}^h - \partial_i F_t{}^h) - F_i{}^t(\partial_t F_j{}^h - \partial_j F_t{}^h). \tag{1.2}$$

The Nijenhuis tensor $N_{\cdot i}{}^h$ satisfies the following identities:

$$N_{jt}{}^t = 0, \tag{1.3}$$

$$N_{ji}{}^h + N_{ij}{}^h = 0, \tag{1.4}$$

$$N_{jt}{}^h F_i{}^t = -N_{ji}{}^t F_t{}^h = -N_{it}{}^h F_j{}^t, \tag{1.5}$$

$$N_{ji}{}^h + F_j{}^t F_i{}^s N_{ts}{}^h = 0, \qquad N_{ji}{}^h - F_i{}^s F_r{}^h N_{js}{}^r = 0. \tag{1.6}$$

The equations (1.6) can be written as

$$*O_{ji}^{ts} N_{ts}{}^h = 0, \qquad O_{ir}^{sh} N_{js}{}^r = 0 \tag{1.7}$$

and show that $N_{ji}{}^h$ is pure in $j$ and $i$ and hybrid in $h$ and $i$.

An almost Hermite space is an almost complex space with a so-called Hermite metric $g_{ji}$, that is, a metric satisfying

$$F_j{}^t F_i{}^s g_{ts} = g_{ji}. \tag{1.8}$$

This means that $g_{ji}$ is hybrid in $j$ and $i$. The tensor

$$F_j{}^t g_{ti} = F_{ji} \tag{1.9}$$

is skew-symmetric. The tensor $F_i{}^h$ is pure in $h$ and $i$ and the tensor $F_{ji}$ is hybrid in $j$ and $i$.

Now the Nijenhuis tensor can be written as

$$N_{ji}{}^h = F_j{}^t(\nabla_t F_i{}^h - \nabla_i F_t{}^h) - F_i{}^t(\nabla_t F_j{}^h - \nabla_j F_t{}^h). \tag{1.10}$$

If we put

$$N_{jih} = N_{ji}{}^s g_{sh} = F_j{}^t(\nabla_t F_{ih} - \nabla_i F_{th}) - F_i{}^t(\nabla_t F_{jh} - \nabla_j F_{th}), \tag{1.11}$$

then we have, from (1.7),

$$^*O_{ji}^{ts} N_{tsh} = 0, \qquad ^*O_{ih}^{sr} N_{jsr} = 0, \tag{1.12}$$

which show that $N_{jih}$ is pure in all its indices.

Now we put

$$F_i = \nabla^j F_{ji}. \tag{1.13}$$

If $F_i = 0$, we call such an almost Hermite space an *almost semi-Kähler space.* (S. Kotō [1]).

We also put

$$F_{jih} = \nabla_j F_{ih} + \nabla_i F_{hj} + \nabla_h F_{ji}, \tag{1.14}$$

$$G_{ji}{}^h = \nabla_j F_i{}^h + \nabla_i F_j{}^h. \tag{1.15}$$

When $F_{jih} = 0$, the space is an almost Kähler space and when $G_{ji}{}^h = 0$ the space is an almost Tachibana space. The following theorem is evident.

THEOREM 1.1. *An almost Kähler space or an almost Tachibana space is an almost semi-Kähler space.*

## § 2. Some properties of the Nijenhuis tensor

The Nijenhuis tensor (1.11) may be written in the from

$$N_{jih} = (F_j{}^t \nabla_t F_{ih} + F_i{}^t \nabla_j F_{th}) - (F_i{}^t \nabla_t F_{jh} + F_j{}^t \nabla_i F_{th}). \tag{2.1}$$

Thus, if we put

$$P_{jih} = F_j{}^t \nabla_t F_{ih} + F_i{}^t \nabla_j F_{th}, \tag{2.2}$$

we have

$$N_{jih}=P_{jih}-P_{ijh}. \tag{2.3}$$

But $F_j^{\ t}\nabla_t F_{ih}$ is skew-symmetric in $i$ and $h$ and also $F_i^{\ t}\nabla_j F_{th}$ is skew-symmetric in $i$ and $h$ by virtue of

$$0=\nabla_j(F_i^{\ t}F_{th})=(\nabla_j F_i^{\ t})F_{th}+F_i^{\ t}\nabla_j F_{th},$$

that is

$$F_i^{\ t}\nabla_j F_{th}=-F_h^{\ t}\nabla_j F_{ti}.$$

Thus $P_{jih}$ is skew-symmetric in $i$ and $h$, and consequently, we have, from (2.3)

$$P_{jih}=\frac{1}{2}(N_{jih}+N_{hji}+N_{hij}). \tag{2.4}$$

From (2.3) and (2.4) we can see that the vanishing of the Nijenhuis tensor is equivalent to the vanishing of the tensor $P_{jih}$, which may be written in the form

$$P_{jih}=2F_j^{\ t}O_{ti}^{sr}\nabla_s F_{rh}. \tag{2.5}$$

Thus, the vanishing of the tensor $P_{jih}$ is equivalent to the vanishing of $O_{ji}^{sr}\nabla_s F_{rh}$. Thus we have

THEOREM 2.1. *A necessary and sufficient condition for an almost Hermitian space to be a Hermite space is that*

$$O_{ji}^{ts}\nabla_t F_s^{\ h}=0, \tag{2.6}$$

*that is, the tensor* $\nabla_j F_i^{\ h}$ *is hybrid in* $j$ *and* $i$. (S. Kotō [1], [2]).

Now, the Nijenhuis tensor $N_{jih}$ satisfies in general

$$N_{jih}+N_{jhi}=-F_{jit}F_h^{\ t}-F_{jht}F_i^{\ t}+2F_j^{\ t}G_{iht} \tag{2.7}$$

and the tensor $G_{jih}$ satisfies

$$2{}^*O_{ji}^{ts}G_{tsh}=-(F_j^{\ t}F_{tsi}+F_i^{\ t}F_{tsj})F_h^{\ s}, \tag{2.8}$$

and consequently we have from (2.7) and (2.8)

$$N_{jih}+N_{jhi}=2F_j^{\ t}O_{ih}^{sr}G_{srt}, \tag{2.9}$$

from which

THEOREM 2.2. *The necessary and sufficient condition for the Nijenhuis tensor* $N_{jih}$ *to be skew-symmetric in all its indices is that*

$$O_{ji}^{ts}G_{ts}{}^{h}=O_{ji}^{ts}(\nabla_t F_s{}^{h}+\nabla_s F_t{}^{h})=0. \tag{2.10}$$

We have already stated that the Nijenhuis tensor is skew-symmetric in all its indices in an almost Tachibana space.

## § 3. Curvature tensors in an almost Hermite space

We denote the curvature tensor of the Hermite metric $g_{ji}$ by

$$K_{kji}{}^{h}=\partial_k\left\{{h \atop j\ i}\right\}-\partial_j\left\{{h \atop k\ i}\right\}+\left\{{h \atop k\ t}\right\}\left\{{t \atop j\ i}\right\}-\left\{{h \atop j\ t}\right\}\left\{{t \atop k\ i}\right\}, \tag{3.1}$$

its covariant components by

$$K_{kjih}=K_{kji}{}^{t}g_{th}, \tag{3.2}$$

and the Ricci tensor and the curvature scalar by

$$K_{ji}=K_{tji}{}^{t} \quad \text{and} \quad K=g^{ji}K_{ji} \tag{3.3}$$

respectively. Moreover we put

$$H_{kj}=\frac{1}{2}K_{kjih}F^{ih}, \tag{3.4}$$

$$K^{*}_{ji}=-H_{jt}F_i{}^{t} \quad \text{or} \quad K^{*}_{jt}F_i{}^{t}=H_{ji} \tag{3.5}$$

and

$$H=-F^{ji}H_{ji}. \tag{3.6}$$

From (3.5), $H_{ji}$ being skew-symmetric, we have

$$K^{*}_{jt}F_i{}^{t}+K^{*}_{it}F_j{}^{t}=0$$

or

$$K^{*}_{ji}=F_i{}^{t}F_j{}^{s}K^{*}_{ts},$$

and consequently, when $K^*_{ji}$ is symmetric, it is hybrid in $j$ and $i$.

Applying the Ricci identity to $F_i^{\ h}$ we find

$$\nabla_k\nabla_j F_i^{\ h}-\nabla_j\nabla_k F_i^{\ h}=K_{kjt}{}^h F_i^{\ t}-K_{kji}{}^t F_t^{\ h}, \tag{3.7}$$

from which, by the contraction with respect to $h$ and $k$,

$$\nabla_t\nabla_j F_i^{\ t}+\nabla_j F_i=K_{jt}F_i^{\ t}-H_{ji}, \tag{3.8}$$

or

$$\nabla_t\nabla_j F_i^{\ t}+\nabla_j F_i=(K_{jt}-K^*_{jt})F_i^{\ t}. \tag{3.9}$$

Thus we have

THEOREM 3.1. *In order that $K_{ji}=K^*_{ji}$ in an almost Hermite space, it is necessary and sufficient that*

$$\nabla_t\nabla_j F_i^{\ t}+\nabla_j F_i=0. \tag{3.10}$$

On the other hand, applying $F^{ts}\nabla_t\nabla_s$ to $F_{ih}$, we find

$$F^{ts}\nabla_t\nabla_s F_{ih}=\frac{1}{2}F^{ts}(\nabla_t\nabla_s-\nabla_s\nabla_t)F_{ih} = -\frac{1}{2}F^{ts}(K_{tsi}{}^r F_{rh}+K_{tsh}{}^r F_{ir})$$

or

$$F^{ts}\nabla_t\nabla_s F_{ih}=H_{ir}F_h^{\ r}+H_{rh}F_i^{\ r}. \tag{3.11}$$

Now, applying $\nabla_j$ to $F^{ji}F_{ih}=-A^j_h$, we find

$$F^i F_{ih}+F^{ji}\nabla_j F_{ih}=0.$$

Applying $\nabla^h$ to this equation and taking account of (3.8), we find

$$(\nabla^h F^i)F_{ih}-F^iF_i+(\nabla^h F^{ji})(\nabla_j F_{ih})+F^{ji}(K_{jt}F_i^{\ t}-H_{ji}-\nabla_j F_i)=0$$

or

$$K-H=(\nabla^h F^{ji})(\nabla_j F_{ih})-F^iF_i-2F^{ji}(\nabla_j F_i). \tag{3.12}$$

From (3.12), we have

THEOREM 3.2. *In an almost semi-Kähler space, we have*

$$K-H=(\nabla^h F^{ji})(\nabla_j F_{ih}). \tag{3.13}$$

(S.Kotō [1]).

Suppose that the space is an almost Kähler space, then we have

$$(\nabla^h F^{ji})(\nabla_j F_{ih}) = \frac{1}{2}(\nabla^h F^{ji})(\nabla_j F_{ih} - \nabla_i F_{jh}) = -\frac{1}{2}(\nabla^h F^{ji})(\nabla_h F_{ji}).$$

Thus, from (3.13), we have

$$K - H = -\frac{1}{2}(\nabla^h F^{ji})(\nabla_h F_{ji}) \leqq 0,$$

from which

THEOREM 3.3. *In an almost Kähler space, we have*

$$K \leqq H, \tag{3.14}$$

*and the equality sign occurs when and only when the space is a Kähler space.*

Suppose next that the space is an almost Tachibana space, then we have

$$(\nabla^h F^{ji})(\nabla_j F_{ih}) = \frac{1}{9} F^{hji} F_{jih} = \frac{1}{9} F^{jih} F_{jih}.$$

Thus, from (3.13), we have

$$K - H = \frac{1}{9} F^{jih} F_{jih} \geqq 0,$$

from which

THEOREM 3.4. *In an almost Tachibana space, we have*

$$K \geqq H \tag{3.15}$$

*and the equality sign occurs when and only when the space is a Kähler space.*

Now suppose that our almost semi-Kähler space is conformally flat, then we have

$$K_{kjih} = \frac{1}{2(n-1)}(g_{kh}K_{ji} - g_{jh}K_{ki} + K_{kh}g_{ji} - K_{jh}g_{ki})$$

$$- \frac{K}{2(n-1)(2n-1)}(g_{kh}g_{ji} - g_{jh}g_{ki}). \tag{3.16}$$

Transvecting (3.16) with $F^{kj}F^{ih}$, we find

$$K=(2n-1)H, \tag{3.17}$$

from which

$$K-H=\frac{2(n-1)}{2n-1}K. \tag{3.18}$$

From Theorem 3.3 and (3.18), we have

THEOREM 3.5. *If an almost Kähler space is conformally flat, then the scalar curvature $K$ satisfies*

$$K\leqq 0, \tag{3.19}$$

*and $K=0$ if and only if the space is a Kähler space.*

From Theorem 3.4 and (3.18), we have

THEOREM 3.6. *If an almost Tachibana space is conformally flat, then the scalar curvature $K$ satisfies*

$$K\geqq 0 \tag{3.20}$$

*and $K=0$ if and only if the space is a Kähler space.*

## § 4. *O-space

S. Kotō [1] dealt with an almost Hermite space in which the condition

$$^*O_{ji}^{ts}\nabla_t F_s{}^h=0 \tag{4.1}$$

is satisfied and he called such a space an almost *O-space.

Since $g^{ji}$ is hybrid in $j$ and $i$, that is

$$g^{ji}O_{ji}^{ts}=g^{ts},$$

transvecting (4.1) with $g^{ji}$, we find

$$g^{ts}\nabla_t F_s{}^h=0,$$

from which

THEOREM 4.1. *An almost *O-space is an almost semi-Kähler space.*

(S. Kotō [1])

In a general almost Hermite space, we have

$$*O_{ih}^{sr}\nabla_j F_{sr}=\frac{1}{2}(\nabla_j F_{ih}+F_i^{\ s}F_h^{\ r}\nabla_j F_{sr})=\frac{1}{2}(\nabla_j F_{ih}+\nabla_j F_{hi}),$$

that is

$$*O_{ih}^{sr}\nabla_j F_{sr}=0,$$

which shows that $\nabla_j F_{ih}$ is pure in $i$ and $h$. Thus

THEOREM 4.2. *In almost *O-space, the* $\nabla_j F_{ih}$ *is pure in all its indices.*

When an almost *$O$-space has vanishing Nijenhuis tensor, it is called an *$O$-space.

In an *$O$-space, we have, from (2.2) and (2.4),

$$P_{jih}=F_j^{\ t}\nabla_t F_{ih}+F_t^{\ i}\nabla_j F_{th}=0,$$

and from (4.1)

$$F_j^{\ t}\nabla_t F_{ih}-F_i^{\ t}\nabla_j F_{th}=0.$$

From these two equations we have

$$\nabla_j F_{th}=0,$$

and consequently the space is a Kähler space.

Thus we have

THEOREM 4.3. *An *O-space is a Kähler space.*

Now, in general, we have

$$*O_{ji}^{ts}\nabla_t F_{sh}=\frac{1}{2}(\nabla_j F_{ih}+F_j^{\ t}F_i^{\ s}\nabla_t F_{sh})$$

$$=-\frac{1}{2}(\nabla_j F_{hi}+F_j^{\ t}F_h^{\ s}\nabla_t F_{si})=-*O_{jh}^{ts}\nabla_t F_{si},$$

which shows that $*O_{ji}^{ts}\nabla_t F_{sh}$ is skew-symmetric in $i$ and $h$. On the other hand, we have in an almost Kähler space,

$$*O_{ji}^{ts}\nabla_t F_{sh}=*O_{ji}^{ts}(-\nabla_s F_{ht}-\nabla_h F_{ts})=*O_{ij}^{ts}\nabla_t F_{sh}$$

by virtue of (4.2), which shows that $*O_{ji}^{ts}\nabla_t F_{sh}$ is symmetric in $j$ and $i$. Thus we have

$$^*O_{ji}^{ts}\nabla_t F_{sh}=0$$

in an almost Kähler space. Thus

THEOREM 4.4. *An almost Kähler space is an almost *O-space.* (S. Kotō [1]).

In an almost Tachibana space, $\nabla_j F_{ih}$ is skew-symmetric in all its indices; thus we have

$$^*O_{ji}^{ts}\nabla_t F_{sh}={}^*O_{ji}^{ts}\nabla_h F_{ts}=0$$

by virtue of (4.2). Thus we have

THEOREM 4.5. *An almost Tachibana space is an almost *O-space.* (S. Kotō [1]).

Now, suppose that an almost *O-space reduces to an almost Tachibana space, then by Theorem 2.2, the Nijenhuis tensor $N_{jih}$ is skew-symmetric in all its indices.

Conversely, suppose that, in an almost *O-space, the Nijenhuis tensor $N_{jih}$ is skew-symmetric in all its indices. Then

$$^*O_{ji}^{ts}\nabla_t F_{sh}=0,$$

from which

$$^*O_{ji}^{ts}(\nabla_t F_s{}^h+\nabla_s F_t{}^h)=0$$

and, from (2.10),

$$O_{ji}^{ts}(\nabla_t F_s{}^h+\nabla_s F_t{}^h)=0.$$

Thus

$$G_{ji}{}^h=\nabla_j F_i{}^h+\nabla_i F_j{}^h$$

being pure and at the same time hybrid in $j$ and $i$, we have

$$G_{ji}{}^h=0,$$

so that the space is an almost Tachibana space. Thus

THEOREM 4.6. *A necessary and sufficient condition for an almost *O-space to be an almost Tachibana space is that the Nijenhuis tensor $N_{jih}$ is skew-symmetric in all its indices.* (S. Kotō [1]; H. Mizusawa [1]).

## § 5. Contravariant almost analytic vectors

In an almost complex space, a contravariant almost analytic vector field is defined as a vector field $v^h$ satisfying

$$\underset{v}{\mathfrak{L}} F_i{}^h = v^t \partial_t F_i{}^h - F_i{}^t \partial_t v^h + F_t{}^h \partial_i v^t = 0, \tag{5.1}$$

which is a tensor equation, where $\underset{v}{\mathfrak{L}}$ denotes the Lie differentiation with respect to the vector field $v^h$.

In an almost Hermite space, the equation (5.1) may be written as

$$\underset{v}{\mathfrak{L}} F_i{}^h = v^t \nabla_t F_i{}^h - F_i{}^t \nabla_t v^h + F_t{}^h \nabla_i v^t = 0, \tag{5.2}$$

from which

$$v^t \nabla_t F_{ih} - F_i{}^t \nabla_t v_h - F_h{}^t \nabla_i v_t = 0, \tag{5.3}$$

and taking the symmetric part of this with respect to $i$ and $h$,

$$O_{ji}^{ts}(\nabla_t v_s + \nabla_s v_t) = 0 \quad \text{or} \quad O_{ji}^{ts}(\underset{v}{\mathfrak{L}} g_{ts}) = 0, \tag{5.4}$$

and also

$$O_{ts}^{ji}(\nabla^t v^s + \nabla^s v^t) = 0 \quad \text{or} \quad O_{ts}^{ji}(\underset{v}{\mathfrak{L}} g^{ts}) = 0, \tag{5.5}$$

where

$$\underset{v}{\mathfrak{L}} g_{ji} = v^t \nabla_t g_{ji} + g_{ti} \nabla_j v^t + g_{jt} \nabla_i v^t = \nabla_j v_i + \nabla_i v_j$$

and

$$\underset{v}{\mathfrak{L}} g^{ji} = v^t \nabla_t g^{ji} - g^{ti} \nabla_t v^j - g^{jt} \nabla_t v^i = -\nabla^j v^i - \nabla^i v^j.$$

Equations (5.4) and (5.5) show that $\underset{v}{\mathfrak{L}} g_{ji}$ and $\underset{v}{\mathfrak{L}} g^{ji}$ are both hybrid for a contravariant almost analytic vector field $v^h$ in an almost Hermite space.

Now by a straightforward calculation, we can prove

$$\frac{1}{2}(F_j{}^r F_{ri}{}^h + F_i{}^r F_{rj}{}^h) - G_{ji}{}^r F_r{}^h$$
$$+ F_j{}^t F_i{}^s \left[\frac{1}{2}(F_t{}^r F_{rs}{}^h + F_s{}^r F_{rt}{}^h) - G_{ts}{}^r F_r{}^h\right] = 0,$$

which shows that the tensor

$$\frac{1}{2}(F_j{}^t F_{ti}{}^h + F_i{}^t F_{tj}{}^h) = G_{ji}{}^t F_t{}^h,$$

is symmetric in $j$ and $i$, pure in $j$ and $i$. Thus, $\underset{v}{\mathfrak{L}}\, g^{ji}$ being hybrid in $j$ and $i$, we have

$$\frac{1}{2}(F_j{}^t F_{ti}{}^h + F_i{}^t F_{tj}{}^h)(\underset{v}{\mathfrak{L}}\, g^{ji}) - G_{ji}{}^t F_t{}^h(\underset{v}{\mathfrak{L}}\, g^{ji}) = 0$$

or

$$\frac{1}{2} F_{ji}{}^h(\underset{v}{\mathfrak{L}}\, F^{ji}) = G_{ji}{}^t F_t{}^h(\nabla^j v^i)$$

by virtue of $\underset{v}{\mathfrak{L}}\, F_i{}^h = 0$ for a contravariant almost analytic vector field $v^h$ in an almost Hermite space.

Now applying the operator $\nabla^i$ to (5.2), we find

$$F_t{}^h[\nabla^i\nabla_i v^t + K_i{}^t v^i - F_i{}^t \underset{v}{\mathfrak{L}}\, F^i - G_{ji}{}^s F_s{}^t(\nabla^j v^i)] = 0 \qquad (5.6)$$

or

$$\nabla^i\nabla_i v^h + K_i{}^h v^i - F_i{}^h \underset{v}{\mathfrak{L}}\, F^i - G_{ji}{}^t F_t{}^h(\nabla^j v^i) = 0, \qquad (5.7)$$

or equivalently

$$\nabla^i\nabla_i v^h + K_i{}^h v^i - F_i{}^h \underset{v}{\mathfrak{L}}\, F^i - \frac{1}{2} F_{ji}{}^h(\underset{v}{\mathfrak{L}}\, F^{ji}) = 0 \qquad (5.8)$$

by virtue of (5.6).

Equations (5.6) and (5.7) or (5.6) and (5.8) are necessary conditions for a vector $v^h$ in a general almost Hermite space to be contravariant almost analytic.

If we put

$$S^{ji}=g^{jt}(\underset{v}{\mathfrak{L}}F_t^{\ i}), \tag{5.9}$$

we have

$$\begin{aligned}\frac{1}{2}S^{ji}S_{ji}=\frac{1}{2}v^tv^s(\nabla_tF^{ji})(\nabla_sF_{ji})\\ -v^t(\nabla_tF^{ji})F_j^{\ s}(\nabla_sv_i)+v^t(\nabla_tF^{ji})F_{si}(\nabla_jv^s)\\ +(\nabla^jv^i)(\nabla_jv_i)-F^{jt}F_{si}(\nabla_tv^i)(\nabla_jv^s)\end{aligned} \tag{5.10}$$

and

$$\nabla^j\{(\underset{v}{\mathfrak{L}}F_j^{\ t})F_t^{\ i}v_i\} \tag{5.11}$$

$$+\left[\nabla^i\nabla_iv^h+K_i^{\ h}v^i-F_i^{\ h}(\underset{v}{\mathfrak{L}}F^i)-\frac{1}{2}F_{ji}^{\ \ h}(\underset{v}{\mathfrak{L}}F^{ji})\right]v_h+\frac{1}{2}S^{ji}S_{ji}=0.$$

Assuming that the Hermite space is compact, we have from (5.11)

$$\int\left[\left\{\nabla^i\nabla_iv^h+K_i^{\ h}v^i-F_i^{\ h}(\underset{v}{\mathfrak{L}}F^i)-\frac{1}{2}F_{ji}^{\ \ h}(\underset{v}{\mathfrak{L}}F^{ji})\right\}v_h +\frac{1}{2}S^{ji}S_{ji}\right]d\sigma=0. \qquad 5.12)$$

From (5.8) and (5.12), we have

THEOREM 5.1. *A necessary and sufficient condition for a vector field $v^h$ in a compact almost Hermite space to be contravariant almost analytic is*

$$\nabla^i\nabla_iv^h+K_i^{\ h}v^i-F_i^{\ h}(\underset{v}{\mathfrak{L}}F^i)-\frac{1}{2}F_{ji}^{\ \ h}(\underset{v}{\mathfrak{L}}F^{ji})=0.$$

(S. Sawaki [2]).

Theorem 4.1 in Chapter VII and Theorem 4.1 in Chapter VIII are corollaries to this theorem.

As another corollary to this theorem, we have

THEOREM 5.2. *A necessary and sufficient condition for a vector field $v^h$ in a compact almost semi-Kähler space to be* *contravariant almost analytic is*

$$\nabla^i \nabla_i v^h + K_i{}^h v^i - \frac{1}{2} F_{ji}{}^h (\underset{v}{\mathfrak{L}} F^{ji}) = 0. \tag{5.13}$$

(S.Kotō [1]).

But, in an almost semi-Kähler space, we have

$$F_{jih} F^{ji} = 2F_t F_h{}^t = 0,$$

from which

$$(\underset{v}{\mathfrak{L}} F_{jih}) F^{ji} + F_{jih} (\underset{v}{\mathfrak{L}} F^{ji}) = 0$$

and consequently (5.13) may be written as

$$\nabla^i \nabla_i v^h + K_i{}^h v^i + \frac{1}{2} (\underset{v}{\mathfrak{L}} F_{jit}) F^{ji} g^{th} = 0. \tag{5.14}$$

Thus

THEOREM 5.3. *If a Killing vector $v^h$ in a compact almost semi-Kähler space satisfies $\underset{v}{\mathfrak{L}} F_{jih} = 0$, or $\underset{v}{\mathfrak{L}} F_{ji}{}^h = 0$ then $v^h$ is contravariant almost analytic.*

Now, in an almost *$O$-space, $\nabla_j F_i{}^h$ is pure in $j$ and $i$, and consequently

$$G_{ji}{}^h = \nabla_j F_i{}^h + \nabla_i F_j{}^h$$

is also pure in $j$ and $i$. On the other hand

$$\underset{v}{\mathfrak{L}} g^{ji} = -\nabla^j v^i - \nabla^i v^j$$

is hybrid in $j$ and $i$ for a contravariant almost analytic vector field $v^h$.

Thus

$$G_{ji}{}^h (\nabla^j v^i) = -\frac{1}{2} G_{ji}{}^h (\underset{v}{\mathfrak{L}} g^{ji}) = 0.$$

Thus we have from (5.6)

$$F_{jih} (\underset{v}{\mathfrak{L}} F^{ji}) = 0.$$

Thus we have

THEOREM 5.4. *A necessary condition for a vector field $v^h$ in a compact almost *O-space to be contravariant almost analytic is that*

$$F_{jih}(\underset{v}{\mathfrak{L}} F^{ji})=0$$

*and*

$$\nabla^i\nabla_i v^h + K_i{}^h v^i = 0$$

*and a sufficient condition is that*

$$\nabla^i\nabla_i v^h + K_i{}^h v^i - \frac{1}{2} F_{jih}(\underset{v}{\mathfrak{L}} F^{ji}) = 0.$$

(S. Kotō [1]).

In exactly the same way as in the case of almost Kähler space, we have

COROLLARY 1. *A Killing vector in a compact almost *O-space satisfying $F_{jih}(\underset{v}{\mathfrak{L}} F^{ji})=0$ is contravariant almost analytic.*

COROLLARY 2. *An infinitesimal almost analytic transformation in a compact almost *O-space which does not change the volume of the space is a motion.*

THEOREM 5.5. *In a compact almost *O-space, we have*

$$\int K_{ji} v^j v^i d\sigma \geqq 0$$

*for a contravariant almost analytic vector field $v^h$ and if the equality sign occurs then $v^h$ is parallel.*

COROLLARY 1. *In a compact almost *O-space with negative definite Ricci curvature, there exists no contravariant almost analytic vector field other than the zero vector.*

*In a compact almost *O-space with negative semi-definite Ricci curvature, a contravariant almost analytic vector field is parallel.*

## § 6. Projective and conformal Killing vectors

For a projective Killing vector $v^h$, we have

$$\nabla^i\nabla_i v^h + K_i{}^h v^i - \frac{2}{2n+1}\nabla^h\nabla_t v^t = 0. \tag{6.1}$$

Thus, if a projective Killing vector is at the same time a contravariant almost analytic vector in a compact almost *$O$-space, combining Theorem 5.4 and the above equation, we find

$$\nabla^h\nabla_t v^t = 0$$

from which

$$\nabla_t v^t = \text{const.} \quad \text{and consequently} \quad \nabla_t v^t = 0$$

by virtue of Green's theorem. Thus $v^h$ is a Killing vector and we have

THEOREM 6.1. *If a projective Killing vector is at the same time a contravariant almost analytic vector in a compact almost *O-space, then the vector is a Killing vector.* (S. Kotō [1]).

For a conformal Killing vector $v^h$, we have

$$\nabla^i\nabla_i v^h + K_i{}^h v^i + \frac{n-1}{n}\nabla^h\nabla_t v^t = 0. \tag{6.2}$$

On the other hand, in a compact almost semi-Kähler space, the formula (5.12) becomes

$$\int\left[\left\{\nabla^i\nabla_i v^h + K_i{}^h v^i + \frac{1}{2}(\underset{v}{\mathfrak{L}} F_{ji}{}^h)F^{ji}\right\} v_h + \frac{1}{2}S^{ji}S_{ji}\right] d\sigma = 0, \tag{6.3}$$

by virtue of

$$F^i = 0 \quad \text{and} \quad F_{ji}{}^h F^{ji} = 0.$$

Thus, if a conformal Killing vector satisfies

$$(\underset{v}{\mathfrak{L}} F_{ji}{}^h)F^{ji} = 0 \tag{6.4}$$

or

$$(\underset{v}{\mathfrak{L}} F_{jih})F^{ji}=0, \tag{6.5}$$

then, substituting (6.2) into (6.3), we find

$$\int\left[\frac{n-1}{n}(\nabla_t v^t)^2+\frac{1}{2}S^{ji}S_{ji}\right]d\sigma=0, \tag{6.6}$$

from which

$$\nabla_t v^t=0, \qquad S_{ji}=0.$$

Thus the conformal Killing vector $v^h$ is a Killing vector and is also a contravariant almost analytic vector. Thus we have

THEOREM 6.2. *If a conformal Killing vector $v^h$ satisfies $(\underset{v}{\mathfrak{L}} F_{jih})F^{ji}=0$ in a compact almost *O-space, then it is a Killing vector and is also a contravariant almost analytic vector:*

## § 7. Covariant almost analytic vectors

In an almost complex space, a covariant almost analytic vector is defined as a vector field $v^h$ which satisfies

$$[(\partial_j F_i{}^h)-(\partial_i F_j{}^h)]w_h - F_j{}^t\partial_t w_i + F_i{}^t\partial_j w_t=0, \tag{7.1}$$

which is easily verified to be a tensor equation.

In an almost Hermite space, the equation (7.1) may be written as

$$[(\nabla_j F_i{}^h)-(\nabla_i F_j{}^h)]w_h - F_j{}^t\nabla_t w_i + F_i{}^t\nabla_j w_t=0, \tag{7.2}$$

from which, taking the symmetric part with respect to $j$ and $i$, we find

$$^*O_{ji}^{ts}(\nabla_t w_s - \nabla_s w_t)=0, \tag{7.3}$$

which shows that $\nabla_j w_i - \nabla_i w_j$ is pure for a covariant almost analytic vector $w_i$ in an almost Hermite space.

Transvecting (7.2) with $\nabla_k F^{ji}$ we find

$$(\nabla_k F^{ji})(\nabla_j F_i^{\ h}) w_h = 0 \tag{7.4}$$

for a covariant almost analytic vector field $w_i$:

Now we define tensors $P_{ji}$ and $Q_{ji}$ by

$$P_{ji} = (\nabla_j F_i^{\ t} - \nabla_i F_j^{\ t}) w_t, \quad Q_{ji} = F_j^{\ t} \nabla_t w_i - F_i^{\ t} \nabla_j w_t \tag{7.5}$$

respectively. Then for a covariant almost analytic vector field $w_i$, we have

$$P_{ji} = Q_{ji}$$

$$P_{ji} P^{ji} = 2(F_{jit} - \nabla_t F_{ji})(\nabla^j F^{ih}) w^t w_h$$

or

$$P_{ji} P^{ji} = 2F_{ji}^{\ \ t}(\nabla^j F^{is}) w_t w_s$$

by virtue of (7.4) and

$$P_{ji} Q^{ji} = F_j^t (\nabla_t w_i + \nabla_i w_t)(G^{jis} - 2\nabla^i F^{js}) w_s .$$

Thus in an almost Kähler space we have $P_{ji} P^{ji} = 0$ for a covariant almost analytic vector $w_i$ from which we obtain

$$P_{ji} = 0, \qquad Q_{ji} = 0.$$

But in an almost Kähler space, we have

$$\nabla_j F_i^{\ t} - \nabla_i F_j^{\ t} = -\nabla^t F_{ji} .$$

Thus $P_{ji} = 0$ gives

$$w^t \nabla_t F_{ji} = 0.$$

On the other hand, in an almost Tachibana space, we have

$$\nabla_j F_{ih} = \frac{1}{3} F_{jih}$$

and consequently (7.4) may be written as

$$F_{jis} F^{jir} w_r = 0,$$

from which

$$(F_{ji}{}^{s}w_s)(F^{jir}w_r)=0$$

and consequently $F_{ji}{}^{r}w_r=0$, that is,

$$(\nabla_j F_i{}^t)w_t=0 \quad \text{or} \quad w^t\nabla_t F_{ji}=0$$

by virtue of $\nabla_j F_{ih}=\nabla_h F_{ji}$.

Thus in an almost Tachibana space, we have $P_{ji}Q^{ji}=0$ and the equations $P_{ji}=Q_{ji}$ and $P_{ji}Q^{ji}=0$ give $P_{ji}=0$ and $Q_{ji}=0$, that is,

$$w^t\nabla_t F_{ji}=0, \qquad Q_{ji}=0$$

for a covariant almost analytic vector in an almost Tachibana space.

Thus we have

THEOREM 7.1. *A necessary and sufficient condition for a vector field $w_i$ in an almost Kähler or in an almost Tachibana space to be covariant almost analytic is that*

$$w^t\nabla_t F_{ji}=0, \tag{7.6}$$

$$F_j{}^t\nabla_t w_i - F_i{}^t\nabla_j w_t=0. \tag{7.7}$$

(S. Sawaki [2]).

For a covariant almost analytic vector field $w_i$ in an almost Hermite space, we have

$$\begin{aligned} N_{ji}{}^{h}w_h &= [F_j{}^t(\nabla_t F_i{}^h-\nabla_i F_t{}^h)-F_i{}^t(\nabla_t F_j{}^h-\nabla_j F_t{}^h)]w_h \\ &= F_j{}^t(F_t{}^s\nabla_s w_i - F_i{}^s\nabla_t w_s)-F_i{}^t(F_t{}^s\nabla_s w_j - F_j{}^s\nabla_t w_s) \\ &= -(\nabla_j w_i - \nabla_i w_j)-F_j{}^tF_i{}^s(\nabla_t w_s-\nabla_s w_t) \\ &= -2{}^*O_{ji}^{ts}(\nabla_t w_s-\nabla_s w_t), \end{aligned}$$

that is,

$$N_{ji}{}^{h}w_h=0 \tag{7.8}$$

by virtue of (7.2) and (7.3).

For such a vector, we have also

$$(\nabla_j F_i{}^r - \nabla_i F_j{}^r)(\nabla^j w^i) F_r{}^h$$

$$= \frac{1}{2}(\nabla_j F_i{}^r - \nabla_i F_j{}^r)(\nabla^j w^i - \nabla^i w^j) F_r{}^h$$

$$= \frac{1}{2}(\nabla_j F_i{}^r - \nabla_i F_j{}^r) O_{ts}^{ji}(\nabla^t w^s - \nabla^s w^t) F_r{}^h$$

$$= \frac{1}{2}[(\nabla_t F_s{}^r - \nabla_s F_t{}^r) - F_t{}^j F_s{}^i(\nabla_j F_i{}^r - \nabla_i F_j{}^r)](\nabla^t w^s) F_r{}^h$$

by virtue of

$$\nabla^j w^i - \nabla^i w^j = O_{ts}^{ji}(\nabla^t w^s - \nabla^s w^t)$$

derived from (7.3). From this we have

$$(\nabla_j F_i{}^t - \nabla_i F_j{}^t)(\nabla^j w^i) F_t{}^h = -\frac{1}{2} N_{ji}{}^h(\nabla^j w^i)$$

and consequently

$$(\nabla_j F_i{}^t - \nabla_i F_j{}^t)(\nabla^j w^i) F_t{}^h w_h = 0 \tag{7.9}$$

by virtue of (7.8) for a covariant almost analytic vector field $w_i$ in an almost Hermite space.

If we suppose that $w_i$ is a contravariant and at the same time covariant almost analytic vector field in an almost Hermite space, then adding

$$w^t \nabla_t F_j{}^h - F_j{}^t \nabla_t w^h + F_t{}^h \nabla_j w^t = 0$$

or

$$w^t \nabla_t F_{ji} - F_j{}^t \nabla_t w_i - F_i{}^t \nabla_j w_t = 0$$

and

$$(\nabla_j F_{it} - \nabla_i F_{jt}) w^t - F_j{}^t \nabla_t w_i + F_i{}^t \nabla_j w_t = 0,$$

we find

$$F_{jit} w^t - 2F_j{}^t \nabla_t w_i = 0. \tag{7.10}$$

In an almost Kähler space, equation (7.10) reduces to

$$F_j{}^t \nabla_t w_i = 0.$$

In an almost Tachibana space, (7.10) is also written as

$$3w^t\nabla_t F_{ji} - 2F_j{}^t\nabla_t w_i = 0$$

or

$$F_j{}^t\nabla_t w_i = 0$$

by virtue of $w^t\nabla_t F_{ji} = 0$ in Theorem 7.1. Thus

THEOREM 7.2. *If, in an almost Kähler or in an almost Tachibana space, $w_i$ is a contravariant and at the same time covariant almost analytic vector field, then it is covariantly constant.*

The equation (7.2) is written as

$$\nabla_j \tilde{w}_i - \nabla_i \tilde{w}_j = F_j{}^t(\nabla_t w_i - \nabla_i w_t), \tag{7.11}$$

where

$$\tilde{w}_i = -F_i{}^t w_t. \tag{7.12}$$

The equation (7.11) may also be written as

$$-(\nabla_j w_i - \nabla_i w_j) = F_j{}^t(\nabla_t \tilde{w}_i - \nabla_i \tilde{w}_t). \tag{7.13}$$

The equations (7.11) and (7.13) give

THEOREM 7.3. *If a vector field $w_i$ in an almost Hermite space is covariant almost analytic, then the vector field $\tilde{w}_i = F_i{}^t w_t$ is also covariant almost analytic.*

If vectors $w_i$ and $\tilde{w}_i$ are both closed, then equations (7.11) is satisfied. Thus we have

THEOREM 7.4. *If vectors $w_i$ and $\tilde{w}_i = F_i{}^t w_t$ in an almost Hermite space are both closed, then they are both covariant almost analytic.*

From (7.11) by transvection with $F^{ji}$ and with $g^{ji}$ we find

$$F^{ji}\nabla_j \tilde{w}_i = 0 \quad \text{and} \quad F^{ji}(\nabla_j w_i - \nabla_i w_j) = 0 \tag{7.14}$$

respectively. Thus applying $g^{ji}\nabla_i$ to $\tilde{w}_j = -F_j{}^t w_t$, we find

$$g^{ji}\nabla_j \tilde{w}_i - F^i w_i = 0. \tag{7.15}$$

If a covariant almost analytic vector field $w_i$ is closed, we have from (7.11)

$$\nabla_j \tilde{w}_i - \nabla_i \tilde{w}_j = 0. \tag{7.16}$$

Thus $\tilde{w}_i$ is covariant almost analytic (Theorem 7.3) and is closed and we have

THEOREM 7.5. *If, in an almost Hermite space, a covariant almost analytic vector* $w_i$ *is closed, then* $\tilde{w}_i$ *is closed.*

Thus from (7.15) and (7.16) we have

THEOREM 7.6. *If, in an almost semi-Kähler space, a covariant almost analytic vector field* $w_i$ *is closed, then it is harmonic.*

Now applying $F_k^{\ j}\nabla^i$ to (7.2) and changing indices we obtain

$$\begin{aligned}\nabla^t\nabla_t w_i-(2K^*_{ji}-K_{ji})w^j+F_i^{\ t}\nabla^s(F_{tsr}w^r)\\ +(\nabla^t w^s)G_{tsr}F_i^{\ r}+F_i^{\ t}(w^s\nabla_s F_t+F_s\nabla^s w_t)=0. \quad (7.17)\end{aligned}$$

For the tensor $S_{ji}$ defined by

$$S_{ji}=(\nabla_j F_i^{\ t}-\nabla_i F_j^{\ t})w_t-F_j^{\ t}\nabla_t w_i+F_i^{\ t}\nabla_j w_t, \quad (7.18)$$

we have the identity

$$\begin{aligned}\nabla^j(T_{ji}F_t^{\ i}w^t)+[\nabla^t\nabla_t w_i-(2K^*_{ji}-K_{ji})w^j\\ +F_i^{\ t}\nabla^s(F_{tsr}w^r)+(\nabla^t w^s)G_{tsr}F_i^{\ r}+F_i^{\ t}(w^s\nabla_s F_t+F_s\nabla^s w_t)\\ -(\nabla_t F_{sr}-\nabla_s F_{tr})(\nabla^t w^s)F_i^{\ r}]w^i+\frac{1}{2}S^{ji}S_{ji}=0. \quad (7.19)\end{aligned}$$

Thus in a compact almost Hermite space, we have

$$\begin{aligned}\int\Big[\{\nabla^t\nabla_t w_i-(2K^*_{ji}-K_{ji})w^j+F_i^{\ t}\nabla^s(F_{tsr}w^r)\\ +(\nabla^t w^s)G_{tsr}F_i^{\ r}+F_i^{\ t}(w^s\nabla_s F_t+F_s\nabla^s w_t)\\ -(\nabla_t F_{sr}-\nabla_s F_{tr})(\nabla^t w^s)F_i^{\ r}\}w^i+\frac{1}{2}S^{ji}S_{ji}\Big]d\sigma=0, \quad (7.20)\end{aligned}$$

and consequently

THEOREM 7.7. *A necessary condition for a vector field* $w_i$ *in an almost Hermite space to be covariant almost analytic is that* (7.9) *and* (7.17) *are satisfied and a sufficient con-*

*dition for $w_i$ in a compact almost Hermite space to be covariant almost analytic is*

$$\begin{aligned}\nabla^t\nabla_t w_i-(2K^*_{ji}-K_{ji})w^j+F_i{}^t\nabla^s(F_{tsr}w^r)\\ +(\nabla^t w^s)G_{tsr}F_i{}^r+F_i{}^t(w^s\nabla_s F_t+F_s\nabla^s w_t)\\ -(\nabla_t F_{sr}-\nabla_s F_{tr})(\nabla^t w^s)F_i{}^r=0.\end{aligned} \tag{7.21}$$

(S. Sawaki [2]; K. Yano and M. Ako [1]).

CHAPTER X

# LOCALLY PRODUCT SPACES

## § 1. Locally product spaces†

Let us consider an $n$-dimensional manifold $M_n$ which is covered by such a system of coordinate neighbourhoods $(\xi^h)$ that in any intersection of two coordinate neighbourhoods $(\xi^h)$ and $(\xi^{h'})$ we have

$$\xi^{a'} = \xi^{a'}(\xi^a), \qquad \xi^{x'} = \xi^{x'}(\xi^x), \tag{1.1}$$

with

$$|\partial_a \xi^{a'}| \neq 0, \qquad |\partial_x \xi^{x'}| \neq 0,$$

where the indices $a$, $b$, $c$, $d$, $e$ run over the range 1, 2,. . .. $p$ and the indices $x$, $y$, $z$ over the range $p+1, \ldots, p+q=n$, Then we say that the manifold $M_n$ admits a locally *product structure* defined by the existence of such a system of coordinate neighbourhoods called *separating coordinate system.* We call *locally product space* a space which admits a locally product structure.

Consider a contravariant vector field $v^h(\xi)$. The transformation law of $v^h$ under a coordinate transformation (1.1) is

$$v^{h'} = A_h^{h'} v^h$$

or

$$v^{a'} = A_a^{a'} v^a\,, \qquad v^{x'} = A_x^{x'} v^x\,, \tag{1.2}$$

† See S. Tachibana [7].

by virtue of

$$A_h^{h'} = \begin{pmatrix} A_a^{a'} & 0 \\ 0 & A_x^{x'} \end{pmatrix}, \tag{1.3}$$

where

$$A_h^{h'} = \partial_h \xi^{h'}.$$

Equations (1.2) show that if $v^h = (v^a, v^x)$ are components of a contravariant vector, then $(v^a, 0)$ and $(0, v^x)$ are also components of contravariant vectors.

Similarly, if $w_i = (w_b, w_y)$ are components of a covariant vector, then $(w_b, 0)$ and $(0, w_y)$ are also components of covariant vectors and also if, for example,

$$T_i^{\ h} = \begin{pmatrix} T_b^{\ a} & T_y^{\ a} \\ T_b^{\ x} & T_y^{\ x} \end{pmatrix}$$

are components of a tensor, then

$$\begin{pmatrix} T_b^{\ a} & 0 \\ 0 & 0 \end{pmatrix}, \quad \begin{pmatrix} 0 & T_y^{\ a} \\ 0 & 0 \end{pmatrix}, \quad \cdots\cdots\cdots$$

are all components of tensor of the same type as $T_i^{\ h}$.

Equations (1.2) also show that if $v^h = (v^a, v^x)$ are components of a contravariant vector, then

$$(v^a, -v^x)$$

are also components of a contravariant vector. This process may be represented by

$$v^h \to F_i^{\ h} v^i$$

where

$$F_i^{\ h} = \begin{pmatrix} \delta_b^a & 0 \\ 0 & -\delta_y^x \end{pmatrix}. \tag{1.4}$$

It is easily seen that $F_i^{\ h}$ having such numerical components in each separating coordinate neighbourhood is a mixed tensor. The tensor $F_i^{\ h}$ satisfies the equation

$$F_j^{\ i} F_i^{\ h} = +A_j^h. \tag{1.5}$$

If we put

$$P_i^{\ h}=\begin{pmatrix}\delta_b^a & 0\\ 0 & 0\end{pmatrix},\quad Q_i^{\ h}=\begin{pmatrix}0 & 0\\ 0 & \delta_y^x\end{pmatrix} \tag{1.6}$$

in a separating coordinate system, then $P_i^{\ h}$ is an operator which projects any vector

$$v^h=(v^a,\ v^x)\quad\text{into}\quad P_i^{\ h}v^i=(v^a,\ 0)$$

and $Q_i^{\ h}$ is an operator which projects

$$v^h=(v^a,\ v^x)\quad\text{into}\quad Q_i^{\ h}v^i=(0,\ v^x).$$

These operators may be expressed in terms of $A_i^h$ and $F_i^{\ h}$ as follows:

$$P_i^{\ h}=\frac{1}{2}(A_i^h+F_i^{\ h}),\qquad Q_i^{\ h}=\frac{1}{2}(A_i^h-F_i^{\ h}) \tag{1.7}$$

from which

$$A_i^h=P_i^{\ h}+Q_i^{\ h},\qquad F_i^{\ h}=P_i^{\ h}-Q_i^{\ h}. \tag{1.8}$$

If a tensor $T_i^{\ h}$ has components of the type

$$T_i^{\ h}=\begin{pmatrix}T_b^{\ a} & 0\\ 0 & T_y^{\ x}\end{pmatrix},$$

we say that $T_i^{\ h}$ is *pure*, and if it has components of the type

$$T_i^{\ h}=\begin{pmatrix}0 & T_y^{\ a}\\ T_b^{\ x} & 0\end{pmatrix},$$

we say that $T_i^{\ h}$ is *hybrid*. Similarly if a tensor $T_{ji}$ has components of the type

$$T_{ji}=\begin{pmatrix}T_{cb} & 0\\ 0 & T_{zy}\end{pmatrix},$$

we say that $T_{ji}$ is *pure*, and if it has components of the type

$$T_{ji}=\begin{pmatrix}0 & T_{zb}\\ T_{cy} & 0\end{pmatrix},$$

we say that $T_{ji}$ is *hybrid.*

If we put

$$O_{ir}^{sh}=\frac{1}{2}(A_i^s A_r^h+F_i{}^s F_r{}^h), \tag{1.9}$$

we have, for a general tensor $T_i{}^h$,

$$O_{ir}^{sh}T_s{}^r=\begin{pmatrix}T_b{}^a & 0\\ 0 & T_y{}^x\end{pmatrix}. \tag{1.10}$$

Thus the operator $O_{ir}^{sh}$ applied to a general tensor $T_i^h$ gives a pure tensor given by (1.10). We call (1.10) the *pure part* of $T_i{}^h$.

If we put

$$*O_{ir}^{sh}=\frac{1}{2}(A_i^s A_r^h-F_i{}^s F_r{}^h), \tag{1.11}$$

then we have, for a general tensor $T_i{}^h$,

$$*O_{ir}^{sh}T_s{}^r=\begin{pmatrix}0 & T_y{}^a\\ T_b{}^x & 0\end{pmatrix}. \tag{1.12}$$

Thus the operator $*O_{ir}^{sh}$ applied to a general tensor $T_i^h$ gives a hybrid tensor given by (1.12). We call (1.12) the *hybrid part* of $T_i{}^h$.

Similarly, we have, for a covariant tensor $T_{ji}$,

$$O_{ji}^{ts}T_{ts}=\begin{pmatrix}T_{cb} & 0\\ 0 & T_{zy}\end{pmatrix} \tag{1.13}$$

and

$$*O_{ji}^{ts}T_{ts}=\begin{pmatrix}0 & T_{zb}\\ T_{cy} & 0\end{pmatrix}. \tag{1.14}$$

We call (1.13) and (1.14) the pure and the hybrid parts of $T_{ji}$.

The operators $O$ and $^*O$ satisfy

$$O+{}^*O=A,\ O\cdot O=O,\ O\cdot{}^*O=0,\ {}^*O\cdot O=0,\ {}^*O\cdot{}^*O={}^*O, \tag{1.15}$$

where $A$ denotes the identity operator.

The condition for $T$ to be pure is given by

$$OT=T \quad \text{or} \quad {}^*OT=0 \tag{1.16}$$

and that for $T$ to be hybrid is given by

$$^*OT=T \quad \text{or} \quad OT=0. \tag{1.17}$$

## § 2. Linear connexions in a locally product space

Let us consider a locally product space $M_n$ covered by a separating coordinate system $(\xi^h)$. Then in an intersection of two coordinate neighbourhoods, we have

$$\xi^{a'}=\xi^{a'}(\xi^a), \qquad \xi^{x'}=\xi^{x'}(\xi^x). \tag{2.1}$$

We denote by $M_p$ the system of subspaces defined by $\xi^x=$const. and by $M_q$ the system of subspaces defined by $\xi^a=$const. Then our space $M_n$ is locally the product $M_p\times M_q$ of two spaces.

Now, suppose that a linear connexion $\Gamma_j{}^h{}_i(\xi)$ is given in the manifold $M_n$. Under a transformation of coordinates (2.1), the linear connexion $\Gamma_j{}^h{}_i$ has the following transformation law:

$$\Gamma_{j'}{}^{h'}{}_{i'}=\frac{\partial\xi^{h'}}{\partial\xi^h}\left(\frac{\partial\xi^j}{\partial\xi^{j'}}\frac{\partial\xi^i}{\partial\xi^{i'}}\Gamma_j{}^h{}_i+\frac{\partial^2\xi^h}{\partial\xi^{j'}\partial\xi^{i'}}\right). \tag{2.2}$$

Since the transformation of coordinates is of the form (2.1), we can see from (2.2) that the quantities

$$\Gamma_c{}^a{}_y,\quad \Gamma_z{}^a{}_b,\quad \Gamma_z{}^a{}_y,\quad \Gamma_c{}^x{}_b,\quad \Gamma_z{}^x{}_b,\quad \Gamma_c{}^x{}_y$$

are all components of tensors.

In exactly the same way as in § 5 of Chapter III, we can prove the following theorems.

THEOREM 2.1. *An arbitrary contravariant vector tangent to* $M_p$, *displaced parallelly along* $M_p$ *is still tangent to* $M_p$, *if and only if*

$$\Gamma_{c\ b}^{\ x}=0. \tag{2.3}$$

If this condition is satisfied, we say that $M_p$ is *totally geodesic* in $M_n$.

THEOREM 2.2. *An arbitrary contravariant vector tangent to* $M_p$, *displaced parallelly along* $M_q$, *is still tangent to* $M_p$ *if and only if*

$$\Gamma_{z\ b}^{\ x}=0. \tag{2.4}$$

THEOREM 2.3. *An arbitrary contravariant vector tangent to* $M_p$ *displaced parallelly in any direction, is still tangent to* $M_p$, *if and only if*

$$\Gamma_{c\ b}^{\ x}=0, \qquad \Gamma_{z\ b}^{\ x}=0. \tag{2.5}$$

If this condition is satisfied, we say that $M_p$ is *parallel* in $M_n$.

THEOREM 2.4. *An arbitrary contravariant vector tangent to* $M_q$, *displaced parallelly along* $M_p$, *is still tangent to* $M_q$, *if and only if*

$$\Gamma_{c\ y}^{\ a}=0. \tag{2.6}$$

THEOREM 2.5. *An arbitrary contravariant vector tangent to* $M_q$, *displaced parallelly along* $M_q$, *is still tangent to* $M_q$, *if and only if*

$$\Gamma_{z\ y}^{\ a}=0. \tag{2.7}$$

In this case, $M_q$ is totally geodesic in $M_n$.

THEOREM 2.6. *An arbitrary contravariant vector tangent to* $M_q$, *displaced parallelly in any direction, is still tangent to* $M_q$, *if and only if*

$$\Gamma_{c\ y}^{\ a}=0, \qquad \Gamma_{z\ y}^{\ a}=0. \tag{2.8}$$

In this case, $M_q$ is parallel in $M_n$.

## § 3. Locally product and locally decomposable Riemannian spaces

Suppose that, in the $n$-dimensional locally product space $M_n$, there is given a positive definite Riemannian metric

$$ds^2 = g_{ji}(\xi) d\xi^j d\xi^i \tag{3.1}$$

which satisfies

$$F_j^{\ t} F_i^{\ s} g_{ts} = g_{ji}. \tag{3.2}$$

The equation (3.2) shows that

$$*O_{ji}^{ts} g_{ts} = 0, \tag{3.3}$$

that is, $g_{ji}$ is pure:

$$g_{ji} = \begin{pmatrix} g_{cb} & 0 \\ 0 & g_{zy} \end{pmatrix} \tag{3.4}$$

and consequently that the metric (3.1) has the form

$$ds^2 = g_{cb}(\xi) d\xi^c d\xi^b + g_{zy}(\xi) d\xi^z d\xi^y, \tag{3.5}$$

or equivalently that the subspaces $M_p$ and $M_q$ are orthogonal. We call such a space a locally product Riemannian space.

It is easily seen that $g^{ih}$ is also pure:

$$g^{ih} = \begin{pmatrix} g^{ba} & 0 \\ 0 & g^{yz} \end{pmatrix}. \tag{3.6}$$

If we put

$$F_j^{\ t} g_{ti} = F_{ji} \tag{3.7}$$

then $F_j^{\ i} F_i^{\ h} = A_j^h$ and this equation shows that

$$F_{ji} = F_{ij} \tag{3.8}$$

and in fact

$$F_{ji} = \begin{pmatrix} g_{cb} & 0 \\ 0 & -g_{zy} \end{pmatrix}. \tag{3.9}$$

The tensor $F_{ji}$ is pure. If we put

$$P_{ji}=P_j^t g_{ti}, \qquad Q_{ji}=Q_j^t g_{ti}, \tag{3.10}$$

we have

$$P_{ji}=\begin{pmatrix} g_{cb} & 0 \\ 0 & 0 \end{pmatrix}, \qquad Q_{ji}=\begin{pmatrix} 0 & 0 \\ 0 & g_{zy} \end{pmatrix} \tag{3.11}$$

and

$$F_{ji}=P_{ji}-Q_{ji}. \tag{3.12}$$

Suppose that the metric of the space (3.5) has the form

$$ds^2=g_{cb}(\xi^a)d\xi^c d\xi^b+g_{zy}(\xi^x)d\xi^z d\xi^y, \tag{3.13}$$

that is, $g_{cb}(\xi)$ are functions of $\xi^a$ only, $g_{cy}=0$, and $g_{zy}(\xi)$ are functions of $\xi^x$ only, then we call the space a *locally decomposable Riemannian space.*

For a locally decomposable Riemannian space, we have

$$\left\{ {x \atop c\ b} \right\}=0, \qquad \left\{ {x \atop z\ b} \right\}=0, \qquad \left\{ {a \atop c\ y} \right\}=0, \qquad \left\{ {a \atop z\ y} \right\}=0. \tag{3.14}$$

Conversely, if (3.14) are satisfied in a locally product Riemannian space, we can easily prove that $g_{cb}(\xi)$ are functions of $\xi^a$ only, $g_{cy}=0$ and $g_{zy}(\xi)$ are functions of $\xi^x$ only. Thus we have

THEOREM 3.1. *A necessary and sufficient condition for a locally product Riemannian space to be a locally decomposable Riemannian space is that* (3.14) *are satisfied in a separating coordinate system.*

Combining the theorems in the previous section and this theorem, we have

THEOREM 3.2. *A necessary and sufficient condition for a locally product Riemannian space* $M_p \times M_q$ *to be a locally decomposable Riemannian space is that* $M_p$ *and* $M_q$ *are both totally geodesic and are parallel.*

Now, for the covariant derivative of the tensor $F_i^{\ h}$, we have

$$\nabla_j F_b^{\ a}=0,\ \nabla_j F_y^{\ a}=-2\begin{Bmatrix} a \\ j\ y \end{Bmatrix},\ \nabla_j F_b^{\ x}=+2\begin{Bmatrix} x \\ j\ b \end{Bmatrix},\ \nabla_j F_y^{\ x}=0. \tag{3.15}$$

Thus, from Theorem **3.1** and (3.15), we have

THEOREM 3.3. *A necessary and sufficient condition for a locally product Riemannian space to be a locally decomposable Riemannian space is that*

$$\nabla_j F_i^{\ h}=0 \tag{3.16}$$

or *equivalently*

$$\nabla_j F_{ih}=0. \tag{3.17}$$

Since, in a locally decomposable Riemannian space, the Christoffel symbols $\begin{Bmatrix} h \\ j\ i \end{Bmatrix}$ are all zero except $\begin{Bmatrix} a \\ c\ b \end{Bmatrix}$ and $\begin{Bmatrix} x \\ z\ y \end{Bmatrix}$, and besides, $\begin{Bmatrix} a \\ c\ b \end{Bmatrix}$ are functions of $\xi^a$ only and $\begin{Bmatrix} x \\ z\ y \end{Bmatrix}$ are functions of $\xi^x$ only, the curvature tensor

$$K_{kji}^{\ \ \ h}=\partial_k\begin{Bmatrix} h \\ j\ i \end{Bmatrix}-\partial_j\begin{Bmatrix} h \\ k\ i \end{Bmatrix}+\begin{Bmatrix} h \\ k\ t \end{Bmatrix}\begin{Bmatrix} t \\ j\ i \end{Bmatrix}-\begin{Bmatrix} h \\ j\ t \end{Bmatrix}\begin{Bmatrix} t \\ k\ i \end{Bmatrix} \tag{3.18}$$

is pure in all its indices and also the successive covariant derivatives of $K_{kji}^{\ \ \ h}$

$$\nabla_l K_{kji}^{\ \ \ h},\qquad \nabla_m\nabla_l K_{kji}^{\ \ \ h},\ \ldots,$$

are all pure in its indices.

We put

$$K_{ji}=K_{tji}^{\ \ \ t}, \tag{3.19}$$

$$K=g^{ji}K_{ji},\qquad K^*=F^{ji}K_{ji}. \tag{3.20}$$

Applying the Ricci formula to $F_i^{\ h}$. we find

$$\nabla_k\nabla_j F_i^{\ h}-\nabla_j\nabla_k F_i^{\ h}=K_{kjt}^{\ \ \ h}F_i^{\ t}-K_{kji}^{\ \ \ t}F_t^{\ h},$$

from which

$$K_{kjt}{}^{h}F_{i}{}^{t}-F_{kji}{}^{t}F_{t}{}^{h}=0. \tag{3.21}$$

Transvecting (3.21) with $g^{ji}$, we find

$$K_{kji}{}^{h}F^{ji}-K_{k}{}^{t}F_{t}{}^{h}=0 \tag{3.22}$$

or

$$K_{k}{}^{h}=F_{k}{}^{t}F^{ji}K_{tji}{}^{h}. \tag{3.23}$$

Now, the tensor $\nabla_{k}K_{ji}$ is pure in all its indices and hence

$$*O_{kj}^{ts}\nabla_{t}K_{si}=0$$

or

$$F_{k}{}^{t}\nabla_{t}K_{ji}-F_{j}{}^{t}\nabla_{k}K_{ti}=0. \tag{3.24}$$

Transvecting this with $g^{ji}$, we find

$$F_{k}{}^{t}\nabla_{t}K=\nabla_{k}K^{*}. \tag{3.25}$$

## § 4. Decomposable vectors and tensors

Consider a contravariant vector field $v^{h}(\xi)$ in a locally product space $M_{n}$. When $v^{a}$ are functions of $\xi^{b}$ only, and $v^{x}$ functions of $\xi^{y}$ only in a separating coordinate system, we call $v^{h}$ a *contravariant decomposable vector field* in $M_{n}$. The condition for $v^{h}(\xi)$ to be decomposable is given by

$$\partial_{y}v^{a}=0, \qquad \partial_{b}v^{x}=0 \tag{4.1}$$

in a separating coordinate system. Equations (4.1) are equivalent to

$$*O_{ir}^{sh}\partial_{s}v^{r}=0 \tag{4.2}$$

or to

$$-F_{i}{}^{t}\partial_{t}v^{h}+F_{r}{}^{h}\partial_{i}v^{r}=0. \tag{4.3}$$

or to

$$\underset{v}{\mathfrak{L}}F_{i}{}^{h}=v^{t}\partial_{t}F_{i}{}^{h}-F_{i}{}^{t}\partial_{t}v^{h}+F_{t}{}^{h}\partial_{i}v^{t}=0 \tag{4.4}$$

which is a tensor equation. Thus we have

THEOREM 4.1. *A necessary and sufficient condition for a contravariant vector field $v^h$ in a locally product space to be decomposable is that the derivative of $F_i{}^h$ with respect to $v^h$ vanishes.*

Consider a covariant vector field $w_i(\xi)$ in a locally product space $M_n$. When $w_b(\xi)$ are functions of $\xi^a$ only and $w_y(\xi)$ functions of $\xi^x$ only in a separating coordinate system, we call $w_i(\xi)$ a *covariant decomposable vector field* in $M_n$. The condition for $w_i(\xi)$ to be decomposable is given by

$$\partial_y w_b = 0, \qquad \partial_c w_y = 0 \tag{4.5}$$

in a separating coordinate system. Equations (4.5) are equivalent to

$$*O_{ji}^{ts}\partial_t w_s = 0 \tag{4.6}$$

or to

$$-F_j{}^t\partial_t w_i + F_i{}^t\partial_j w_t = 0 \tag{4.7}$$

or to

$$(\partial_j F_i{}^t - \partial_i F_j{}^t)w_t - F_j{}^t\partial_t w_i + F_i{}^t\partial_j w_t = 0, \tag{4.8}$$

which is a tensor equation. Thus we have

THEOREM 4.2. *A necessary and sufficient condition for a covariant vector field $w_i$ in a locally product space to be decomposable is* (4.8).

Consider next a tensor field, say $T_{ji}{}^h(\xi)$, in a locally product space $M_n$. When $T_{ji}{}^h$ is pure in all its indices, that is, when all the components other than $T_{cb}{}^a(\xi)$ and $T_{zy}{}^x(\xi)$ are zero, and moreover, $T_{cb}{}^a(\xi)$ are functions of $\xi^a$ only and $T_{zy}{}^x(\xi)$ functions of $\xi^x$ only in a separating coordinate system, we call $T_{ji}{}^h(\xi)$ a *decomposable tensor field* in $M_n$. The condition for $T_{ji}{}^h(\xi)$ to be decomposable is given by

$$\partial_y T_{cb}{}^a = 0, \qquad \partial_b T_{zy}{}^x = 0 \tag{4.9}$$

in a separating coordinate system. Equations (4.9) are

equivalent to

$$*O_{kj}^{ts}\partial_t T_{si}{}^h=0 \tag{4.10}$$

or to

$$-F_k{}^t\partial_t T_{ji}{}^h+F^t\partial_k T_{ti}{}^h=0 \tag{4.11}$$

or to

$$\begin{aligned}F_k{}^t\partial_t T_{ji}{}^h-\partial_k(T_{ji}{}^t F_t{}^h)-T_{ti}{}^h\partial_j F_k{}^t-T_{jt}{}^h\partial_i F_k{}^t\\ -(\partial_k F_t{}^h-\partial_t F_k{}^h)T_{ji}{}^t=0,\end{aligned} \tag{4.12}$$

which is a tensor equation.

Since $T_{ji}{}^h$ is pure, we have

$$T_{ti}{}^h F_j{}^t=T_{ji}{}^t F_t{}^h=T_{jt}{}^h F_i{}^t$$

and consequently (4.12) may be written also in the form

$$\begin{aligned}F_k{}^t\nabla_t T_{ji}{}^h-\nabla_k(T_{jt}{}^h F_i{}^t)-T_{ti}{}^h\nabla_j F_k{}^t-T_{jt}{}^h\nabla_i F_k{}^t\\ -(\nabla_k F_t{}^h-\nabla_t F_k{}^h)T_{ji}{}^t=0.\end{aligned} \tag{4.13}$$

Thus we have

THEOREM 4.3. *A necessary and sufficient condition for a pure tensor field, say* $T_{ji}{}^h$ *in a locally product space to be decomposable is* (4.12) or (4.13).

## § 5. Necessary and sufficient conditions for vectors and tensors to be decomposable

We consider a compact locally decomposable Riemannian space and study the necessary and sufficient conditions for vectors and tensors in it to be decomposable. In such a space, we have

$$\nabla_j F_i{}^h=0, \qquad \nabla_j F_{ih}=0. \tag{5.1}$$

Now, the condition (4.4) for a contravariant vector $v^h$ to be decomposable is written as

$$\underset{v}{\mathfrak{L}} F_i{}^h=v^t\nabla_t F_i{}^h-F_i{}^t\nabla_t v^h+F_t{}^h\nabla_i v^t=0,$$

or

$$-F_i^{\ t}\nabla_t v^h + F_t^{\ h}\nabla_i v^t = 0, \tag{5.2}$$

applying $\nabla^i$ to (5.2), we find

$$g^{ji}\nabla_j\nabla_i v^h - (F^{ji}\nabla_j\nabla_i v^t)F_t^{\ h} = 0. \tag{5.3}$$

On the other hand, putting

$$S_{ji} = -F_j^{\ t}\nabla_t v_i + F_i^{\ t}\nabla_j v_t,$$

we have the identity

$$[g^{ji}\nabla_j\nabla_i v^h - (F^{ji}\nabla_j\nabla_i v^t)F_t^{\ h}]v_h + \frac{1}{2}S^{ji}S_{ji}$$

$$= \nabla^j[v^i(\nabla_j v_i) - F_j^{\ t}F_i^{\ s}(\nabla_t v^i)v_s],$$

from which

$$\int\left[\{g^{ji}\nabla_j\nabla_i v^h - (F^{ji}\nabla_j\nabla_i v^t)F_t^{\ h}\}v_h + \frac{1}{2}S^{ji}S_{ji}\right]d\sigma = 0. \tag{5.4}$$

Thus we have

THEOREM 5.1. *A necessary and sufficient condition for a contravariant vector field $v^h$ in a compact orientable locally decomposable Riemannian space to be decomposable is*

$$g^{ji}\nabla_j\nabla_i v^h - (F^{ji}\nabla_j\nabla_i v^t)F_t^{\ h} = 0.$$

The condition (4.8) for a covariant vector $w_i$ to be decomposable is written as

$$-F_j^{\ t}\nabla_t w_i + F_i^{\ t}\nabla_j w_t = 0$$

or

$$-F_j^{\ t}\nabla_t w_h + F_h^{\ t}\nabla_j w_t = 0. \tag{5.5}$$

Comparing (5.2) and (5.5) we have

THEOREM 5.2. *If a contravariant vector $v^h$ in a locally decomposable Riemannian space is decomposable, then the covariant vector $v_i = g_{ih}v^h$ is also decomposable, and vice versa.*

Now, consider a pure covariant tensor $w_{i_1 i_2 \dots i_p}$. Then the condition for this tensor to be decomposable is given by

$$F_j{}^t\partial_t w_{i_1i_2\ldots i_p} - \partial_j(w_{i_1i_2\ldots t}F_{i_p}{}^t) \\ - w_{ti_2\ldots i_p}\partial_{i_1}F_j{}^t - \ldots - w_{i_1\ldots i_{p-1}t}\partial_{i_p}F_j{}^t = 0$$

or

$$F_j{}^t\nabla_t w_{i_1\ldots i_p} - \nabla_j(w_{i_1\ldots i_{p-1}t}F_{i_p}{}^t) = 0. \tag{5.6}$$

Applying $\nabla^j$ to (5.6), we find

$$g^{ji}\nabla_j\nabla_i w_{i_1\ldots i_p} - (F^{ji}\nabla_j\nabla_i w_{i\ldots i_{p-1}t})F_{i_p}{}^t = 0.$$

On the other hand, putting

$$S_{ji_1\ldots i_p} = F_j{}^t\nabla_t w_{i_1\ldots i_p} - \nabla_j(w_{i_1\ldots i_{p-1}t}F_{i_p}{}^t)$$

we have

$$[g^{ji}\nabla_j\nabla_i w_{i_1\ldots i_p} - (F^{ji}\nabla_j\nabla_i w_{i_1\ldots i_{p-1}t})F_{i_p}{}^t]w^{i_1\ldots i_p} \\ + \frac{1}{2}S_{ji_1\ldots i_p}S^{ji_1\ldots i_p} = \nabla_t(w_{i_1\ldots i_p}\nabla^t w^{i_1\ldots i_p}) \\ - \nabla_t(F_j{}^tF_s{}^{i_p}w_{i\ldots i_p}\nabla^j w^{i_1\ldots i_{p-1}s}),$$

from which

$$\int\left[\{g^{ji}\nabla_j\nabla_i w_{i_1\ldots i_p} - (F^{ji}\nabla_j\nabla_i w_{i_1\ldots i_{p-1}t})F_{i_p}{}^t\}w^{i_1\ldots i_p} \right. \\ \left. + \frac{1}{2}S_{ji_1\ldots i_p}S^{ji_1\ldots i_p}\right]d\sigma = 0. \tag{5.7}$$

Thus we have

THEOREM 5.3. *A necessary and sufficient condition for a pure tensor field $w_{i_1\ldots i_p}$ in a compact orientable locally decomposable Riemannian space to be decomposable is*

$$g^{ji}\nabla_j\nabla_i w_{i_1\ldots i_p} - F^{ji}\nabla_j\nabla_i w_{i_1\ldots i_{p-1}t}F_{i_p}{}^t = 0.$$

Now, a necessary and sufficient condition for a skew-symmetric covariant tensor in a compact orientable Riemannian space to be harmonic is

$$g^{ji}\nabla_j\nabla_i w_{i_1\ldots i_p} - \textstyle\sum_{s=1}^p K_{i_s}{}^u w_{i_1\ldots u\ldots i_p} \\ + \textstyle\sum_{r<s} K_{i_ri_s}{}^{rs} w_{i_1\ldots r\ldots s\ldots i_p} = 0. \tag{5.8}$$

If we assume $w_{i_1\ldots i_p}$ to be pure, we have

$$w_{i_1\ldots i_s\ldots i_t\ldots i_p}-w_{i_1\ldots r\ldots q\ldots i_p}F_{i_s}{}^{r}F_{i_t}{}^{q}=0.$$

or

$$w_{i_1\ldots i_s\ldots r\ldots i_p}F_{i_t}{}^{r}-w_{i_1\ldots r\ldots i_t\ldots i_p}F_{i_s}{}^{r}=0.$$

Thus

$$w_{i_1\ldots i_{p-1}t}F_{i_p}{}^{t}$$

is also a skew-symmetric tensor.

Now we can easily see that, if a pure harmonic tensor $w_{i_1\ldots i_p}$ satisfies (5.8), then the above skew-symmetric tensor satisfies the same equation. Thus we have

THEOREM 5.4. *In a compact orientable locally decomposable Riemannian space, if* $w_{i_1\ldots i_p}$ *is a pure harmonic tensor, then so is* $w_{i_1\ldots i_{p-1}t}F_{i_p}{}^{t}$.

Suppose that $w_{i_1\ldots i_p}$ is pure and harmonic, then we have

$$\nabla_j w_{i_1\ldots i_p}=\nabla_{i_1}w_{ji_2\ldots i_p}+\ldots+\nabla_{i_p}w_{i_1\ldots i_{p-1}j}$$

and also, $w_{i_1\ldots i_{p-1}t}F_{i_p}{}^{t}$ being harmonic,

$$\nabla_j w_{i_1\ldots i_{p-1}t}F_{i_p}{}^{t}=\nabla_{i_1}w_{ji_2\ldots i_{p-1}t}F_{i_p}{}^{t}+\ldots+\nabla_{i_p}w_{i_1\ldots i_{p-1}t}F_{j}{}^{t}.$$

Thus

$$\begin{aligned}
&F^{kj}\nabla_k\nabla_j w_{i_1\ldots i_{p-1}}F_{i_p}{}^{t}\\
&\quad=F^{kj}\nabla_k[\nabla_{i_1}w_{ji_2\ldots i_{p-1}t}F_{i_p}{}^{t}+\ldots+\nabla_{i_p}w_{i_1\ldots i_{p-1}t}F_{j}{}^{t}]\\
&\quad=\nabla^{k}[\nabla_{i_1}w_{ji_2\ldots i_{p-1}t}F_{k}{}^{j}F_{i_p}{}^{t}+\ldots+\nabla_{i_{p-1}}w_{i_1\ldots jt}F_{k}{}^{j}F_{i_p}{}^{t}\\
&\qquad\qquad+\nabla_{i_p}w_{i_1\ldots i_{p-1}k}]\\
&\quad=\nabla^{k}[\nabla_{i_1}w_{ki_2\ldots i_{p-1}i_p}+\ldots+\nabla_{i_{p-1}}w_{i_1\ldots ki_p}\\
&\qquad\qquad+\nabla_{i_p}w_{i_1\ldots i_{p-1}k}]\\
&\quad=\nabla^{k}\nabla_k w_{i_1\ldots i_p},
\end{aligned}$$

that is,

$$\nabla^{k}\nabla_k w_{i_1\ldots i_p}-F^{kj}\nabla_k\nabla_j w_{i_1\ldots i_{p-1}t}F_{i_p}{}^{t}=0.$$

Thus, we have

THEOREM 5.5. *A pure harmonic tensor in a compact orientable decomposable space is decomposable.*

## § 6. Projective and conformal Killing vectors

We consider a contravariant vector field $v^h$. The Lie derivative of the Christoffel symbols $\left\{ {h \atop j\ i} \right\}$ is given by

$$\underset{v}{\mathfrak{L}} \left\{ {h \atop j\ i} \right\} = \nabla_j \nabla_i v^h + K_{kji}{}^h v^k, \tag{6.1}$$

from which, by transvection with $g^{ji}$,

$$g^{ji} \underset{v}{\mathfrak{L}} \left\{ {h \atop j\ i} \right\} = g^{ji} \nabla_j \nabla_i v^h + K_i{}^h v^i \tag{6.2}$$

and, by transvection with $F^{ji}$

$$F^{ji} \underset{v}{\mathfrak{L}} \left\{ {h \atop j\ i} \right\} = F^{ji} \nabla_j \nabla_i v^h + K_i{}^t F_t{}^h v^i \tag{6.3}$$

by virtue of (3.22).

Equation (6.3) can be also written as

$$F_t{}^h F^{ji} \underset{v}{\mathfrak{L}} \left\{ {t \atop j\ i} \right\} = F^{ji} \nabla_j \nabla_i (F_t{}^h v^t) + K_i{}^h v^i . \tag{6.4}$$

From (6.2) and (6.4), we find

$$g^{ji} \underset{v}{\mathfrak{L}} \left\{ {h \atop j\ i} \right\} - F_t{}^h F^{ji} \underset{v}{\mathfrak{L}} \left\{ {t \atop j\ i} \right\} = g^{ji} \nabla_j \nabla_i v^h - F^{ji} \nabla_j \nabla_i (F_t{}^h v^t). \tag{6.5}$$

For a Killing vector $v^h$ or an affine Killing vector $v^h$, we have

$$\underset{v}{\mathfrak{L}} \left\{ {h \atop j\ i} \right\} = 0$$

and consequently we have

$$g^{ji} \nabla_j \nabla_i v^h - F^{ji} \nabla_j \nabla_i (F_t{}^h v^t) = 0$$

for such a vector field $v^h$. But this is a necessary and sufficient condition that the vector field $v^h$ in a compact orientable locally Riemannian space be decomposable.

Thus we have

THEOREM 6.1. *A Killing vector or an affine Killing vector in a compact orientable locally decomposable Riemannian space is decomposable.*

For a projective Killing vector $v^h$, we have

$$\underset{v}{\mathfrak{L}}\left\{{h \atop j\ i}\right\} = A_j^h p_i + A_i^h p_j, \tag{6.6}$$

where $p_i$ is a certain gradient vector field, and consequently

$$g \underset{v}{\mathfrak{L}}\left\{{h \atop j\ i}\right\} - F_t^{\ h} F^{ji} \underset{v}{\mathfrak{L}}\left\{{t \atop j\ i}\right\}$$
$$= g^{ji}(A_j^h p_i + A_i^h p_j) - F_t^{\ h} F^{ji}(A_j^t p_i + A_i^t p_j) = 2p^h - 2p^h = 0$$

and consequently, from (6.5)

$$g^{ji}\nabla_j\nabla_i v^h - F^{ji}\nabla_j\nabla_i F_t^{\ h} v^t = 0.$$

Thus we have

THEOREM 6.2. *A projective Killing vector field $v^h$ in a compact orientable locally decomposable Riemannian space is decomposable. Thus such a vector satisfies*

$$\underset{v}{\mathfrak{L}}\, F_i^{\ h} = 0.$$

On the other hand, we have

$$\underset{v}{\mathfrak{L}}\nabla_j F_i^{\ h} - \nabla_j \underset{v}{\mathfrak{L}}\, F_i^{\ h} = \left(\underset{v}{\mathfrak{L}}\left\{{h \atop j\ t}\right\}\right) F_i^{\ t} - \left(\underset{v}{\mathfrak{L}}\left\{{t \atop j\ i}\right\}\right) F_t^{\ h}.$$

Substituting (6.6) and (6.7) in this equation we find

$$(A_j^h p_t + A_t^h p_j) F_i^{\ t} - (A_j^t p_i + A_i^t p_j) F_t^{\ h} = 0,$$

or

$$A_j^h p_t F_i^{\ t} - F_j^{\ h} p_i = 0,$$

from which, by contraction with respect to $h$ and $j$,

$$(p+q) p_t F_i^{\ t} - (p-q) p_i = 0$$

and consequently

$$(p-q) p_t F_i^{\ t} - (p+q) p_i = 0.$$

Thus

$$p_i=0,$$

and we have

THEOREM 6.3. *A projective Killing vector field in a compact orientable locally decomposable Riemannian space is an affine Killing vector and consequently a Killing vector field.*

We next consider a conformal Killing vector $v^h$. Such a vector field satisfies

$$\underset{v}{\mathfrak{L}} g_{ji}=\nabla_j v_i+\nabla_i v_j=2\Phi g_{ji}, \tag{6.7}$$

$\Phi$ being a scalar, from which

$$g^{ji}\nabla_j v_i=\nabla_j v^j=n\Phi, \tag{6.8}$$

and

$$F^{ji}\nabla_j v_i=(p-q)\Phi. \tag{6.9}$$

From (6.7), we obtain

$$\underset{v}{\mathfrak{L}}\left\{ {h \atop j\ i} \right\}=\nabla_j\nabla_i v^h+K_{kji}{}^h v^k=A_j^h\Phi_i+A_i^h\Phi_j-\Phi^h g_{ji}, \tag{6.10}$$

where $\Phi_i=\nabla_i\Phi$ and $\Phi^h=\Phi_i g^{ih}$.

From (6.10), we find

$$g^{ji}\nabla_j\nabla_i v^h+K_i{}^h v^i=-(n-2)\Phi^h$$

and

$$F_t{}^h F^{ji}\nabla_j\nabla_i v^t+K_i{}^h v^i=2\Phi^h-(p-q)F_t{}^h\Phi^t,$$

and from these two equations

$$g^{ji}\nabla_j\nabla_i v^h-F^{ji}\nabla_j\nabla_i F_t{}^h v^t=-n\Phi^h+(p-q)F_t{}^h\Phi^t.$$

Substituting this into (5.4), we find

$$\int\left[-n\Phi^h v_h+(p-q)F^{ji}\Phi_j v_i+\frac{1}{2}S^{ji}S_{ji}\right]d\sigma=0. \tag{6.11}$$

On the other hand,

$$\int(-n\Phi^h v_h)d\sigma=n\int\Phi\nabla^h v_h d\sigma=n^2\int\Phi^2 d\sigma$$

by virtue of (6.8), and

$$\int (p-q)F^{ji}\Phi_j v_i d\sigma = -(p-q)\int \Phi F^{ji}\nabla_j v_i d\sigma = -(p-q)^2 \int \Phi^2 d\sigma$$

by virtue of (6.9). Thus equation (6.11) gives

$$\int \left[ \{n^2-(p-q)^2\}\Phi^2 + \frac{1}{2} S^{ji}S_{ji} \right] d\sigma = 0,$$

from which

$$\Phi = 0, \qquad S_{ji} = 0.$$

Thus we have

THEOREM 6.4. *A conformal Killing vector in a compact orientable locally decomposable Riemannian space is a Killing vector and is decomposable.*

## § 7. Almost Einstein spaces and spaces of almost constant curvature

Suppose that our Riemannian space is locally decomposable and that we have chosen a coordinate system in which the first fundamental form is of the form

$$ds^2 = g_{cb}(\xi^a)d\xi^c d\xi^b + g_{zy}(\xi^x)d\xi^z d\xi^y \; (p > 2, \; q > 2): \qquad (7.1)$$

We know already that the Ricci tensor $K_{ji}$ is pure in $j$ and $i$.

Now suppose that $g_{cb}d\xi^c d\xi^b$ and $g_{zy}d\xi^z d\xi^y$ are both Einstein spaces, then we have

$$K_{cb} = \lambda g_{cb}, \qquad K_{zy} = \mu g_{zy} \qquad (7.2)$$

for certain constants $\lambda$ and $\mu$.

The equations (7.2) may also be written in the form

$$K_{ji} = \frac{1}{2}(\lambda+\mu)g_{ji} + \frac{1}{2}(\lambda-\mu)F_{ji}: \qquad (7.3)$$

Conversely suppose that the Ricci curvature of a locally decomposable Riemannian space has the form

$$K_{ji} = ag_{ji} + bF_{ji}. \tag{7.4}$$

If we choose a coordinate system in which the first fundamental form takes the form (7.1), equation (7.4) gives

$$K_{cb} = (a+b)g_{cb}, \qquad K_{zy} = (a-b)g_{zy}: \tag{7.5}$$

Thus, if $p > 2$ and $q > 2$, then $g_{cb}d\xi^c d\xi^b$ and $g_{zy}d\xi^z d\xi^y$ are both metrics of Einstein spaces, and

$$a+b = \lambda, \qquad a-b = \mu$$

are both constants. Thus we have

THEOREM 7.1. *In a locally decomposable Riemannian space where $p > 2$, $q > 2$, a necessary and sufficient condition that the two components are both Einstein spaces is that the Ricci tensor of the space has the form*

$$K_{ji} = ag_{ji} + bF_{ji},$$

*a and b being necessarily constant.*

We call such a space an almost Einstein space.

We next suppose that $g_{cb}d\xi^c d\xi^b$ and $g_{zy}d\xi^z d\xi^y$ are both spaces of constant curvature, then we have

$$K_{dcba} = \lambda(g_{da}g_{cb} - g_{ca}g_{db}), \quad K_{zyxw} = \mu(g_{zw}g_{yx} - g_{yw}g_{zx}) \tag{7.6}$$

for certain constants $\lambda$ and $\mu$.

The equations may also be written in the form

$$K_{kjih} = \frac{1}{4}(\lambda+\mu)[(g_{kh}g_{ji} - g_{jh}g_{ki}) + (F_{kh}F_{ji} - F_{jh}F_{ki})]$$

$$+\frac{1}{4}(\lambda-\mu)[(F_{kh}g_{ji} - F_{jh}g_{ki}) + (g_{kh}F_{ji} - g_{jh}F_{ki})]. \tag{7.7}$$

Conversely suppose that the curvature tensor of a locally decomposable Riemannian space has the form

$$\begin{aligned} K_{kjih} = {} & a[(g_{kh}g_{ji} - g_{jh}g_{ki}) + (F_{kh}F_{ji} - F_{jh}F_{ki})] \\ & + b[(F_{kh}g_{ji} - F_{jh}g_{ki}) + (g_{kh}F_{ji} - g_{jh}F_{ki})]. \end{aligned} \tag{7.8}$$

If we choose a coordinate system in which the first fundamental form takes the form (7.1), equation (7.8) gives

$$K_{dcba} = 2(a+b)(g_{da}g_{cb} - g_{ca}g_{db}), \qquad (7.9)$$
$$K_{zyxw} = 2(a-b)(g_{zw}g_{yx} - g_{yw}g_{zx}).$$

Thus if $p > 2$ and $q > 2$ then $g_{cb}d\xi^c d\xi^b$ and $g_{zy}d\xi^z d\xi^y$ are both metrics of spaces of constant curvature, and

$$2(a+b) = \lambda \quad \text{and} \quad 2(a-b) = \mu$$

are both constants. Thus we have

THEOREM 7.2. *In a locally decomposable Riemannian space where $p > 2$, $q > 2$, a necessary and sufficient condition that the two components are both spaces of constant curvature is that the curvature tensor of the space has the form*

$$K_{kjih} = a[(g_{kh}g_{ji} - g_{jh}g_{ki}) + (F_{kh}F_{ji} - F_{jh}F_{ki})]$$
$$+ b[(F_{kh}g_{ji} - F_{jh}g_{ki}) + (g_{kh}F_{ji} - g_{jh}F_{ki})],$$

*$a$ and $b$ being necessarily constants.*

We call such a space a space of almost constant curvature.

CHAPTER XI

# ALMOST PRODUCT SPACES

## § 1. Almost product spaces†

Let $M_n$ be an $n$-dimensional manifold of differentiability class $C^\infty$ and let there be given two globally complementary distributions $P$ and $Q$ of dimensions $p$ and $q$ respectively, where $p+q=n$ and $p \geq 1$, $q \geq 1$. Here we mean by a distribution of dimension $p$ a $p$-dimensional plane in the $n$-dimensional tangent plane at each point of the manifold.

When there is given only one distribution $P$ in $M_n$, we can construct globally a complementary distribution $Q$ in the following way. Since the manifold $M_n$ is of differentiability class, $C^\infty$, we can introduce a global Riemannian metric of class $C^\infty$ in $M_n$. Then we have only to define $Q$ as a distribution which is always orthogonal to the distribution $P$ with respect to the introduced Riemannian metric. The distribution $Q$ is thus globally defined.

We take $p$ linearly independent contravariant vectors $B_b{}^h$ $(a, b, c, \ldots = 1, 2, \ldots, p;\ x, y, z = p+1,\ p+2,\ \ldots,\ p+ +q=n)$ in $P$ and $q$ linearly independent contravariant vectors $C_y{}^h$ in $Q$. Then $n$ vectors $B_b{}^h$ and $C_y{}^h$ being linearly independent, we construct the inverse of the matrix $(B_b{}^h, C_y{}^h)$ which we denote by $(B^a{}_i, C^x{}_i)$.
Then we have the identities

$$B_b{}^h B^a{}_h = \delta_b^a, \quad B_b{}^h C^x{}_h = 0, \quad C_y{}^h B^a{}_h = 0, \quad C_y{}^h C^x{}_h = \delta_y^x \qquad (1.1)$$

† See A. Aragnol [1]; T. Fukami [3]; T. Nagano [2]; A. G. Walker [1]; T. J. Willmore [1], [3]; K. Yano [15].

and

$$B_a^{\ h} B^a_{\ i} + C_x^{\ h} C^x_{\ i} = A_i^h. \tag{1.2}$$

We use the notations

$$A_\beta^{\ h} = (B_b^{\ h},\ C_y^{\ h}), \qquad A^\alpha_{\ h} = (B^a_{\ h},\ C^x_{\ h}) \tag{1.3}$$

$(\alpha,\ \beta,\ \gamma\ \ldots = 1, 2, \cdots, p,\ p+1, \ldots, p+q = n)$, and write (1.1) and (1.2) in the following form:

$$A_\beta^{\ h} A^\alpha_{\ h} = \delta^\alpha_\beta, \qquad A_\alpha^{\ h} A^\alpha_{\ i} = A_i^h. \tag{1.4}$$

We call the set $A_\beta^{\ h} = (B_b^{\ h},\ C_y^{\ h})$ the non-holonomic frame.

If we put

$$B_a^{\ h} B^a_{\ i} = P_i^{\ h}, \qquad C_x^{\ h} C^x_{\ i} = Q_i^{\ h}, \tag{1.5}$$

then we have from (1.1) and (1.2)

$$P_j^{\ i} P_i^{\ h} = P_j^{\ h}, \quad P_j^{\ i} Q_i^{\ h} = 0, \quad Q_j^{\ i} P_i^{\ h} = 0, \quad Q_j^{\ i} Q_i^{\ h} = Q_j^{\ h} \tag{1.6}$$

and

$$A_i^h = P_i^{\ h} + Q_i^{\ h}. \tag{1.7}$$

It will be easily verified that the tensors $P_i^{\ h}$ and $Q_i^{\ h}$ do not depend on the choice of $B_b^{\ h}$ in $P$ and $C_y^{\ h}$ in $Q$. These are projection tensors on $P$ and $Q$ respectively, that is, an arbitrary vector $v^h$ in the tangent space is decomposed into

$$v^h = P_i^{\ h} v^i + Q_i^{\ h} v^i,$$

$P_i^{\ h} v^i$ and $Q_i^{\ h} v^i$ being respectively in $P$ and $Q$.

A vector in $P$ is characterized by

$$v^h = P_i^{\ h} v^i \quad \text{or} \quad Q_i^{\ h} v^i = 0$$

and a vector in $Q$ by

$$v^h = Q_i^{\ h} v^i \quad \text{or} \quad P_i^{\ h} v^i = 0.$$

We now define the tensor $F_i^{\ h}$ by

$$F_i^{\ h} = P_i^{\ h} - Q_i^{\ h}, \tag{1.8}$$

then (1.7) and (1.8) give

$$P_i^{\ h} = \frac{1}{2}(A_i^h + F_i^{\ h}), \quad Q_i^{\ h} = \frac{1}{2}(A_i^h - F_i^{\ h}). \tag{1.9}$$

Taking account of (1.6) we can easily see that

$$F_j^{\ i}F_i^{\ h} = +A_j^h. \tag{1.10}$$

We notice here that if the distributions $P$ and $Q$ are given globally, then the tensor $F_i^{\ h}$ of rank $n$ will also be defined globally.

Conversely if, in the manifold $M_n$, a tensor $F_i^{\ h}$ satisfying equation (1.10) is given globally, then we define $P_i^{\ h}$ and $Q_i^{\ h}$ by (1.9) and we can easily see that these $P_i^{\ h}$ and $Q_i^{\ h}$ satisfy (1.6) and (1.7). Thus $P_i^{\ h}$ and $Q_i^{\ h}$ define two complementary distributions $P$ and $Q$ globally.

Let $a$ be an eigenvalue of the matrix $F_i^{\ h}$ and $v^h$ the corresponding eigenvector, then we have

$$F_j^{\ i}v^j = av^i, \tag{1.11}$$

from which, transvecting with $F_i^{\ h}$,

$$F_i^{\ h}F_j^{\ i}v^j = aF_i^{\ h}v^i$$

or

$$v^h = a^2 v^h,$$

because of (1.10) and (1.11). Thus we have $a^2 = 1$, which shows that the eigenvalues of the matrix $F_i^{\ h}$ are $+1$ or $-1$. For an eigenvector $v^h$ corresponding to the eigenvalue $+1$, we have

$$F_i^{\ h}v^i = v^h \quad \text{or} \quad Q_i^{\ h}v^i = \frac{1}{2}(A_i^h - F_i^{\ h})v^i = 0,$$

which shows that $v^h$ is in $P$. An eigenvector corresponding to the eigenvalue $-1$, is in $Q$.

Thus, if $F_i^{\ h}$ has eigenvalue $+1$ of multiplicity $p$ and eigenvalue $-1$ of multiplicity $q$, then the dimension of $P$ is $p$ and that of $Q$ is $q$.

An $n$-dimensional manifold $M_n$ in which a tensor field $F_i^{\ h}(\neq A_i^h)$ satisfying (1.10) is given is called an *almost product space*.

## § 2. Integrability conditions

We put

$$\Omega_{\gamma\beta}{}^{\alpha} = \frac{1}{2}(\partial_\gamma A_\beta{}^h - \partial_\beta A_\gamma{}^h) A^\alpha{}_h \tag{2.1}$$

and call $\Omega_{\beta\gamma}{}^{\alpha}$ the non-holonomic object, where $\partial_\gamma$ denotes the non-holonomic or Pfaffian derivative with respect to $A_\gamma{}^h$, that is,

$$\partial_\gamma = A_\gamma{}^h \partial_h = A_\gamma{}^h \frac{\partial}{\partial \xi^h}, \tag{2.2}$$

$\xi^h$ being the local coordinates. Thus

$$\partial_b = B_b{}^h \partial_h \quad \text{and} \quad \partial_y = C_y{}^h \partial_h. \tag{2.3}$$

The equation (2.1) can also be written in the form

$$\Omega_{\gamma\beta}{}^{\alpha} = -\frac{1}{2} A_\gamma^j A_\beta^i (\partial_j A^\alpha_i - \partial_i A^\alpha{}_j), \tag{2.4}$$

which shows that $\Omega_{\gamma\beta}{}^{\alpha}$ are scalars under a transformation of the local coordinates. But if we effect the transformation of the non-holonomic frame:

$$A_{\beta'} = A^\beta_{\beta'} A_\beta{}^h \; (|A^\beta_{\beta'}| \neq 0), \tag{2.5}$$

then the non-holonomic object $\Omega_{\gamma\beta}{}^{\alpha}$ undergoes the transformation

$$\Omega_{\gamma'\beta'}{}^{\alpha'} = A^\gamma_{\gamma'} A^\beta_{\beta'} A^{\alpha'}_\alpha \Omega_{\gamma\beta}{}^{\alpha} + \frac{1}{2}(\partial_{\gamma'} A^\alpha_{\beta'} - \partial_{\beta'} A^\alpha_{\gamma'}) A^{\alpha'}_\alpha, \tag{2.6}$$

where $A^{\alpha'}_\alpha$ is the inverse matrix of $A^\alpha_{\alpha'}$. Equation (2.6) shows that the non-holonomic object $\Omega_{\gamma\beta}{}^{\alpha}$ is not a tensor under the transformation of non-holonomic frame.

Since we are considering a non-holonomic frame whose first $p$ vectors are in the distribution $P$ and whose second $q$ vectors are in the distribution $Q$, the transformation (2.5) of the non-holonomic frame will split into

$$B_{b'}{}^h = A^b_{b'} B_b{}^h, \qquad C_{y'}{}^h = A^y_{y'} C_y{}^h, \tag{2.7}$$

which implies

$$A^{\alpha}_{\alpha'} = \begin{pmatrix} A^b_{b'} & 0 \\ 0 & A^y_{y'} \end{pmatrix}. \tag{2.8}$$

Thus equation (2.6) gives

$$\Omega_{c'b'}{}^{x'} = A^c_{c'} A^b_{b'} A^{x'}_x \Omega_{cb}{}^x \qquad \Omega_{z'y'}{}^{a'} = A^z_{z'} A^y_{y'} A^{a'}_a \Omega_{zy}{}^a , \tag{2.9}$$

which show that $\Omega_{cb}{}^{x'}$ and $\Omega_{zy}{}^a$ are tensors under a transformation of the non-holonomic frame.

We shall now consider the integrability condition of the distribution $P$.

An arbitrary contravariant vector $d\xi^h$ in the tangent space at a point of the manifold can be written in the form

$$d\xi^h = B_a{}^h (d\xi)^a + C_y{}^h (d\xi)^y \tag{2.10}$$

because of (1.7), where

$$(d\xi)^a = B^a{}_i d\xi^i , \qquad (d\xi)^x = C^x{}_i d\xi^i . \tag{2.11}$$

Thus the distribution $P$ is defined by

$$(d\xi)^x = C^x{}_i d\xi^i = 0. \tag{2.12}$$

The condition for the distribution $P$ to be completely integrable is then that

$$(\partial_j C^x{}_i - \partial_i C^x{}_j) d\xi^j \wedge d\xi^i = 0$$

be satisfied by any $d\xi^h$ satisfying (2.12), that is,

$$\Omega_{cb}{}^x = -\frac{1}{2} B_c{}^j B_b{}^i (\partial_j C^x{}_i - \partial_i C^x{}_j) = 0. \tag{2.13}$$

This is the condition for the distribution $P$ to be completely integrable.

The same condition may be found also in the following way. An arbitrary contravariant vector $d\xi^h$ in the tangent space at a point of the manifold can be written in the form:

$$d\xi^h = P_i{}^h d\xi^i + Q_i{}^h d\xi^i , \tag{2.14}$$

because of (1.7). Thus the distribution $P$ is defined by

$$Q_i{}^h d\xi^i = 0. \tag{2.15}$$

The condition for $P$ to be completely integrable is then that

$$(\partial_j Q_i{}^h - \partial_i Q_j{}^h) d\xi^j \wedge d\xi^i = 0$$

be satisfied by any $d\xi^h$ satisfying (2.15), that is, by any vector satisfying $P_i{}^h d\xi^i = d\xi^h$. Thus we have

$$-\frac{1}{2} P_j{}^m P_i{}^l (\partial_m Q_l{}^h - \partial_l Q_m{}^h) = 0 \tag{2.16}$$

as the condition for $P$ to be completely integrable.

By a straightforward calculation, we can show that

$$B^c{}_j B^b{}_i C_x{}^h \Omega_{cb}{}^x = -\frac{1}{2} P_j{}^m P_i{}^l (\partial_m Q_l{}^h - \partial_l Q_m{}^h), \tag{2.17}$$

and the equivalence of (2.13) and (2.16) is evident.

Now substituting (1.9) in the left hand member of equation (2.16), we find

$$-\frac{1}{2} P_j{}^m P_i{}^l (\partial_m Q_l{}^h - \partial_l Q_m{}^h) = \frac{1}{16} (N_{ji}{}^h - N_{ji}{}^l F_l{}^h), \tag{2.18}$$

where $N_{ji}{}^h$ is the so-called Nijenhuis tensor formed with $F_i{}^h$:

$$N_{ji}{}^h = F_j{}^l (\partial_l F_i{}^h - \partial_i F_l{}^h) - F_i{}^l (\partial_l F_j{}^h - \partial_j F_l{}^h). \tag{2.19}$$

The equation (2.18) shows that the conditions for $P$ to be completely integrable is expressed also in the form:

$$N_{ji}{}^h - N_{ji}{}^l F_l{}^h = 0. \tag{2.20}$$

Similarly we can find that the condition for $Q$ to be completely integrable is

$$\Omega_{zy}{}^a = 0 \tag{2.21}$$

or

$$-\frac{1}{2} Q_j{}^m Q_i{}^l (\partial_m P_l{}^h - \partial_l P_m{}^h) = 0 \tag{2.22}$$

or

$$N_{ji}{}^h + N_{ji}{}^l F_l{}^h = 0. \tag{2.23}$$

Gathering the above results, we have

THEOREM 2.1.† *In order that the distribution $P$ ($Q$) be completely integrable, it is necessary and sufficient that*

$$\Omega_{cb}{}^{x} = 0, \qquad (\Omega_{zy}{}^{a} = 0)$$

*or equivalently*

$$-\frac{1}{2} P_j^m P_i^l (\partial_m Q_l^h - \partial_l Q_m{}^h) = 0 \quad \left( -\frac{1}{2} Q_j^m Q_i^l (\partial_m P_l^h - \partial_l P_m{}^h) = 0 \right)$$

*or equivalently*

$$N_{ji}{}^h - N_{ji}{}^l F_l{}^h = 0, \qquad (N_{ji}{}^h + N_{ji}{}^l F_l{}^h = 0).$$

*Consequently, in order that both of the distributions $P$ and $Q$ be completely integrable, it is necessary and sufficient that*

$$\Omega_{cb}{}^{x} = 0, \qquad \Omega_{zy}{}^{a} = 0$$

*or equivalently*

$$-\frac{1}{2} P_j^m P_i^l (\partial_m Q_l^h - \partial_l Q_m{}^h) = 0, \quad -\frac{1}{2} Q_j^m Q_i^l (\partial_m P_l^h - \partial_l P_m{}^h) = 0$$

*or equivalently*

$$N_{ji} = 0.$$

## § 3. Affine connexions and distributions

Let $\Gamma_j{}^h{}_i$ be components of an affine connexion in $M_n$ and the covariant differentiation of a contravariant vector $v^h$ be denoted by

$$\nabla_j v^h = \partial_j v^h + \Gamma_j{}^h{}_i v^i. \tag{3.1}$$

If we put

$$\Gamma_\gamma{}^\alpha{}_\beta = (\partial_\gamma A_\beta{}^h + A_\gamma{}^j A_\beta{}^i \Gamma_j{}^h{}_i) A^\alpha{}_h, \tag{3.2}$$

† See also T. Fukami [3]; A. Lichnerowicz [3], [7], [17]; G. de Rham [3]; A. G. Walker [1]; T. J. Willmore [1]; K. Yano [15].

then the components $\nabla_\gamma v^\alpha$ of $\nabla_j v^h$ with respect to the non-holonomic frame $(A_\alpha{}^h)$ are given by

$$\nabla_\gamma v^\alpha = \partial_\gamma v^\alpha + \Gamma_\gamma{}^\alpha{}_\beta v^\beta . \tag{3.3}$$

From (3.2) we have

$$\frac{1}{2}(\Gamma_\gamma{}^\alpha{}_\beta - \Gamma_\beta{}^\alpha{}_\gamma) = S_{\gamma\beta}{}^\alpha + \Omega_{\gamma\beta}{}^\alpha , \tag{3.4}$$

where

$$S_{\gamma\beta}{}^\alpha = A_\gamma{}^j A_\beta{}^i A_h{}^\alpha S_{ji}{}^h \tag{3.5}$$

and

$$S_{ji}{}^h = \frac{1}{2}(\Gamma_j{}^h{}_i - \Gamma_i{}^h{}_j) \tag{3.6}$$

is the torsion tensor for the affine connexion $\Gamma_j{}^h{}_i$.

From (3.4) we can see that the $\Gamma_\gamma{}^\alpha{}_\beta$ is not necessarily symmetric with respect to $\gamma$ and $\beta$ even if the affine connexion $\Gamma_j{}^h{}_i$ is symmetric.

Now if we effect a transformation (2.5) of the non-holonomic frame, the components $\Gamma_\gamma{}^\alpha{}_\beta$ of the affine connexion undergo the transformation

$$\Gamma_{\gamma'}{}^{\alpha'}{}_{\beta'} = (\partial_{\gamma'} A^\alpha_{\beta'} + A^\gamma_{\gamma'} A^\beta_{\beta'} \Gamma_\gamma{}^\alpha{}_\beta) A^{\alpha'}_\alpha . \tag{3.7}$$

Since the matrix $(A^\alpha_{\alpha'})$ has the form (2.8) the equation (3.7) gives

$$\left.\begin{aligned} \Gamma_{c'}{}^{x'}{}_{b'} &= A^c_{c'} A^b_{b'} A^{x'}_x \Gamma_c{}^x{}_b , & \Gamma_{z'}{}^{x'}{}_{b'} &= A^z_{z'} A^b_{b'} A^{x'}_x \Gamma_z{}^x{}_{b'} , \\ \Gamma_{z'}{}^{a'}{}_{y'} &= A^z_{z'} A^y_{y'} A^{a'}_a \Gamma_z{}^a{}_y , & \Gamma_{c'}{}^{a'}{}_{y'} &= A^c_{c'} A^y_{y'} A^{a'}_a \Gamma_c{}^a{}_y , \end{aligned}\right\} \tag{3.8}$$

which show that $\Gamma_c{}^x{}_b$, $\Gamma_z{}^x{}_b$, $\Gamma_z{}^a{}_y$ and $\Gamma_c{}^a{}_y$ are all components of tensors with respect to the transformations of the non-holonomic frame $A_{\alpha'}{}^h = A^\alpha_{\alpha'} A_\alpha{}^h$ having a special form (2.8). For example, the vanishing of one of these tensors should have a geometrical meaning independent of the choice of the vectors $B_a{}^h$ in $P$ and $C_x{}^h$ in $Q$.

Now equation (3.2) can be written also in the form:

$$\partial_\gamma A_\beta{}^h + A_\gamma{}^j A_\beta{}^i \Gamma_j{}^h{}_i = \Gamma_\gamma{}^\alpha{}_\beta A_\alpha{}^h . \tag{3.9}$$

If we put $\gamma=c$ and $\beta=b$ in (3.9), we get

$$\partial_c B_b{}^h + B_c{}^j B_b{}^i \Gamma_j{}^h{}_i = \Gamma_c{}^a{}_b B_a{}^h + \Gamma_c{}^x{}_b C_x{}^h$$

or

$$\partial_c B_b{}^h + B_c{}^j B_b{}^i \Gamma_j{}^h{}_i - \Gamma_c{}^a{}_b B_a{}^h = \Gamma_c{}^x{}_b C_x{}^h.$$

We denote by $\nabla_c B_b{}^h$ the left hand member of the above equation:

$$\nabla_c B_b{}^h = \partial_c B_b{}^h + B_c{}^j B_b{}^i \Gamma_j{}^h{}_i - \Gamma_c{}^a{}_b B_a{}^h = \Gamma_c{}^x{}_b C_x{}^h. \quad (3.10)$$

Here $\nabla_c B_b{}^h$ is the so-called van der Waerden–Bortolotti derivative of $B_b{}^h$ along the distribution $P$ and consequently equation (3.10) reduces to that of Gauss when $P$ is integrable.

If we put $\gamma=c$ and $\beta=y$ in (3.9), we get

$$\nabla_c C_y{}^h = \partial_c C_y{}^h + B_c{}^j C_y{}^i \Gamma_j{}^h{}_i = \Gamma_c{}^a{}_y B_a{}^h + \Gamma_c{}^x{}_y C_x{}^h, \quad (3.11)$$

which reduces to the equation of Weingarten when $P$ is integrable.

Similarly we get, from (3.9),

$$\nabla_z B_b{}^h = \partial_z B_b{}^h + C_z{}^j B_b{}^i \Gamma_j{}^h{}_i = \Gamma_z{}^a{}_b B_a{}^h + \Gamma_z{}^x{}_b C_x{}^h \quad (3.12)$$

and

$$\nabla_z C_y{}^h = \partial_z C_y{}^h + C_z{}^j C_y{}^i \Gamma_j{}^h{}_i - \Gamma_z{}^x{}_y C_x{}^h = \Gamma_z{}^a{}_y B_a{}^h, \quad (3.13)$$

which reduce respectively to the equation of Weingarten and that of Gauss when the distribution $Q$ is integrable.

We shall now consider various conditions which we can put on the distribution $P$ and $Q$.

(i) The condition for $P$ to be flat.

Let us consider a vector field $v^h$. If the vector is parallel when we displace in any direction contained in $P$, we say that the vector is parallel along $P$. We can use the same terminology also for the distribution, that is, if a distribution is parallel when we displace in any direction contained in $P$, we say that the distribution is parallel along $P$

Now when we displace a vector contained in $P$ parallelly along $P$, if the displaced vector is always contained in the distribution $P$, we say that the distribution $P$ is a *flat* distribution. Walker [1] calls such a distribution a semi-parallel distribution.

The equation (3.10) shows that the condition for the distribution $P$ to be flat is

$$\Gamma_{c\ b}^{\ x} = 0. \tag{3.14}$$

(ii) The condition for $P$ to be geodesic.

Take a point $\xi^h$ and a direction $v^h$ at $\xi^h$ which is contained in $P$. The auto-parallel curve or path with respect to the affine connexion under consideration is uniquely determined by the initial point $\xi^h$ and the initial direction $v^h$. If the tangent to the path thus determined is always contained in $P$ for any initial point and for any initial direction contained in $P$, we say that the distribution is *geodesic*. Walker calls such a distribution a *path-parallel* distribution.

The condition for the distribution $P$ to be geodesic is then that, if the equation

$$C_i^x \frac{d\xi^i}{ds} = 0$$

is satisfied at the initial point $\xi_0^h$ and for the initial direction $\left(\frac{d\xi^h}{ds}\right)_0$ at $\xi_0^h$, it should always be satisfied along the path:

$$\frac{d^2\xi^h}{ds^2} + \Gamma_{j\ i}^{\ h} \frac{d\xi^j}{ds} \frac{d\xi^i}{ds} = 0$$

having $\xi_0^h$ as the initial point and $\left(\frac{d\xi^h}{ds}\right)_0$ as the initial direction at $\xi_0^h$.

Thus differentiating $C_i^x \frac{d\xi^i}{ds} = 0$ along the path, we have

$$(\partial_j C_i^x - C_h^x \Gamma_{j\ i}^{\ h}) \frac{d\xi^j}{ds} \frac{d\xi^i}{ds} = 0,$$

or $\dfrac{d\xi^i}{ds}$ being always contained in $P$,

$$B_{(c}{}^j B_{b)}{}^i(\partial_j C_i^x - C_h^x \Gamma_j{}^h{}_i) = 0$$

or

$$(\partial_{(c} B_{b)} + B_{(c}{}^j B_{b)}{}^i \Gamma_j{}^h{}_i) = 0,$$

which is equivalent to

$$\Gamma_{(c}^x{}_{b)} = 0. \tag{3.15}$$

This is the condition for $P$ to be geodesic.

Thus we see that a flat distribution is always geodesic, but a geodesic distribution is not necessarily flat. The distinction between flat distributions and geodesic distributions goes back to Hayden [1].

(iii) The condition for $P$ to be parallel along $Q$.

The equation (3.12) shows that the condition for the distribution $P$ to be parallel along $Q$ is

$$\Gamma_z{}^x{}_b = 0. \tag{3.16}$$

(iv) The condition for $Q$ to be parallel along $P$.

$$\Gamma_c{}^a{}_y = 0. \tag{3.17}$$

(v) The condition for $Q$ to be flat.

$$\Gamma_z{}^a{}_y = 0. \tag{3.18}$$

(vi) The condition for $Q$ to be geodesic.

$$\Gamma_{(z}{}^a{}_{y)} = 0. \tag{3.19}$$

(vii) The condition for $P$ to be parallel.

From (i) and (iii) we have

$$\Gamma_c{}^x{}_b = 0 \quad \text{and} \quad \Gamma_z{}^x{}_b = 0$$

or

$$\Gamma_\gamma{}^x{}_b = 0. \tag{3.20}$$

(viii) The condition for $Q$ to be parallel.

From (iv) and (v) we have

$$\Gamma_{c\ y}^{\ a} = 0 \quad \text{and} \quad \Gamma_{z\ y}^{\ a} = 0$$

or

$$\Gamma_{\gamma\ y}^{\ a} = 0. \tag{3.21}$$

Suppose that there is given a symmetric connexion with respect to which the distribution $P$ is flat, then we have

$$\Gamma_{c\ b}^{\ x} = 0.$$

On the other hand, from (3.4), we find

$$\frac{1}{2}(\Gamma_{c\ b}^{\ x} - \Gamma_{b\ c}^{\ x}) = \Omega_{cb}^{\ \ x},$$

the torsion tensor $S_{\gamma\beta}^{\ \ \alpha}$ being zero. This shows that when the distribution $P$ is flat with respect to a symmetric affine connexion, $P$ is integrable. The same is true of course for the distribution $Q$.

## § 4. The determination of affine connexions

Let us consider an almost product space $M_n$ of class $C^\infty$ in which two complementary distributions $P$ and $Q$ of class $C^\infty$ are given globally. Walker [1] studied the existence of global affine connexion with respect to which the given distributions are flat, geodesic or parallel and which are symmetric whenever possible. We shall study the same problems with the use of the existing theory of non-holonomic subspaces and of the Nijenhuis tensor which is related closely to the integrability conditions of the distributions.

Following Walker, we first choose a symmetric affine connexion $\mathring{\Gamma}_{j\ i}^{\ h}$ defined globally in the almost product space $M_n$. Since the space is of class $C^\infty$, we can introduce a global Riemannian metric of class $C^\infty$ in $M_n$ and construct the Levi–Civita affine connexion which can be taken as our $\mathring{\Gamma}_{j\ i}^{h}$.

Then for any global affine connexion $\Gamma_{j\ i}^{h}$, if we put

$$\Gamma_{j\ i}^{h}=\mathring{\Gamma}_{j\ i}^{h}+T_{ji}{}^{h}, \tag{4.1}$$

$T_{ji}{}^{h}$ is a tensor field defined globally in $M_n$.

The problem of determination of a global affine connexion $\Gamma_{j\ i}^{h}$ satisfying certain conditions is then reduced to that of the global tensor $T_{ji}{}^{h}$ satisfying certain conditions.

Substituting (4.1) into (3.2), we find

$$\Gamma_{\gamma\ \beta}^{\ \alpha}=\mathring{\Gamma}_{\gamma\ \beta}^{\ \alpha}+T_{\gamma\beta}{}^{\alpha}, \tag{4.2}$$

where

$$\mathring{\Gamma}_{\gamma\ \beta}^{\ \alpha}=(\partial_\gamma A_\beta{}^h+A_\gamma{}^j A_\beta{}^i \mathring{\Gamma}_{j\ i}^{h})A^\alpha{}_h \tag{4.3}$$

are components of the affine connexion $\mathring{\Gamma}_{j\ i}^{h}$ with respect to the non-holonomic frame $(A_\alpha{}^h)$, and

$$T_{\gamma\beta}{}^{\alpha}=A_\gamma{}^j A_\beta{}^i A^\alpha{}_h T_{ji}{}^h \tag{4.4}$$

are those of the tensor $T_{ji}{}^{h}$ with respect to the same frame.

The affine connexion $\mathring{\Gamma}_{j\ i}^{h}$ being symmetric, we have from (4.3)

$$\frac{1}{2}(\mathring{\Gamma}_{\gamma\ \beta}^{\ \alpha}-\mathring{\Gamma}_{\beta\ \gamma}^{\ \alpha})=\Omega_{\gamma\beta}{}^{\alpha}. \tag{4.5}$$

We shall study the existence of global affine connexions with respect to which the given distributions are flat, geodesic or parallel and which are symmetric whenever possible.

The condition for $P$ to be flat is

$$\Gamma_{c\ b}^{\ x}=0.$$

Thus we have from (4.2)

$$0=\mathring{\Gamma}_{c\ b}^{\ x}+T_{cb}{}^{x}.$$

Thus the distribution $P$ is flat with respect to the affine connexion $\Gamma_{j\ i}^{h}=\mathring{\Gamma}_{j\ i}^{h}+T_{ji}^{\ \ h}$ if and only if the tensor $T_{ji}^{\ \ h}$ satisfies

$$T_{cb}^{\ \ x}=-\mathring{\Gamma}_{c\ b}^{\ x}. \tag{4.6}$$

To get the simplest $T_{ji}^{\ \ h}$ which satisfies this condition, we define $T_{ji}^{\ \ h}$ by requiring that all the components $T_{\gamma\beta}^{\ \ \alpha}$ of $T_{ji}^{\ \ h}$ with respect to the non-holonomic frame other than $T_{cb}^{\ \ x}$ given by (4.6) are zero. Such a $T_{ji}^{\ \ h}$ is given by the formula

$$T_{ji}^{\ \ h}=-B_{j}^{c}B_{i}^{b}C_{x}^{\ h}\mathring{\Gamma}_{c\ b}^{\ x}. \tag{4.7}$$

As we remarked at the beginning of Section 3, the $\mathring{\Gamma}_{c\ b}^{\ x}$ are components of a tensor with respect to the transformation of the non-holonomic frame, and the $T_{ji}^{\ \ h}$ defined here does not depend on the choice of the vectors $B_{a}^{\ h}$ in $P$ and $C_{x}^{\ h}$ in $Q$. Thus we can see that the tensor $T_{ji}^{\ \ h}$ is determined uniquely by the distributions $P$ and $Q$ and the connexion $\mathring{\Gamma}_{j\ i}^{h}$. But these are given globally and consequently the $T_{ji}^{\ \ h}$ is defined globally and we can conclude that the affine connexion

$$\Gamma_{j\ i}^{h}=\mathring{\Gamma}_{j\ i}^{h}+T_{ji}^{\ \ h}$$

is also defined globally.

We can further require that the affine connexion $\Gamma_{j\ i}^{h}$ is symmetric whenever the distribution $P$ is integrable. In order that this may be the case, we must have

$$S_{\gamma\beta}^{\ \ \alpha}=\frac{1}{2}(T_{\gamma\beta}^{\ \ \alpha}-T_{\beta\gamma}^{\ \ \alpha})=0,$$

whenever

$$\Omega_{cb}^{\ \ x}=0.$$

On the other hand, we have from (4.6),

$$S_{cb}^{\ \ x}=-\Omega_{cb}^{\ \ x}.$$

Thus the distribution $P$ is flat with respect to the affine connexion $\Gamma_{j\,i}^{h}=\mathring{\Gamma}_{j\,i}^{h}+T_{ji}{}^{h}$ and the affine connexion $\Gamma_{j\,i}^{h}$ is symmetric whenever the distribution $P$ is integrable, if, and only if the tensor $T_{ji}{}^{h}$ is given by

$$T_{cb}{}^{x}=-\mathring{\Gamma}_{c\,b}^{x}, \text{ the other } T\text{'s satisfying } T_{\gamma\beta}{}^{\alpha}-T_{\beta\gamma}{}^{\alpha}=0. \quad (4.8)$$

The simplest $T_{ji}{}^{h}$ which satisfies these conditions is given by (4.7).

From (4.7), we have

$$T_{ji}{}^{h}=-B_{j}{}^{c}B_{i}{}^{b}C_{x}{}^{h}\mathring{\Gamma}_{c\,b}^{x}=-B_{j}{}^{c}B_{i}{}^{b}C_{x}{}^{h}(\partial_{c}B_{b}{}^{k}+B_{c}{}^{m}B_{b}{}^{l}\mathring{\Gamma}_{m\,l}^{k})C_{k}{}^{x}$$

from which

$$T_{ji}{}^{h}=P_{j}{}^{m}P_{i}{}^{l}(\mathring{\nabla}_{m}Q_{l}{}^{h}). \quad (4.9)$$

Substituting (1.9) into (4.9) we find

$$T_{ji}{}^{h}=-\frac{1}{8}[\mathring{\nabla}_{j}F_{i}{}^{h}+F_{j}{}^{l}(\mathring{\nabla}_{l}F_{i}{}^{h})+F_{i}{}^{l}(\mathring{\nabla}_{l}F_{j}{}^{h})+F_{j}{}^{m}F_{i}{}^{l}(\mathring{\nabla}_{m}F_{l}{}^{h})], \quad (4.10)$$

from which

$$T_{ji}{}^{h}-T_{ij}{}^{h}=-\frac{1}{8}(N_{ji}{}^{h}-N_{ji}{}^{l}F_{l}{}^{h}). \quad (4.11)$$

Thus we have

THEOREM 4.1. *For any distribution $P$ given globally, there exists a global affine connexion $\Gamma_{j\,i}^{h}$ with respect to which the given distribution $P$ is flat and which is symmetric whenever the distribution is integrable. One such affine connexion is given by* (4.10).

In exactly the same way, we can prove the following theorems.

THEOREM 4.2. *For any distribution given globally, there exists a global symmetric connexion with respect to which the distribution is geodesic, and one such connexion is given by* (4.12)

$$\begin{aligned}\Gamma_{j\,i}^{h}=\mathring{\Gamma}_{j\,i}^{h}&-\frac{1}{16}[F_{j}{}^{l}(\mathring{\nabla}_{l}F_{i}{}^{h})+F_{i}{}^{l}(\mathring{\nabla}_{j}F_{l}{}^{h})\\&+F_{i}{}^{l}(\mathring{\nabla}_{l}F_{j}{}^{h})+F_{j}{}^{l}(\mathring{\nabla}_{i}F_{l}{}^{h})]\\&+\frac{1}{16}[F_{j}{}^{l}(\mathring{\nabla}_{l}F_{i}{}^{k})+F_{i}{}^{l}(\mathring{\nabla}_{j}F_{l}{}^{k})\\&+F_{i}{}^{l}(\mathring{\nabla}_{l}F_{j}{}^{k})+F_{j}{}^{l}(\mathring{\nabla}_{l}F_{j}{}^{k})]F_{k}{}^{h}.\end{aligned} \quad (4.12)$$

THEOREM 4.3. *For any complementary distributions given globally, there exists a global affine connexion with respect to which one of the distributions is parallel along the other. One such connexion is given by*

$$\Gamma_j{}^h{}_i = \mathring{\Gamma}_j{}^h{}_i - \frac{1}{8}[\mathring{\nabla}_j F_i{}^h - F_j{}^l(\mathring{\nabla}_l F_i{}^h) + F_i{}^l(\mathring{\nabla}_j F_l{}^h) - F_j{}^m F_i(\mathring{\nabla}_m F_l{}^h)]. \quad (4.13)$$

THEOREM 4.4. *For any complementary distributions P and Q given globally, there exists a global affine connexion with respect to which both of the distributions are flat and which is symmetric whenever both of the distributions are integrable. One such connexion is given by*

$$\Gamma_j{}^h{}_i = \mathring{\Gamma}_j{}^h{}_i - \frac{1}{4}[F_j{}^l(\mathring{\nabla}_l F_i{}^h) + F_i{}^l(\mathring{\nabla}_j F_l{}^h)]. \quad (4.14)$$

THEOREM 4.5. *For any two complementary distributions given globally in* $M_n$, *there exists a global symmetric affine connexion with respect to which both of the distributions are geodesic. One such affine connexion is given by*

$$\Gamma_j{}^h{}_i = \mathring{\Gamma}_j{}^h{}_i - \frac{1}{8}[F_j{}^l(\mathring{\nabla}_l F_i{}^h) + F_i{}^l(\mathring{\nabla}_j F_l{}^h) + F_i{}^l(\mathring{\nabla}_l F_j{}^h) + F_j{}^l(\mathring{\nabla}_i F_l{}^h)]. \quad (4.15)$$

THEOREM 4.6. *For any two complementary distributions given globally, there exists a global symmetric affine connexion with respect to which both of the distributions are geodesic and one of the distributions is parallel along the other. One such affine connexion is given by*

$$\Gamma_j{}^h{}_i = \mathring{\Gamma}_j{}^h{}_i - \frac{1}{8}[F_j{}^l(\mathring{\nabla}_l F_i{}^h) + F_i{}^l(\mathring{\nabla}_j F_l{}^h) + F_i{}^l(\mathring{\nabla}_l F_j{}^h) + F_j{}^l(\mathring{\nabla}_i F_l{}^h)]$$
$$+ \frac{1}{4}[F_j{}^l(\mathring{\nabla}_l F_i{}^h) - F_i{}^l(\mathring{\nabla}_j F_l{}^h)] + \frac{1}{4}[F_i{}^l(\mathring{\nabla}_j F_l{}^h) - F_j{}^l(\mathring{\nabla}_i F_l{}^h)]. \quad (4.16)$$

THEOREM 4.7. *For any distribution given globally, there exists a global affine connexion with respect to which the given distribution is parallel and which is symmetric whenever the distribution is integrable. One such affine connexion is given by*

$$\Gamma_{j\ i}^{\ h} = \mathring{\Gamma}_{j\ i}^{\ h} - \frac{1}{4}[(\mathring{\nabla}_j F_i^{\ h}) + (\mathring{\nabla}_i F_j^{\ h}) + F_i^{\ l}(\mathring{\nabla}_j F_l^{\ h})$$

$$+ F_j^{\ l}(\mathring{\nabla}_i F_l^{\ h})] + \frac{1}{8}[\mathring{\nabla}_i F_j^{\ h}) + F_j^{\ l}(\mathring{\nabla}_i F_l^{\ h})$$

$$+ F_i^{\ l}(\mathring{\nabla}_l F_j^{\ h}) + F_j^{\ m} F_i^{\ l}(\mathring{\nabla}_l F_m^{\ h})]. \tag{4.17}$$

THEOREM 4.8. *For any two complementary distributions given globally, there always exists a global affine connexion with respect to which both of the given distributions are parallel and which is symmetric whenever both of the distributions are integrable. One such affine connexion is given by*

$$\Gamma_{j\ i}^{\ h} = \mathring{\Gamma}_{j\ i}^{\ h} - \frac{1}{4}[F_j^{\ l}(\mathring{\nabla}_i F_l^{\ h}) - F_i^{\ l}(\mathring{\nabla}_l F_j^{\ h})] + \frac{1}{2}(\mathring{\nabla}_j F_i^{\ l}) F_l^{\ h}. \tag{4.18}$$

## § 5. Parallel distributions

We shall give here the proof of Theorem 4.8 of the last section.

The distributions $P$ and $Q$ are both parallel with respect to the affine connexion $\Gamma_{j\ i}^{\ h} = \mathring{\Gamma}_{j\ i}^{\ h} + T_{ji}^{\ \ h}$ if and only if

$$T_{cb}^{\ \ x} = -\mathring{\Gamma}_{c\ b}^{\ x},\ T_{zb}^{\ \ x} = -\mathring{\Gamma}_{z\ b}^{\ x},\ T_{cy}^{\ \ a} = -\mathring{\Gamma}_{c\ y}^{\ a},\ T_{zy}^{\ \ a} = -\mathring{\Gamma}_{z\ y}^{\ a}. \tag{5.1}$$

The simplest $T_{ji}^{\ \ h}$ satisfying these conditions is given by

$$T_{ji}^{\ \ h} = -B_j^{\ c} B_i^{\ b} C_x^{\ h} \mathring{\Gamma}_{c\ b}^{\ x} - C_j^{\ z} B_i^{\ b} C_x^{\ h} \mathring{\Gamma}_{z\ b}^{\ x}$$

$$- B_j^{\ c} C_i^{\ y} B_a^{\ h} \mathring{\Gamma}_{c\ y}^{\ a} - C_j^{\ z} C_i^{\ y} B_a^{\ h} \mathring{\Gamma}_{z\ y}^{\ a}. \tag{5.2}$$

We can further require that the affine connexion $\Gamma_{j\ i}^{\ h}$ is symmetric whenever the given distribution is integrable.

The distributions $P$ and $Q$ are both parallel with respect to the affine connexion $\Gamma_{j\ i}^{\ h} = \mathring{\Gamma}_{j\ i}^{\ h} + T_{ji}^{\ \ h}$ and the affine connexion is symmetric whenever the distributions $P$ and $Q$ are both integrable if and only if the tensor $T_{ji}^{\ \ h}$ satisfies

$$T_{cb}^{\ \ x} = -\mathring{\Gamma}_{c\ b}^{\ x}, \quad T_{zb}^{\ \ x} = T_{bz}^{\ \ x} = -\mathring{\Gamma}_{z\ b}^{\ x},$$
$$T_{cy}^{\ \ a} = T_{yc}^{\ \ a} = -\mathring{\Gamma}_{c\ y}^{\ a}, \quad T_{zy}^{\ \ a} = -\mathring{\Gamma}_{z\ y}^{\ a} \tag{5.3}$$

all the other $T_{\gamma\beta}^{\ \ \alpha}$ being symmetric in $\gamma$ and $\beta$.

The simplest $T_{ji}^{\ \ h}$ satisfying these conditions is given by

$$T_{ji}^{\ \ h} = -B_j^{\ c} B_i^{\ b} C_x^{\ h} \mathring{\Gamma}_{c\ b}^{\ x} - C_j^{\ z} B_i^{\ b} C_x^{\ h} \mathring{\Gamma}_{z\ b}^{\ x}$$
$$- B_j^{\ b} C_i^{\ z} C_x^{\ h} \mathring{\Gamma}_{z\ b}^{\ x} - B_j^{\ c} C_i^{\ y} B_a^{\ h} \mathring{\Gamma}_{c\ y}^{\ a} - C_j^{\ y} B_i^{\ c} B_a^{\ h} \mathring{\Gamma}_{c\ y}^{\ a} - C_j^{\ z} C_i^{\ y} B_a^{\ h} \mathring{\Gamma}_{z\ y}^{\ a}, \tag{5.4}$$

from which

$$T_{ji}^{\ \ h} = -\frac{1}{4}\left[F_j^{\ l}(\mathring{\nabla}_i F_l^{\ h}) - F_i^{\ l}(\mathring{\nabla}_l F_j^{\ h})\right] + \frac{1}{2}(\mathring{\nabla}_j F_i^{\ l}) F_l^{\ h} \tag{5.5}$$

and

$$T_{ji}^{\ \ h} - T_{ij}^{\ \ h} = -\frac{1}{4} N_{ji}^{\ \ h}. \tag{5.6}$$

We can prove the same result in the following way. Since the distribution $P$ is parallel,

$$Q_i^{\ h} v^i = 0$$

and

$$(\delta Q_i^{\ h}) v^i + Q_i^{\ h}(\delta v^i) = 0$$

should be satisfied whenever $v^i$ satisfies

$$Q_i^{\ h} v^i = 0 \quad \text{and} \quad \delta v^i = 0,$$

from which

$$(\delta Q_i^{\ h}) v^i = 0$$

for any vector $v^i$ satisfying

$$Q_i^{\ h} v^i = 0 \quad \text{or} \quad P_i^{\ h} v^i = v^h,$$

that is

$$(\nabla_j Q_i^{\ h}) P_k^{\ i} = 0. \tag{5.7}$$

In a similar way, since the distribution $Q$ is parallel, we have

$$(\nabla_j P_i^{\ h}) Q_k^{\ i} = 0. \tag{5.8}$$

But since

$$P_i^{\ h} + Q_i^{\ h} = A_i^h,$$

$$\nabla_j P_i^{\ h} = -\nabla_j Q_i^{\ h}. \tag{5.9}$$

Thus (5.8) becomes

$$(\nabla_j Q_i^{\ h}) Q_k^{\ i} = 0. \tag{5.10}$$

Adding (5.7) and (5.10) and taking account of $P_k^{\ i} + Q_k^{\ i} = A_k^i$ we find

$$\nabla_j Q_i^{\ h} = 0 \tag{5.11}$$

and consequently

$$\nabla_j P_i^{\ h} = 0. \tag{5.12}$$

Thus, from (5.11) and (5.12), we have

$$\nabla_j F_i^{\ h} = 0. \tag{5.13}$$

Conversely, if we have (5.13), then from

$$\nabla_j A_i^h = \nabla_j P_i^{\ h} + \nabla_j Q_i^{\ h} = 0,$$

$$\nabla_j F_i^{\ h} = \nabla_j P_i^{\ h} - \nabla_j Q_i^{\ h} = 0,$$

we get

$$\nabla_j P_i^{\ h} = 0, \qquad \nabla_j Q_i^{\ h} = 0, \tag{5.14}$$

which show that the distributions $P$ and $Q$ are parallel.

Now suppose that there is given in $M_n$ a symmetric affine connection $\mathring{\Gamma}_{j\ i}^{\ h}$, then any affine connexion $\Gamma_{j\ i}^{\ h}$ in $M_n$ is expressed in the form

$$\Gamma_{j\ i}^{\ h} = \mathring{\Gamma}_{j\ i}^{\ h} + T_{ji}^{\ \ h}, \tag{5.15}$$

where $T$ is a tensor. Denoting by $\nabla_j$ and $\mathring{\nabla}_j$ the covariant differentiations with respect to $\Gamma_{j\,i}^{\,h}$ and $\mathring{\Gamma}_{j\,i}^{\,h}$ respectively, we have

$$\nabla_j F_i^{\,h} = \mathring{\nabla}_j F_i^{\,h} + T_{jl}^{\;\;h} F_i^{\,l} - T_{ji}^{\;\;l} F_l^{\,h}. \tag{5.16}$$

We have to find an affine connexion $\Gamma_{j\,i}^{\,h}$ such that

$$\nabla_j F_i^{\,h} = 0. \tag{5.17}$$

From (5.16) and (5.17) we find

$$(\mathring{\nabla}_j F_l^{\,h}) F_i^{\,l} + T_{ji}^{\;\;h} - F_i^{\,m} F_l^{\,h} T_{jm}^{\;\;\;l} = 0,$$

from which

$$*O_{il}^{mh} T_{jm}^{\;\;\;l} = \frac{1}{2} (\mathring{\nabla}_j F_i^{\,l}) F_l^{\,h}. \tag{5.18}$$

Denoting by $V$ the right hand member of the equation (5.18) we can easily verify that

$$*OV = 0$$

and consequently, by the Lemma (p. 132), equation (5.18) admits a solution and a special solution is given by

$$T_{ji}^{\;\;h} = \frac{1}{2} (\mathring{\nabla}_j F_i^{\,l}) F_l^{\,h}$$

and the general solution by

$$T_{ji}^{\;\;h} = \frac{1}{2} (\mathring{\nabla}_j F_i^{\,l}) F_l^{\,h} + *O_{il}^{mh} W_{jm}^{\;\;\;l}, \tag{5.19}$$

$W$ being an arbitrary tensor. Thus if we take as $W$

$$W_{ji}^{\;\;h} = \frac{1}{4} (\mathring{\nabla}_i F_j^{\,l}) F_l^{\,h} + \frac{1}{4} (\mathring{\nabla}_l F_j^{\,h}) F_i^{\,l}, \tag{5.20}$$

then it will be easily verified that

$$*O_{il}^{mh} W_{jm}^{\;\;\;l} = 0$$

and consequently

$$*O_{il}^{mh} W_{jm}{}^{l} = W_{ji}{}^{h}.$$

Thus (5.19) gives

$$T_{ji}{}^{h} = \frac{1}{4}(\mathring{\nabla}_j F_i{}^l + \mathring{\nabla}_i F_j{}^l) F_l{}^h - \frac{1}{4}(\mathring{\nabla}_j F_l{}^h - \mathring{\nabla}_l F_j{}^h) F_i{}^l, \quad (5.21)$$

from which

$$T_{ji}{}^{h} - T_{ij}{}^{h} = -\frac{1}{4} N_{ji}{}^{h}. \quad (5.22)$$

Thus we see that the affine connexion

$$\Gamma_{j\,i}^{h} = \mathring{\Gamma}_{j\,i}^{h} + \frac{1}{4}(\mathring{\nabla}_j F_i{}^l + \mathring{\nabla}_i F_j{}^l) F_l{}^h - \frac{1}{4}(\mathring{\nabla}_j F_l{}^h - \mathring{\nabla}_l F_j{}^h) F_i{}^l \quad (5.23)$$

satisfies

$$\nabla_j F_i{}^h = 0, \qquad S_{ji}{}^{h} = -\frac{1}{4} N_{ji}{}^{h}, \quad (5.24)$$

where

$$S_{ji}{}^{h} = \Gamma_{j\,i}^{h} - \Gamma_{i\,j}^{h} \quad (5.25)$$

is the torsion tensor of the affine connection $\Gamma_{j\,i}^{h}$.

Now, it is easily seen that if there exists a symmetric affine connexion $\Gamma_{j\,i}^{h}$ such that $\nabla_j F_i{}^h = 0$, then

$$N_{ji}{}^{h} = F_j{}^l(\nabla_l F_i{}^h - \nabla_i F_l{}^h) - F_i{}^l(\nabla_l F_j{}^h - \nabla_j F_l{}^h) = 0.$$

Conversely if $N_{ji}{}^{h} = 0$, then the affine connexion defined by (5.23) is symmetric and gives $\nabla_j F_i{}^h = 0$. Thus we have

THEOREM 5.1. *In order that an almost product space be integrable, it is necessary and sufficient that it is possible to introduce a symmetric affine connection with respect to which the structure tensor is covariantly constant.*

This theorem is due to A. G. Walker [1].

CHAPTER XII

# *H*-PROJECTIVE TRANSFORMATIONS

## § 1. *F*-connexions†

In an almost complex space, an affine connexion is called an *F-connexion* if the almost complex structure $F_i{}^h$ is covariant constant with respect to this connection. Given an arbitrary affine connection $\mathring{\Gamma}_j{}^h{}_i$, the connexion

$$\Gamma_j{}^h{}_i = \mathring{\Gamma}_j{}^h{}_i - \frac{1}{2}(\mathring{\nabla}_j F_i{}^r) F_r{}^h \tag{1.1}$$

is an $F$-connexion, where $\mathring{\nabla}$ denotes the covariant differentiation with respect to $\mathring{\Gamma}_j{}^h{}_i$.

THEOREM 1.1. *Let* $\mathring{\Gamma}_j{}^h{}_i$ *be a symmetric affine connexion and* $S_{ji}{}^h$ *be the torsion of the F-connexion* $\Gamma_j{}^h{}_i$ *given by* (1.1). *Then we have*

$$N_{ji}{}^h = 8 O_{ji}^{ts} S_{ts}{}^h. \tag{1.2}$$

*Proof*. From (1.1), we find

$$S_{ji}{}^h = -\frac{1}{2}(\mathring{\nabla}_{[j} F_{i]}{}^s) F_s{}^h, \tag{1.3}$$

and then

$$8 O_{ji}^{ts} S_{ts}{}^h = 2(-F_{[j}{}^r \mathring{\nabla}_{i]} F_r{}^h + F_{[j}{}^r \mathring{\nabla}_{|r|} F_{i]}{}^h) = N_{ji}{}^h, \tag{1.4}$$

which is the required relation.

---

† See S. Ishihara [2]; M. Obata [1].

Let $\Gamma_{j\ i}^{\ h}$ be an $F$-connexion and $S_{ji}^{\ \ h}$ be its torsion tensor. If we put

$$\overset{*}{\Gamma}{}_{j\ i}^{\ h}=\Gamma_{j\ i}^{\ h}-2{}^{*}O_{ji}^{tk}O_{kr}^{sh}S_{ts}^{\ \ r}, \tag{1.5}$$

then $\overset{*}{\Gamma}{}_{j\ i}^{\ h}$ is obviously an $F$-connexion. The torsion $\overset{*}{S}{}_{ji}^{\ \ h}$ of $\overset{*}{\Gamma}{}_{j\ i}^{\ h}$ is given by

$$\overset{*}{S}{}_{ji}^{\ \ h}=O_{ji}^{ts}S_{ts}^{\ \ h}. \tag{1.6}$$

Thus we obtain

THEOREM 1.2. *Given any $F$-connexion $\Gamma_{j\ i}^{\ h}$ with the torsion tensor $S_{ji}^{\ \ h}$, then there exists an $F$-connexion $\overset{*}{\Gamma}{}_{j\ i}^{\ h}$ with the torsion tensor $\overset{*}{S}{}_{ji}^{\ \ h}$ given by* (1.6).

Next, we shall prove

THEOREM 1.3. *In an almost complex space there always exists an $F$-connexion $\overset{*}{\Gamma}{}_{j\ i}^{\ h}$ with the torsion tensor $\overset{*}{S}{}_{ji}^{\ \ h}$ satisfying*

$$8\overset{*}{S}{}_{ji}^{\ \ h}=N_{ji}^{\ \ h}. \tag{1.7}$$

*Proof*. We know that there exists a symmetric affine connexion $\overset{\circ}{\Gamma}{}_{j\ i}^{\ h}$. Then by means of Theorem 1.1, the affine connexion $\Gamma_{j\ i}^{\ h}$ defined by (1.1) has the torsion tensor $S_{ji}^{\ \ h}$ satisfying (1.2). According to Theorem 1.2, there always exists an $F$-connexion $\overset{*}{\Gamma}{}_{j\ i}^{\ h}$ with the torsion tensor $\overset{*}{S}{}_{ji}^{\ \ h}=O_{ji}^{ts}S_{ts}^{\ \ h}$. Thus $\overset{*}{\Gamma}{}_{j\ i}^{\ h}$ is an $F$-connexion with the required property.

For any $F$-connexion $\Gamma_{j\ i}^{\ h}$, we have

$$\begin{aligned}2N_{ji}^{\ \ h}&=F_{[j}^{\ \ t}\nabla_{|t|}F_{i]}^{\ \ h}-F_{j}^{\ t}F_{i}^{\ s}\Gamma_{[ts]}^{\ \ h}+F_{[j}^{\ \ t}\Gamma_{|t|i]}^{\ \ r}F_{r}^{\ h}\\&\quad-F_{[j}^{\ \ t}\nabla_{i]}F_{t}^{\ h}+F_{[j}^{\ \ t}F_{|t|}^{\ \ s}\Gamma_{i]\ s}^{\ h}-F_{[j}^{\ \ t}\Gamma_{i]\ t}^{\ r}F_{r}^{\ h}\\&=S_{ji}^{\ \ h}+F_{i}^{\ s}S_{js}^{\ \ r}F_{r}^{\ h}+F_{j}^{\ t}S_{ti}^{\ \ r}F_{r}^{\ h}-F_{j}^{\ t}F_{i}^{\ s}S_{ts}^{\ \ h}.\\&=4O_{ji}^{tk}{}^{*}O_{kr}^{sh}S_{ts}^{\ \ r},\end{aligned} \tag{1.8}$$

where $S_{ji}{}^h$ is the torsion tensor of $\Gamma_j{}^h{}_i$. Thus we have the following

THEOREM 1.4. *Let $\Gamma_j{}^h{}_i$ be any F-connexion and $S_{ji}{}^h$ be its torsion tensor. Then we have*

$$N_{ji}{}^h = 2O_{ji}^{tk} {}^*O_{kr}^{sh} S_{ts}{}^r. \tag{1.9}$$

Combining Theorems 1.3 and 1.4, we obtain

THEOREM 1.5. *In an almost complex space, there exists a symmetric F-connection, if and only if the tensor $N_{ji}{}^h$ vanishes identically.*

An $F$-connexion $\Gamma_j{}^h{}_i$ is called *half-symmetric* if its torsion tensor $S_{ji}{}^h$ satisfies

$$O_{ji}^{tk} O_{kr}^{sh} S_{ts}{}^r = 0. \tag{1.10}$$

Let $\overset{1}{\Gamma}{}_j{}^h{}_i$ be an $F$-connexion and $\overset{1}{S}{}_{ji}{}^h$ be its torsion tensor. Then, if we consider an $F$-connexion

$$\Gamma_j{}^h{}_i = \overset{1}{\Gamma}{}_j{}^h{}_i - O_{ji}^{tk} O_{kr}^{sh} \overset{1}{S}{}_{ts}{}^r, \tag{1.11}$$

the torsion tensor $S_{ji}{}^h$ of $\Gamma_j{}^h{}_i$ is given by

$$S_{ji}{}^h = \overset{1}{S}{}_{ji}{}^h - O_{ji}^{tk} O_{kr}^{sh} \overset{1}{S}{}_{ts}{}^r, \tag{1.12}$$

which implies that $S_{ji}{}^h$ satisfies (1.10). Thus, we have the

THEOREM 1.6. *In any almost complex space, there exists always a half-symmetric F-connexion.*

Let $\overset{\circ}{\Gamma}{}_j{}^h{}_i$ be a symmetric connexion. Then the connexion $\Gamma_j{}^h{}_i$ given by (1.1) is an $F$-connexion. It is easily seen that the torsion tensor $S_{ji}{}^h$ of the connexion $\Gamma_j{}^h{}_i$ satisfies (1.10). Thus we obtain

THEOREM 1.7. *Let $\overset{\circ}{\Gamma}{}_j{}^h{}_i$ be a symmetric connexion. Then the connexion $\Gamma_j{}^h{}_i$ given by* (1.1) *is a half-symmetric F-connexion.*

We consider an $F$-connexion $\Gamma_j{}^h{}_i$ whose torsion tensor $S_{ji}{}^h$ is given by

$$S_{ji}{}^h = \frac{4}{n} {}^*O_{ji}^{ts} S_{[t} A_{s]}^h \qquad S_j = S_{jr}{}^r. \tag{1.13}$$

We call such an $F$-connexion a *semi-symmetric* one. The torsion tensor $S_{ji}{}^h$ given by (1.13) satisfies obviously (1.10). Hence *any semi-symmetric $F$-connexion is half-symmetric*.

Now we shall prove

THEOREM 1.8. *In order that, in an almost complex space, there exists a semi-symmetric $F$-connexion, it is necessary and sufficient that the Nijenhuis tensor $N_{ji}{}^h$ vanishes identically.*

*Proof*. We suppose that there exists a semi-symmetric $F$-connexion. By virtue of $O \cdot {}^*O = 0$, the torsion tensor $S_{ji}{}^h$ of the $F$-connexion satisfies

$$O_{ji}^{th} {}^*O_{kr}^{sh} S_{ts}{}^r = 0, \tag{1.14}$$

which implies together with (1.9) $N_{jr}{}^h = 0$.

Conversely, if we suppose $N_{ji}{}^h = 0$, then there exists at least one symmetric $F$-connexion which is obviously semi-symmetric.

## § 2. *H*-projective transformations†

In an almost complex space with an $F$-connexion $\Gamma_j{}^h{}_i$, we consider a curve $\xi^h(t)$ satisfying differential equations

$$\frac{d^2\xi^h}{dt^2} + \Gamma_j{}^h{}_i \frac{d\xi^j}{dt}\frac{d\xi^i}{dt} = \alpha(t)\frac{d\xi^h}{dt} + \beta(t) F_r{}^h \frac{d\xi^r}{dt}, \tag{2.1}$$

where $\alpha(t)$ and $\beta(t)$ are certain functions of the parameter $t$. We call such a curve a *holomorphically planar curve*. We see directly from (2.1) that a curve is holomorphically planar if and only if the holomorphic sections determined by its tangent vectors are parallel along the curve itself.

Given two $F$-connexions $\Gamma_j{}^h{}_i$ and $\bar{\Gamma}_j{}^h{}_i$, we suppose that they have all holomorphically planar curves in common. Then we say that these two $F$-connexions are *H-projectively related* to each other.

† S. Ishihara [2].

Now we have

THEOREM 2.1. *Let* $\Gamma_{j\,i}^{h}$ *and* $\bar{\Gamma}_{j\,i}^{h}$ *be two half-symmetric F-connexions. Then these two F-connexions are H-projectively related to each other if and only if*

$$\bar{\Gamma}_{j\,i}^{h}=\Gamma_{j\,i}^{h}+P_{(j}A_{i)}^{h}-P_{t}F_{(j}{}^{t}F_{i)}{}^{h}+Q_{j}A_{i}^{h}+Q_{t}F_{j}{}^{t}F_{i}{}^{h} \tag{2.2}$$

*holds for certain vector field* $P_i$ *and* $Q_i$.

*Proof*. When $\bar{\Gamma}_{j\,i}^{h}$ is given by (2.2), the two connexions $\Gamma_{j\,i}^{h}$ and $\bar{\Gamma}_{j\,i}^{h}$ have all holomorphically planar curves in common. Conversely, we suppose that $\Gamma_{j\,i}^{h}$ and $\bar{\Gamma}_{j\,i}^{h}$ have all holomorphically planar curves in common. On putting

$$A_{ji}{}^{h}=\bar{\Gamma}_{j\,i}^{h}-\Gamma_{j\,i}^{h}, \tag{2.3}$$

we see† that at any point

$$A_{ji}{}^{h}v^{j}v^{i}=av^{h}+bF_{r}{}^{h}v^{r} \tag{2.4}$$

must hold for any vector $v^h$, where $a$ and $b$ depend on both $v^h$ and the point. Therefore we have

$$A_{ji}{}^{h}=U_{(j}A_{i)}^{h}+V_{(j}F_{i)}{}^{h}+P_{ji}{}^{h} \tag{2.5}$$

for certain vectors $U_i$ and $V_i$, where $P_{ji}{}^{h}=A_{[ji]}{}^{h}$.

Since both $\Gamma_{j\,i}^{h}$ and $\bar{\Gamma}_{j\,i}^{h}$ are half-symmetric $F$-connexions, we have

$$*O_{ir}^{sh}A_{js}{}^{r}=0, \qquad O_{ji}^{tk}O_{kr}^{sh}P_{ts}{}^{r}=0. \tag{2.6}$$

Now, by making use of the identity

$$O_{ji}^{tk}*O_{kr}^{sh}+*O_{ji}^{tk}O_{kr}^{sh}=*O_{jr}^{th}A_{i}^{s}, \tag{2.7}$$

according to $P_{(ji)}{}^{k}=0$, we find

$$*O_{ji}^{tk}O_{kr}^{sh}P_{ts}{}^{r}=-*O_{jr}^{th}P_{it}{}^{r}-O_{ji}^{tk}*O_{kr}^{sh}P_{ts}{}^{r}. \tag{2.8}$$

This, together with (2.6), implies

$$P_{ji}{}^{h}=*O_{ir}^{sh}P_{js}{}^{r}-*O_{jr}^{th}P_{it}{}^{r}-O_{ji}^{tk}*O_{kr}^{sh}P_{ts}{}^{r}. \tag{2.9}$$

† S. Tachibana and S. Ishihara [1], appendix.

On the other hand, applying $^*O^{il}_{mh}$ to the two sides of (2.5) and changing indices, we have, by means of (2.6),

$$^*O^{sh}_{ir}P_{js}{}^r = -\frac{1}{2}\{A^h_j(U_i - F_i{}^sV_s) + F_j{}^h(F_i{}^sU_s + V_i)\}. \quad (2.10)$$

If we substitute (2.10) in (2.9), we find

$$P_{ji}{}^h = U_{[j}A^h_{i]} + U_tF_{[j}{}^tF_{i]}^h - V_tF_{[j}{}^tA^h_{i]} - V_{[j}F_{i]}{}^h. \quad (2.11)$$

Then, from (2.5) it follows

$$A_{ji}{}^h = P_{(j}A^h_{i)} - P_tF_{(j}{}^tF_{i)}{}^h + Q_jA^h_i + Q_tF_j{}^tF_i{}^h, \quad (2.12)$$

where we have put $P_i = U_i + V_sF_i{}^s$, $Q_i = U_i - V_sF_i{}^s$. Thus we get Theorem 2.1.

Theorem 2.1 implies immediately

THEOREM 2.2. *Two symmetric F-connexions* $\Gamma^h_{j\,i}$ *and* $\bar{\Gamma}^h_{j\,i}$ *are H-projectively related to each other when and only when*

$$\bar{\Gamma}^h_{j\,i} = \Gamma^h_{j\,i} + P_{(j}A^h_{i)} - P_sF_{(j}{}^sF_{i)}{}^h \quad (2.13)$$

*holds for a certain vector field* $P_i$.

Let $\Gamma^h_{j\,i}$ and $\bar{\Gamma}^h_{j\,i}$ be two half-symmetric $F$-connexions satisfying (2.2) for certain vector fields $P_i$ and $Q_i$. Then the correspondence $\Gamma^h_{j\,i} \to \bar{\Gamma}^h_{j\,i}$ is called a *holomorphically projective transformation*, or shortly a *H-projective transformation* of $\Gamma^h_{j\,i}$.

If we take the skew-symmetric part of the both sides of (2.2) with respect to the covariant indices, we find

$$\bar{S}_{ji}{}^h = S_{ji}{}^h + Q_{[j}A^h_{i]} + Q_tF_{[j}{}^tF_{i]}{}^h, \quad (2.14)$$

where $S_{ji}{}^h$ and $\bar{S}_{ji}{}^h$ denote the torsion tensors of $\Gamma^h_{j\,i}$ and $\bar{\Gamma}^h_{j\,i}$ respectively. Contracting $h$ and $i$, we obtain

$$Q_j = \frac{2}{n}(\bar{S}_{jr}{}^r - S_{jr}{}^r). \quad (2.15)$$

If we substitute (2.15) in (2.14), we obtain easily

$$\bar{S}_{ji}{}^{h}-\frac{4}{n}{}^{*}O_{ji}^{ts}\bar{S}_{[t}A_{s]}^{h}=S_{ji}{}^{h}-\frac{4}{n}{}^{*}O_{ji}^{ts}S_{[t}A_{s]}^{h}, \tag{2.16}$$

where we have put

$$S_{j}=S_{jr}{}^{r}, \qquad \bar{S}_{j}=\bar{S}_{jr}{}^{r}. \tag{2.17}$$

Therefore we have

THEOREM 2.3. *Let* $S_{ji}{}^{h}$ *be the torsion tensor of a half-symmetric F-connexion. Then the tensor*

$$S_{ji}{}^{h}-\frac{4}{n}{}^{*}O_{ji}^{ts}S_{[t}A_{s]}^{h}$$

*is invariant under any H-projective transformation of* $\Gamma_{j\,i}^{h}$.

Here we have

THEOREM 2.4. *In order that a half-symmetric F-connexion be semi-symmetric, it is necessary and sufficient that it be H-projectively related to a symmetric F-connexion.*

*Proof.* First, we assume the semi-symmetry of the given connexion $\Gamma_{j\,i}^{h}$. Denoting by $S_{ji}{}^{h}$ the torsion tensor of $\Gamma_{j\,i}^{h}$, we consider an $H$-projectively related $F$-connexion

$$\Lambda_{ji}^{h}=\Gamma_{j\,i}^{h}-\frac{4}{n}{}^{*}O_{ji}^{ts}S_{[t}A_{s]}^{h}, \tag{2.18}$$

where $S_{j}=S_{jr}{}^{r}$. If we denote by $T_{ji}{}^{h}$ the torsion tensor of $\Lambda_{ji}^{h}$, we find easily

$$T_{ji}{}^{h}=S_{ji}{}^{h}-\frac{4}{n}{}^{*}O_{ji}^{ts}S_{[t}A_{s]}^{h}, \tag{2.19}$$

which implies $T_{ji}{}^{h}=0$ by virtue of the semi-symmetry of $\Gamma_{j\,i}^{h}$.

Conversely, if $\Gamma_{j\,i}^{h}$ is $H$-projectively related to a symmetric $F$-connexion, it is easily seen from Theorem 2.3 that $\Gamma_{j\,i}^{h}$ is semi-symmetric.

Given a half-symmetric $F$-connexion $\Gamma_{j}{}^{h}{}_{i}$, we define a quantity $\Pi^{h}_{ji}$ by

$$\Pi^{h}_{ji}=\Gamma_{j}{}^{h}{}_{i}-\frac{2}{n+2}(\Gamma^{r}_{r(j}A^{h}_{i)}-\Gamma^{r}_{rs}F_{(j}{}^{s}F_{i)}{}^{h})$$

$$-\frac{2}{n}(\Gamma^{r}_{r[j}A^{h}_{i]}+\Gamma^{r}_{[rs]}F_{j}{}^{s}F_{i}{}^{h}), \tag{2.20}$$

which is not an affine connexion. Then we easily prove

THEOREM 2.5. *In order that two half-symmetric $F$-connexions be $H$-projectively related to each other, it is necessary and sufficient that the quantities $\Pi$ corresponding to these $F$-connexions coincide.*

## § 3. *H*-projective flatness and the *H*-projective curvature tensor†

Let $\Gamma_{j}{}^{h}{}_{i}$ be a half-symmetric $F$-connexion in an almost complex space. We assume that for any point in the space there exists at least a neighbourhood of the point in which $\Gamma_{j}{}^{h}{}_{i}$ is $H$-projectively related to a flat $F$-connexion. Then the half-symmetric $F$-connexion is said to be *$H$-projectively flat.*

Let a half-symmetric $F$-connexion $\Gamma_{j}{}^{h}{}_{i}$ be $H$-projectively flat. Then the connexion $\Lambda^{h}_{ji}$ defined by (2.18) is necessarily symmetric. In fact, by virtue of Theorem 2.3, we see that the torsion tensor $T_{ji}{}^{h}$ given by (2.19) vanishes because of the $H$-projective flatness of $\Gamma_{j}{}^{h}{}_{i}$. Further, the symmetric $F$-connexion $\Lambda^{h}_{ji}$ is also $H$-projectively flat. Thus we obtain

THEOREM 3.1. *In order that a half-symmetric $F$-connexion $\Gamma_{j}{}^{h}{}_{i}$ be $H$-projectively flat, it is necessary and sufficient that there exist an $H$-projectively flat, symmetric $F$-connexion which is $H$-projectively related to the given connexion.*

† See S. Ishihara [2]; T. Ōtsuki and Y. Tashiro [1]; Y. Tashiro [1].

Theorem 3.1 implies together with Theorem 1.8 that the Nijenhuis tensor of an almost complex space vanishes if it admits a half-symmetric $F$-connexion which is $H$-projectively flat.

Now, supposing that the space is of dimension $n>2$, we define the *H-projective curvature tensor* $P_{kji}{}^h$ of a symmetric $F$-connexion $\Gamma_{j}{}^{h}{}_{i}$ by

$$P_{kji}{}^h=R_{kji}{}^h+A^h_{[k}P_{j]i}-P_{[kj]}A^h_i-F_{[k}{}^hP_{j]s}F_i{}^s+P_{[k|t|}F_{j]}{}^tF_i{}^h, \quad (3.1)$$

$P_{ji}$ being defined by

$$P_{ji}=-\frac{2}{n+2}\left\{R_{ji}+\frac{2}{n-2}O^{ts}_{ji}(R_{ts}+R_{st})\right\}, \quad (3.2)$$

where $R_{kji}{}^h$ and $R_{ji}$ are respectively the curvature tensor and the Ricci tensor of $\Gamma_{j}{}^{h}{}_{i}$. (See also Mizusawa and S. Kotō [1]). It is easily verified that

$$P_{rji}{}^r=0. \quad (3.3)$$

THEOREM 3.2. *Any two symmetric F-connexions, which are H-projectively related to each other, have the H-projective curvature tensor in common.*

*Proof.* We suppose that a symmetric $F$-connexion $\bar\Gamma_{j}{}^{h}{}_{i}$ is $H$-projectively related to another symmetric $F$-connexion $\Gamma_{j}{}^{h}{}_{i}$. Then we may put

$$\bar\Gamma_{j}{}^{h}{}_{i}=\Gamma_{j}{}^{h}{}_{i}+P_{(j}A^h_{i)}-P_tF_{(j}{}^tF_{i)}{}^h, \quad (3.4)$$

$P_i$ being a certain vector field. Denoting by $R_{kji}{}^h$ and $\bar R_{kji}{}^h$ the curvature tensor of $\Gamma_{j}{}^{h}{}_{i}$ and $\bar\Gamma_{j}{}^{h}{}_{i}$ respectively, we have

$$\bar R_{kji}{}^h=R_{kji}{}^h-A^h_{[k}M_{j]i}+M_{[kj]}A^h_i+F_{[k}{}^hM_{j]s}F_i{}^s-M_{[k|t|}F_{j]}{}^tF_i{}^h, \quad (3.5)$$

where

$$M_{ji}=\nabla_jP_i-O^{ts}_{ji}P_tP_s. \quad (3.6)$$

Contracting $h$ and $k$ in (3.5), we find

$$\bar R_{ji}=R_{ji}-\frac{n+2}{2}M_{ji}+2O^{ts}_{ji}M_{ts}, \quad (3.7)$$

where $\bar{R}_{ji}$ is the Ricci tensor of $\bar{\Gamma}_{j\,i}^{\,h}$. If we contract both sides with $O_{ml}^{ji}$, change indices and take the symmetric parts of the both sides, we obtain

$$O_{ji}^{ts}M_{ts} = -\frac{2}{n-2}\,O_{ji}^{ts}(\bar{R}_{(ts)} - R_{(ts)}). \tag{3.8}$$

Substituting (3.8) in (3.7), we find

$$M_{ji} = \bar{P}_{ji} - P_{ji}, \tag{3.9}$$

where $\bar{P}_{ji}$ is defined by

$$\bar{P}_{ji} = -\frac{2}{n+2}\left\{\bar{R}_{ji} + \frac{2}{n-2}\,O_{ji}^{ts}(\bar{R}_{ts} + \bar{R}_{st})\right\}.$$

If we substitute (3.9) in (3.5), we have $\bar{P}_{kji}{}^{h} = P_{kji}{}^{h}$, where $\bar{P}_{kji}{}^{h}$ is the H-projective curvature tensor of $\bar{\Gamma}_{j\,i}^{\,h}$.

Next, we shall prove

THEOREM 3.3. *In an almost complex space of dimension $n>2$, a symmetric F-connexion is H-projectively flat if and only if its H-projective curvature tensor $P_{kji}{}^{h}$ vanishes identically.*

*Proof.* When a symmetric $F$-connexion $\Gamma_{j\,i}^{\,h}$ is $H$-projectively flat, $P_{kji}{}^{h}$ obviously vanishes. Conversely, if we suppose that $P_{kji}{}^{h}$ vanishes identically, we have from (3.1)

$$R_{kji}{}^{h} = -A_{[k}^{h}P_{j]i} + P_{[kj]}A_{i}^{h} + F_{[k}{}^{h}P_{j]s}F_{i}{}^{s} - P_{[k|s|}F_{j]}{}^{s}F_{i}{}^{h}. \tag{3.10}$$

By means of (3.5), in order to prove that the $F$-connexion $\Gamma_{j\,i}^{\,h}$ is H-projectively flat, it is sufficient to show that in any neighbourhood there exists a local vector field $P_j$ such that

$$\nabla_j P_i = P_{ji} + O_{ji}^{ts}P_tP_s; \tag{3.11}$$

The integrability condition of the differential equations (3.11) is given by

$$-R_{kji}{}^{h}P_h = 2\{\nabla_{[k}P_{j]i} - O_{[j|i|}^{ts}\nabla_{k]}P_tP_s - O_{[j|i}^{ts}P_{t|}\nabla_{k]}P_s\}. \tag{3.12}$$

If we take account of (3.10) and (3.11), we find that (3.12) is equivalent to

$$\nabla_{[k} P_{j]i} = 0. \tag{3.13}$$

Now we shall show that the condition (3.10) implies the integrability condition (3.10) of (3.11). By means of the Bianchi identity, it follows from (3.10) that

$$A^h_{[k} \nabla_l P_{j]i} - \nabla_{[l} P_{kj]} A^h_i - F_{[k}{}^h \nabla_l P_{j]s} F_i{}^s + \nabla_{[l} P_{k|s|} F_{j]}{}^s F_i{}^h = 0 \tag{3.14}$$

which implies (3.13). Thus, the differential equation (3.11) is completely integrable under the condition (3.10). Therefore, Theorem 3.3 is proved.

Now we consider a Kähler space. The Riemannian connexion $\left\{ {h \atop j\ i} \right\}$ of the space is obviously a symmetric $F$-connexion. We know that

$$O^{ts}_{ji} K_{ts} = 0 \tag{3.15}$$

holds good, in which case the $H$-projective curvature tensor $P_{kji}{}^h$ of $\left\{ {h \atop j\ i} \right\}$ becomes:

$$P_{kji}{}^h = K_{kji}{}^h - \frac{2}{n+2} [A^h_{[k} K_{j]i} - F_{[k}{}^h K_{j]s} F_i{}^s + K_{[k|s|} F_{j]}{}^s F_i{}^h]. \tag{3.16}$$

If the space is $H$-projectively flat, i.e. if $P_{kji}{}^h = 0$, then contracting with $g_{ji}$, we have

$$K_{ji} = \frac{K}{n} g_{ji}. \tag{3.17}$$

Hence $K$ is a constant. Then, if we put

$$K = \frac{n(n+2)}{4} k, \tag{3.18}$$

we have

$$K_{kjih} = \frac{k}{4} \{(g_{kh} g_{ji} - g_{jh} g_{ki}) + (F_{kh} F_{ji} - F_{jh} F_{ki}) - 2 F_{kj} F_{ih}\}. \tag{3.19}$$

so that the space is of constant holomorphic sectional curvature.

Conversely, if the space is of constant holomorphic sectional curvature, by virtue of (3.19) the tensor $P_{kji}{}^h$ vanishes identically. Summing up, we obtain

THEOREM 3.4. *A Kähler space is H-projectively flat if and only if it is of constant holomorphic sectional curvature.*

## § 4. *H*-projective vectors†

In a space we consider a tensor valued function $V$ depending not only on a point $\xi^h$ of the space but also on $k$ vectors $u^h_{(1)}, u^h_{(2)}, \ldots, u^h_{(k)}$ at the point $\xi^h$. We denote such a function $V$ by $V(\xi, u_{(1)}, u_{(2)}, \ldots, u_{(k)})$. Assuming the space to be affinely connected, we take an arbitrary curve $C$: $\xi^h = \xi^h(t)$ and denote its successive covariant derivatives by

$$\frac{d\xi^h}{dt}, \quad \frac{\delta^2\xi^h}{dt^2}, \quad \ldots, \quad \frac{\delta^k\xi^h}{dt^k}. \tag{4.1}$$

Then, if we substitute (4.1) in the function $V$ instead of $u^h_{(1)}, u^h_{(2)}, \ldots, u^h_{(k)}$, we have a family of tensors

$$V(C) = V\left(\xi, \frac{d\xi}{dt}, \ldots, \frac{\delta^k\xi}{dt^k}\right)$$

along the curve $C$.

Let $v^h$ be a vector field and $T_\varepsilon$: $'\xi^h = \xi^h + \varepsilon v^h$ be the infinitesimal point transformation determined by $v^h$, $\varepsilon$ being an arbitrary infinitesimal constant. Given a curve $C$: $\xi^h = \xi^h(t)$, the image $'C$ of $C$ is expressed by $'\xi^h = \xi^h(t) + \varepsilon v^h(t))$. We call the limiting value

$$\underset{v}{\mathfrak{L}} V(C) = \lim_{\varepsilon \to 0} \frac{T_\varepsilon^{-1} V('C) - V(C)}{\varepsilon} \tag{4.2}$$

† See S. Ishihara [1]; S. Tachibana and S. Ishihara [1].

the *Lie derivative* of $V(C)$ with respect to $v^h$, where $T_\varepsilon^{-1}V('C)$ is the inverse image of the tensor $V('C)$ by $T_\varepsilon$.

In an almost complex space with a half-symmetric $F$-connexion $\Gamma_{j\,i}^{h}$, given vector field $v^h$, we assume that for any $\varepsilon$ the infinitesimal point-transformation $T_\varepsilon$ maps any holomorphically planar curve into another one. We then say that $v^h$ *preserves the holomorphically planar curves*. For such a vector $v$, taking account of (2.1), we have

$$\underset{v}{\mathfrak{L}}\left[\frac{d^2\xi^h}{dt^2}+\Gamma_{ji}^{h}\frac{d\xi^j}{dt}\frac{d\xi^i}{dt}-\alpha\frac{d\xi^h}{dt}-\beta F_r{}^h\frac{d\xi^r}{dt}\right]$$

$$=\gamma\frac{d\xi^h}{dt}+\delta F_r{}^h\frac{d\xi^r}{dt} \tag{4.3}$$

along any holomorphically planar curve defined by (2.1), where $\gamma$ and $\delta$ are certain functions of $t$.

On putting

$$t_{ji}{}^h=\underset{v}{\mathfrak{L}}\,\Gamma_{j\,i}^{h}, \qquad a_i{}^h=\underset{v}{\mathfrak{L}}\,F_i{}'^h, \tag{4.4}$$

we have from (4.3)

$$t_{ji}{}^h\dot\xi^j\dot\xi^i+a\dot\xi^h+bF_r{}^h\dot\xi^r-\beta a_r{}^h\dot\xi^r=0, \tag{4.5}$$

where we have put

$$a=-(\gamma+\underset{v}{\mathfrak{L}}\alpha), \qquad b=-(\delta+\underset{v}{\mathfrak{L}}\beta), \qquad \dot\xi^h=d\xi/dt.$$

Since (4.5) is established along any holomorphically planar curve, it must hold identically for any values of $\xi^h$ and $\dot\xi^h$. By the definition of holomorphically planar curves the identity (4.5) holds for any value of the coefficient $\beta$. Taking account of these facts we can easily see that the relations

$$a_r{}^h\dot\xi^r=f\dot\xi^h+gF_r{}^h\dot\xi^r, \tag{4.6}$$

$$t_{ji}{}^h\dot\xi^j\dot\xi^i=p\dot\xi^h+qF_r{}^h\dot\xi^r \tag{4.7}$$

hold for any values of $\xi^h$ and $\dot\xi^h$, where $f$, $g$, $p$ and $q$ are certain functions of $\xi^h$ and $\dot\xi^h$.

Now, if we take account of $F_i{}^r a_r{}^h + a_i{}^r F_r{}^h = 0$, and of the identity (4.6) we can find $a_i{}^h = 0$. Next, if we take account of $O_{ir}^{sh} t_{js}{}^r = 0$, $O_{ji}^{tk} O_{kr}^{sh}(t_{ts}{}^r - t_{st}{}^r) = 0$, and of the identity (4.7) we can see that

$$t_{ji}{}^h = 2\varrho_{(j} A_{i)}^h - 2\varrho_t F_{(j}{}^t F_{i)}{}^h + \sigma_j A_i^h + \sigma_t F_j{}^t F_i{}^h \tag{4.8}$$

holds for certain vectors $\varrho_i$ and $\sigma_i$.

When a vector field $v^h$ has the property

$$\underset{v}{\mathfrak{L}}\, \Gamma_j{}^h{}_i = 2\varrho_{(j} A_{i)}^h - 2\varrho_t F_{(j}{}^t F_{i)}{}^h + \sigma_j A_i^h + \sigma_t F_j{}^t F_i{}^h, \tag{4.9}$$

we call it an *H-projective vector* (with respect to the half-symmetric $F$-connexion $\Gamma_j{}^h{}_i$). We then have

THEOREM 4.1. *In an almost complex space with a half-symmetric F-connexion a vector field $v^h$ preserves holomorphically planar curves, if and only if $v^h$ is both contravariant almost analytic and H-projective.*

When the $F$-connexion $\Gamma_j{}^h{}_i$ is symmetric, the formula (4.9) becomes:

$$\underset{v}{\mathfrak{L}}\, \Gamma_j{}^h{}_i = 2\varrho_{(j} A_{i)}^h - 2\varrho_t F_{(j}{}^t F_{i)}{}^h. \tag{4.10}$$

Thus we have

THEOREM 4.2. *If $v^h$ is an H-projective vector of a symmetric F-connexion $\Gamma_j{}^h{}_i$, then we have* (4.10) *for a certain vector field $\varrho_i$.*

Let $v^h$ be an almost analytic $H$-projective vector. Then $v^h$ satisfies (4.9). For the given connexion $\Gamma_j{}^h{}_i$ we consider the half-symmetric $F$-connexion $\Lambda_{ji}^h$ defined by (2.18). Then we have by means of (4.9)

$$\underset{v}{\mathfrak{L}}\, \Lambda_{ji}^h = 2\varrho_{(j} A_{i)}^h - 2\varrho_t F_{(j}{}^t F_{i)}{}^h. \tag{4.11}$$

Now, we call the vector $\varrho_i$ the *associated vector* of $v^h$.

Suppose that there are given two almost analytic projective vectors $u^h$ and $v^h$. Denote by $w = [u, v]$ their product, i. e.

$$w^h = u^r \partial_r v^h - v^r \partial_r u^h . \tag{4.12}$$

Then, by virtue of the well-known formula (K. Yano [1])

$$\underset{w}{\mathfrak{L}} \Lambda_{ji}^h = \underset{u}{\mathfrak{L}} \underset{v}{\mathfrak{L}} \Lambda_{ji}^h - \underset{v}{\mathfrak{L}} \underset{u}{\mathfrak{L}} \Lambda_{ji}^h , \tag{4.13}$$

we find easily

$$\underset{w}{\mathfrak{L}} \Lambda_{ji}^h = 2\zeta_{(j} A_{i)}^h - 2\zeta_r F_{(j}{}^r F_{i)}{}^h , \qquad \zeta_i = \underset{u}{\mathfrak{L}} \varrho_i - \underset{v}{\mathfrak{L}} \mu_i , \tag{4.14}$$

where $\varrho_i$ and $\mu_i$ are the associated vectors of $v^h$ and $u^h$ respectively. This means that $w^h$ is also $H$-projective. Thus we have

THEOREM 4.3. *In an almost complex space with a half-symmetric F-connexion, the set of all almost analytic H-projective vectors forms a Lie algebra.*

Let $v^h$ be an almost analytic $H$-projective vector of a half-symmetric $F$-connexion $\Gamma_j{}^h{}_i$. By contracting $h$ and $i$ in (4.9) we obtain

$$\underset{v}{\mathfrak{L}} \Gamma_{jr}^r = \frac{n+2}{2} \varrho_j + \sigma_j . \tag{4.15}$$

Contracting $h$ and $j$ also in (4.9), we get

$$\underset{v}{\mathfrak{L}} \Gamma_{ri}^r = \frac{n+2}{2} \varrho_i . \tag{4.16}$$

These equations imply

$$\varrho_j = \frac{2}{n+2} \underset{v}{\mathfrak{L}} \Gamma_r{}^r{}_j , \qquad \sigma_j = \frac{2}{n} \underset{v}{\mathfrak{L}} \Gamma_{[jr]}^r . \tag{4.17}$$

Substituting (4.17) in (4.9), we find

$$\underset{v}{\mathfrak{L}} \Pi_{ji}^h = 0, \tag{4.18}$$

where $\Pi_{ji}^h$ is the quantity defined by (2.20). By means of (4.18) we obtain

THEOREM 4.4. *Let $v^h$ be an almost analytic H-projective vector of a half-symmetric F-connexion $\Gamma_{j\ i}^{\ h}$. Suppose that $v^h$ does not vanish at a point p. Then in a certain neighbourhood of p there exists a system of coordinates $(\xi^h)$ such that $\partial_l \Pi_{j\ i}^{\ h}=0$.*

Next, taking account of the definition (2.20) of $\Pi_{j\ i}^{\ h}$, we have easily

THEOREM 4.5. *For a half-symmetric F-connexion $\Gamma_{j\ i}^{\ h}$ an almost contravariant analytic vector field $v^h$ is H-projective, if and only if $\underset{v}{\mathfrak{L}}\, \Pi_{j\ i}^{\ h}=0$.*

For a vector field $v^h$ we get

$$\underset{v}{\mathfrak{L}}\, \nabla_j F_i^{\ h} - \nabla_j \underset{v}{\mathfrak{L}}\, F_i^{\ h} = F_i^{\ s} \underset{v}{\mathfrak{L}}\, \Gamma_{j\ s}^{\ h} - F_r^{\ h} \underset{v}{\mathfrak{L}}\, \Gamma_{ji}^{r}, \tag{4.19}$$

which implies, because of $\nabla_j F_i^{\ h}=0$,

$$\nabla_j \underset{v}{\mathfrak{L}}\, F_i^{\ h} = F_i^{\ s} \underset{v}{\mathfrak{L}}\, \Gamma_{j\ s}^{\ h} - F_r^{\ h} \underset{v}{\mathfrak{L}}\, \Gamma_{j\ i}^{\ r}. \tag{4.20}$$

If the vector $v^h$ is $H$-projective, it is easily verified that the right hand of (4.20) vanishes. Thus we obtain the following theorems by virtue of Obata's theorem, [1], [2], [3].

THEOREM 4.6. *If an almost complex space is irreducible with respect to a half-symmetric F-connexion $\Gamma_{j\ i}^{\ h}$ and if $\Gamma_{j\ i}^{\ h}$ does not admit any covariant constant quaternion structure, then any H-projective vector is almost contravariant analytic.*

THEOREM 4.7. *In an almost complex space having a half-symmetric F-connexion whose Ricci tensor does not vanish, any H-projective vector is almost contravariant analytic.*

## § 5. Some properties of *H*-projective vectors†

Keeping notations as in the previous section, we consider the $F$-connexion $\Lambda_{j\ i}^{\ h}$ defined by (2.18). Denoting by $R_{kji}^{\ \ \ h}$ and $T_{ji}^{\ \ h}$ the curvature tensor and the torsion tensor

† See S. Ishihara [2].

of $\Lambda_j{}^h{}_i$ respectively, we find

$$\underset{v}{\mathfrak{L}} R_{kji}{}^h = \nabla_k(\underset{v}{\mathfrak{L}} \Lambda_j{}^h{}_i) - \nabla_j(\underset{v}{\mathfrak{L}} \Lambda_k{}^h{}_i) - 2T_{kj}{}^r(\underset{v}{\mathfrak{L}} \Lambda_r{}^h{}_i), \tag{5.1}$$

where $\nabla_j$ denote the covariant differentiation with respect to $\Lambda_j{}^h{}_i$. If we substitute (4.11) in (5.1), we get

$$\underset{v}{\mathfrak{L}} R_{kji}{}^h = -A^h_{[k}\nabla_{j]}\varrho_i + \nabla_{[k}\varrho_{j]}A^h_i + F_{[k}{}^h\nabla_{j]}\varrho_s F_i{}^s - \nabla_{[k}\varrho_{|t|}F_{j]}{}^t F_i{}^h$$
$$- 2T_{kj}{}^r(\varrho_{(r}A^h_{i)} - \varrho_s F_{(r}{}^s F_{i)}{}^h). \tag{5.2}$$

By contraction with respect to $h$ and $k$, we have

$$\underset{v}{\mathfrak{L}} R_{ji} = O^{ts}_{ji}(\nabla_t\varrho_s + \nabla_s\varrho_t) - \frac{n+2}{2}\nabla_j\varrho_i - O^{sh}_{jr}T_{is}{}^k F_k{}^r, \tag{5.3}$$

where $R_{ji} = R_{rji}{}^r$. From (5.3) follows

$$\nabla_j\varrho_i = \underset{v}{\mathfrak{L}} P_{ji} + \frac{2}{n+2} O^{sh}_{jr}T_{is}{}^k F_k{}^r. \tag{5.4}$$

By virtue of (4.11) and (5.4), if we put $v_i{}^h = \nabla_i v^h$, we have a system of differential equations

$$\left.\begin{aligned}
&\nabla_i v^h = v_i{}^h, \\
&\nabla_j v_i{}^h = 2v^r\nabla_j T_{ri}{}^h + 2v_j{}^r T_{ri}{}^h - v^r R_{rji}{}^h - \varrho_{(j}A^h_{i)} + \varrho_r F_{(j}{}^r F_{i)}{}^h, \\
&\nabla_j\varrho_i = v^r\nabla_r P_{ji} + P_{ri}v_j{}^r + P_{jr}v_i{}^r + \frac{n}{n+2}\varrho_k O^{sk}_{ir}T_{js}{}^r,
\end{aligned}\right\} \tag{5.5}$$

with respect to the unknowns $v^h$, $\varrho_i$, $v_i{}^h$. Thus, we have

THEOREM 5.1. *Let $v^h$ be an almost analytic H-projective vector of a half-symmetric F-connexion and $\varrho_i$ the vector field associated with $v^h$. If the space is connected and of dimension $n>2$, and if all of $v^h$, $\varrho_j$ and $\nabla_i v^h$ vanish at a point, then $v^h$ vanishes identically.*

In an almost complex space with a half-symmetric $F$-connexion $\Gamma_j{}^h{}_i$, let $\mathfrak{G}$ be the Lie algebra of all almost analytic $H$-projective vectors and $\mathfrak{G}_0$ be the subalgebra of $\mathfrak{G}$ which consists of all elements of $\mathfrak{G}$ vanishing at a point $P_0$. We call $\mathfrak{G}_0$ the *isotropy algebra* of $\mathfrak{G}$ at $P_0$. Let $T_0$ be the tangent space at $P_0$. Then there exists in $T_0$ a system of complex Cartesian coordinates $(\xi^\varkappa, \xi^{\bar\varkappa})$ in which the values $(F_i^{\ h})$ of the structure $F_i^{\ h}$ have the components

$$(F_i^{\ h})_0 = \begin{pmatrix} \sqrt{-1}\delta_\lambda^\varkappa & 0 \\ 0 & -\sqrt{-1}\delta_{\bar\lambda}^{\bar\varkappa} \end{pmatrix} \tag{5.6}$$

For any element $v^h$ of $\mathfrak{G}_0$, by virtue of its analyticity, we have

$$(F_j^{\ t})_0(\nabla_t v^h)_0 = (\nabla_j v^t)_0 (F_t^{\ h})_0 . \tag{5.7}$$

Thus, denoting by $(\nabla_i v^h)_0$ the value of $\nabla_i v^h$ at $P_0$, we see that $(\nabla_i v^h)_0$ has self-adjoint components

$$(\nabla_i v^h)_0 = \begin{pmatrix} a_\lambda^\varkappa(v) & 0 \\ 0 & a_{\bar\lambda}^{\bar\varkappa}(v) \end{pmatrix} \tag{5.8}$$

in $(\xi^\varkappa, \xi^{\bar\varkappa})$. Let $\varrho_i$ be the associated vector of $v^h$ and $(a_\lambda(v), \alpha_{\bar\lambda}(v))$ be its components in $(\xi^\varkappa, \xi^{\bar\varkappa})$. Then we associate with $v^h$ a complex $(m+1, m+1)$-matrix

$$\alpha(v) = \begin{pmatrix} 0 & a_\lambda(v) \\ 0 & a_\lambda^\varkappa(v) \end{pmatrix}. \tag{5.9}$$

It is easily seen that for any element $v^h$, $u^h$ of $\mathfrak{G}_0$ we have

$$\alpha([u, v]) = [\alpha(u), \alpha(v)]. \tag{5.10}$$

This means that there exists a homomorphism $\alpha$ of $\mathfrak{G}_0$ into the algebra $\mathfrak{L}$ of all complex $(m+1, m+1)$-matrices of the form (5.9). Howewer, it follows from Theorem 5.1 that $\alpha$ is an isomorphism. Thus we have

THEOREM 5.2. *The isotropy algebra $\mathfrak{G}_0$ of $\mathfrak{G}$ is isomorphic to a subalgebra of $\mathfrak{L}$ if the space is of dimension $n > 2$.*

We define for an element $v^h$ of $\mathfrak{G}_0$ a complex $(m,m)$-matrix $\beta(v)$ by

$$\beta(v) = (a_\lambda^\varkappa(v)). \tag{5.11}$$

Then from (5.10) we see easily that the correspondence

$v^h \to \beta(v)$ defines a homomorphism of $\mathfrak{G}_0$ into the algebra $\mathfrak{L}_m$ of all complex $(m,m)$-matrices. The image of $\mathfrak{G}_0$ by $\beta$ is called the *linear isotropy algebra* $\beta(\mathfrak{G}_0)$ of $\mathfrak{G}_0$ at the point $P_0$.

Let us consider an almost complex space with a half-symmetric $F$-connexion $\Gamma_{j\,i}^{\,h}$ which is $H$-projectively flat. Then for each point there exists a neighbourhood $U$ of the point in which there exists a flat $P$-connexion $\bar{\Gamma}_{j\,i}^{\,h}$ $H$-projectively related to $\Gamma_{j\,i}^{\,h}$. Let $v^h$ be an almost analytic $H$-projective vector of $\Gamma_{j\,i}^{\,h}$. Then, by Theorems 2.5 and 4.5, we see that $v^h$ is also $H$-projective with respect to $\bar{\Gamma}_{j\,i}^{\,h}$. Thus, we may assume that the given $\bar{\Gamma}$-connexion $\Gamma_{j\,i}^{\,h}$ is flat in $U$.

Since $\Gamma_{j\,i}^{\,h}$ is flat in $U$, we have in $U$

$$\nabla_j \varrho_i = 0, \tag{5.12}$$

$\varrho_i$ being the associated vector of $v^h$, if $n > 2$.

Take another almost analytic $H$-projective vector $u^h$ whose associated vector is $u_i$. We put $w = [u, v]$. If we denote by $\tau_i$ the associated vector of $w^h$, then from (4.15) we find in $U$

$$\tau_i = \varrho_r \sigma_i u^r - \mu_r \nabla_i v^r. \tag{5.13}$$

Since $\Gamma_{j\,i}^{\,h}$ is flat in $U$, we obtain from (4.9)

$$\nabla_j \nabla_i v^h = \varrho_{(j} A_{i)}^h - \varrho_t F_{(j}{}^t F_{i)}{}^h. \tag{5.14}$$

Taking account of (5.14), we see further that

$$\begin{aligned} \nabla_i w^h &= \nabla_i u^r \nabla_r v^h - \nabla_i v^r \nabla_r u^h \\ &+ u^r(\varrho_{(i} A_{r)}^h - \varrho_s F_{(i}{}^s F_{r)}{}^h) - v^r(\mu_{(i} A_{r)}^h - \mu_s F_{(i}{}^s F_{r)}{}^h). \end{aligned} \tag{5.15}$$

Finally, we have from the definition

$$w^h = u^r \nabla_r v^h - v^r \nabla_r u^h. \tag{5.16}$$

At a point $P_0 \varepsilon U$, we consider for $v^h$ a complex $(m+1, m+1)$-matrix

$$B(v) = \begin{pmatrix} b_0^0(v) & b_\lambda^0(v) \\ b_0^\varkappa(v) & b_\lambda^\varkappa(v) \end{pmatrix}, \tag{5.17}$$

where

$$b_0^0(v) = -\frac{1}{n+1} a_\alpha^\alpha(v), \qquad b_\lambda^0(v) = a_\lambda(v),$$

$$b_0^\varkappa(v) = -a^\varkappa(v), \qquad b_\lambda^\varkappa(v) = a_\lambda^\varkappa(v) - \frac{1}{n+1} a_\alpha^\alpha(v)\delta_\lambda^\varkappa,$$

$a^\varkappa(v)$, $a_\lambda(v)$ and $a_\lambda^\varkappa(v)$ denoting respectively the values of $v^h$, $\varrho_i$ and $\nabla_i v^h$ at the point $P_0$. It is easily verified that the trace of $B(v)$ vanishes. Then it follows easily from (5.13), (5.15) and (5.16) that the correspondence $v^h \to B(v)$ defines a homomorphism $B$ of $\mathfrak{G}$ into the Lie algebra $S\mathfrak{L}_{m+1}$ of all complex $(m+1, m+1)$-matrices whose trace vanishes. However, if we take account of Theorem 5.1, we see that $B$ is necessarily an isomorphism if $n > 2$. Summing up, we obtain

THEOREM 5.3. *Let $\mathfrak{G}$ be the algebra of all almost analytic $H$-projective vectors of a $H$-projectively flat, half-symmetric $F$-connexion in an almost complex space of dimension $n > 2$. Then, if the space is connected, $\mathfrak{G}$ is isomorphic to a subalgebra of the algebra $S\mathfrak{L}_{m+1}$.*

## § 6. The Lie algebra of *H*-projective vectors of sufficiently high order†

We shall determine the space admitting a Lie algebra of almost analytic $H$-projective vectors of sufficiently high order. Let $\mathfrak{G}$ be the Lie algebra of all almost analytic $H$-projective vectors in an almost complex space with a half-symmetric $F$-connexion $\Gamma_{j\,i}^{\,h}$. Denote by $\mathfrak{J}_0$ the isotropy Lie algebra of $\mathfrak{G}$ at a point $P_0$. The linear isotropy Lie algebra is denoted by $\mathfrak{J}_0$.

† See S. Ishihara [2].

Let $\Lambda_j{}^h{}_i$ be the half-symmetric $F$-connexion determined by (2.18). Then it follows from (4.9) that

$$\varrho_j = \frac{2}{n+2} \underset{v}{\mathfrak{L}} \Gamma^r_{rj} \tag{6.1}$$

holds for any almost analytic $H$-projective vector $v^h$. If we substitute (6.1) in (4.11), we have

$$\nabla_j \varrho_i = \underset{v}{\mathfrak{L}} \Pi_{ji}, \tag{6.2}$$

the quantity $\Pi_{ji}$ being defined by

$$\Pi_{ji} = P_{ji} + \frac{4}{(n+2)} \Lambda^r_{hs} O^{ks}_{ir} T_{jk}{}^h, \tag{6.3}$$

where $P_{ji}$ is the tensor defined by (3.2) and $T_{ji}{}^h$ is the torsion tensor of $\Lambda_j{}^h{}_i$. The quantity $\Pi_{ji}$ is not a tensor but $\underset{v}{\mathfrak{L}} \Pi_{ji}$ is a tensor.

If we substitute (6.2) in (5.2), we find

$$\underset{v}{\mathfrak{L}} \Pi_{kji}{}^h = -T_{kj}{}^r \varrho_s (A^s_{(r} A^h_{i)} - F_{(r}{}^s F_{i)}{}^h), \tag{6.4}$$

where $\Pi_{kji}{}^h$ is defined by

$$\Pi_{kji}{}^h = R_{kji}{}^h + A^h_{[k} \Pi_{j]i} - \Pi_{[kj]} A^h_i - F_{[k}{}^h \Pi_{j]r} F_i{}^r + \Pi_{[k|t|} F_{j]}{}^t F_i{}^h. \tag{6.5}$$

The quantity $\Pi_{kji}{}^h$ is not a tensor but $\underset{v}{\mathfrak{L}} \Pi_{kji}{}^h$ is a tensor.

THEOREM 6.1. *If the kernel of the homomorphism* $\beta: \mathfrak{G}_0 \to \mathfrak{H}_0$ *is not the trivial subalgebra* $\{0\}$, *then the torsion tensor* $T_{ji}{}^h$ *of* $\Lambda^h_{ji}$ *vanishes at* $P_0$.

*Proof*. From the condition of the theorem it follows that there exists an almost analytic $H$-projective vector $v^h$ such that $v^h$ and $\nabla_i v^h$ vanish at $P_0$ but the associated vector $\varrho_i$ of $v^h$ does not vanish at $P_0$. Then, if we note the fact that $\underset{v}{\mathfrak{L}} \Pi_{kji}{}^h$ is a linear combination of $v^h$ and $\nabla_i v^h$, we see that $\underset{v}{\mathfrak{L}} \Pi_{kji}{}^h$ vanishes at $P_0$. Therefore, from (6.4) we have

$$T_{kj}{}^r \varrho_s (A^s_{(r} A^h_{i)} - F_{(r}{}^s F_{i)}{}^h) = 0, \tag{6.6}$$

which implies $T_{ji}{}^h = 0$ at $P_0$ because of $\varrho_i \neq 0$. This proves Theorem 6.1.

Now we consider a semi-symmetric $F$-connexion $\Gamma_j{}^h{}_i$. Then the corresponding $P$-connexion $\Lambda_j{}^h{}_i$ becomes a symmetric one, i. e. $T_{ji}{}^h = 0$. For an almost analytic $H$-projective vector $v^h$ of $\Gamma_j{}^h{}_i$ we have from (5.2) and (5.4)

$$\underset{v}{\mathfrak{L}} P_{kji}{}^h = 0, \tag{6.7}$$

where $P_{kji}{}^h$ is the $H$-projective curvature tensor of $\Lambda_j{}^h{}_i$. The integrability condition of (5.4) is nothing but

$$\underset{v}{\mathfrak{L}} P_{kji} = P_{kji}{}^h \varrho_h, \tag{6.8}$$

where $P_{kji}$ is defined by

$$P_{kji} = 2\nabla_{[k} P_{j]i}. \tag{6.9}$$

By means of (6.8) we obtain

THEOREM 6.2. *Let $\Gamma_j{}^h{}_i$ be a semi-symmetric $F$-connexion. If the kernel of the homomorphism $\beta: \mathfrak{G}_0 \to \mathfrak{H}_0$ is of the maximum dimension $n$, then the $H$-projective curvature tensor $P_{kji}{}^h$ of $\Lambda_j{}^h{}_i$ vanishes at $P_0$.*

Now we give the following theorem for later use. We consider the following subalgebra of the Lie algebra $\mathfrak{L}_m$ of complex $(m,m)$-matrices $(a^\varkappa_\lambda)$.

$$\begin{aligned}
&S\,\mathfrak{L}_m = \{(a^\varkappa_\lambda) \mid a^\alpha_\alpha = 0\},\\
&\quad \mathfrak{M} = \{(a^\varkappa_\lambda) \mid a^p_1 = 0,\ p = 2,\ 3,\ \ldots,\ m\},\\
&\quad \mathfrak{M}' = \{(a^\varkappa_\lambda) \mid a^1_q = 0,\ q = 2,\ 3,\ \ldots,\ m\},\\
&J(A) = \{(a^\varkappa_\lambda) \mid a^\varkappa_\lambda = At\delta^\varkappa_\lambda\},\ A \text{ being a non-vanishing complex number,}
\end{aligned} \tag{6.10}$$

where $t$ is a real variable.

It is easily seen that

$$\begin{aligned}
\dim S\mathfrak{L}_m &= 2(m^2 - 1),\\
\dim \mathfrak{M} &= \dim \mathfrak{M}' = 2(m^2 - m + 1),\\
\dim J(A) &= 1.
\end{aligned} \tag{6.11}$$

We denote by $J(A)\times S\mathfrak{L}_m$ the Lie algebra generated by $J(A)$ and $S\mathfrak{L}_m$.

THEOREM 6.3. *Each subalgebra of* $\mathfrak{L}_m$ *is, if its dimension is not less than* $2(m^2-m+1)$, *conjugate to one of the Lie algebras:*

$$\mathfrak{L}_m,\quad J(A)\times S\mathfrak{L}_m,\quad S\mathfrak{L}_m,\quad \mathfrak{M},\quad \mathfrak{M}'.$$

By means of Theorem 6.3 we obtain

THEOREM 6.4. *Let* $\mathfrak{H}$ *be a subalgebra of the Lie algebra* $\tilde{\mathfrak{L}}$ *of all complex* $(m+1,m+1)$*-matrices of the form* (5.9). *If dim* $\mathfrak{G}\geqq 2(m^2+1)$, *then the kernel of* $\beta$ *in* $\mathfrak{H}$ *is of dimension* $2m$ *and the image* $\beta(\mathfrak{H})$ *is conjugate to one of the algebras indicated in Theorem* 6.3.

Finally, we consider an almost complex space of dimension $n=2m>2$ with a half-symmetric $F$-connexion $\Gamma_{j\,i}^{\,h}$. We assume that in the space the Lie algebra $\mathfrak{G}$ is of dimension not less than $2(m^2+m+1)$. Then, taking an arbitrary point $P_0$ of the space, we see easily that $\dim\mathfrak{G}_0\geqq\dim\mathfrak{G}-2m$, that is

$$\dim\,\mathfrak{G}_0\geqq 2(m^2+1). \tag{6.12}$$

Thus, Theorems 5.2 and 6.4 imply that the kernel of $\beta$ in $\mathfrak{G}_0$ is of dimension $2m$ and the linear isotropy Lie algebra $\mathfrak{H}_0=\beta(\mathfrak{G}_0)$ is conjugate to one of the Lie algebras indicated in Theorem 6.4.

Since the kernel of $\beta$ in $\mathfrak{G}_0$ is of dimension $2m$, the point $P_0$ being taken arbitrarily, Theorem 6.1 implies that the $F$-connexion $\Gamma_{j\,i}^{\,h}$ is semi-symmetric. Thus, from Theorem 1.8 it follows that the Nijenhuis tensor $N_{ji}{}^h$ of the space vanishes identically if $\dim\mathfrak{G}\geqq 2(m^2+m+1)$. Furthermore, by virtue of Theorem 6.2 the given $F$-connexion $\Gamma_{j\,i}^{\,h}$ is $H$-projectively flat. Hence, taking account of Theorem 5.3, we see that the Lie algebra $\mathfrak{G}$ is isomorphic to a subalgebra of $S\mathfrak{L}_{m+1}$. On the other hand, we have supposed $\dim\mathfrak{G}\geqq 2(m^2+m+1)$. From Theorem 6.3 it follows thereby that $\mathfrak{G}$ is isomorphic to $S\mathfrak{L}_{m+1}$ itself. Summing up, we obtain

THEOREM 6.5. *If an almost complex space of dimension* $n=2m>2$ *with a half-symmetric F-connexion* $\Gamma_{j\ i}^{\ h}$ *admits a Lie algebra* $\mathfrak{G}$ *of almost analytic H-projective vectors of dimension not less than* $2(m^2+m+1)$, *then the space is H-projectively flat and* $\mathfrak{G}$ *is of the maximum dimension* $2(m^2+m)$, *that is,* $\mathfrak{G}$ *is isomorphic to the Lie algebra* $S\mathfrak{L}_{m+1}$.

## § 7. Groups of *H*-projective transformations†

In an almost complex space, if a transformation $T$ of the space preserves the almost complex structure, then we say that $T$ is an *almost analytic one.* Given an $F$-connexion $\Gamma_{j\ i}^{\ h}$ in the space, if a transformation $T$ of the space carries any holomorphically planar curve of $\Gamma_{j\ i}^{\ h}$ into another one, then we call $T$ an *H-projective transformation.* Assuming that $\Gamma_{j\ i}^{\ h}$ is half-symmetric, if a transformation $T$ is both almost analytic and $H$-projective, then we can see that $\Gamma_{j\ i}^{\ h}$ is $H$-projectively related to the connexion $\bar{\Gamma}_{j\ i}^{\ h}$ induced by $T$ from $\Gamma_{j\ i}^{\ h}$, where $\bar{\Gamma}_{j\ i}^{\ h}$ is also an $F$-connexion. When the $F$-connexion $\Gamma_{j\ i}^{\ h}$ is half-symmetric, so is the induced connexion $\bar{\Gamma}_{j\ i}^{\ h}$. Therefore, if $\Gamma_{j\ i}^{\ h}$ is half-symmetric the two connexions $\Gamma_{j\ i}^{\ h}$ and $\bar{\Gamma}_{j\ i}^{\ h}$ are related to each other by the formula (2.2). Combining Theorems 2.5, 4.4 and 4.5, we have

THEOREM 7.1. *In an almost complex space with a half-symmetric F-connexion, a one-parameter group of transformations is one of the almost analytic, H-projective transformations, if and only if the vector field induced by the group is almost analytic and H-projective.*

Now, by virtue of Theorem 6.5, we obtain

THEOREM 7.2. *Let G be an effective, connected, Lie group*

† See S. Ishihara [2].

*of almost analytic H-projective transformations in an almost complex space of dimension* $n=2m>2$ *with a half-symmetric F-connexion. Suppose that the group G is of dimension not less that* $2(m^2+m+1)$. *Then the connexion is H-projectively flat, and the group is of the maximum dimension* $2(m^2+m)$. *If the space is moreover connected, the group G is transitive in the space.*

Keeping the conditions as in Theorem 6.5, we see that the algebra $\mathfrak{G}_0$ is isomorphic to $S\mathfrak{L}_{m+1}$. If we denote by $\mathfrak{U}$ the subalgebra of $S\mathfrak{L}_{m+1}$ such that

$$\mathfrak{U}=\{(b^p_q)\,|\,b^p_p=0,\quad \bar{b}^q_p+b^p_q=0\},\quad (p,\ q=0,\ 1,\ \ldots,\ m)$$

then there exists in $\mathfrak{G}$ a subalgebra $\mathfrak{G}'$ corresponding to $\mathfrak{U}$. Since $\mathfrak{U}$ is the Lie algebra of the unimodular unitary group which is compact and semi-simple, then $\mathfrak{G}'$ generates a compact subgroup $G'$ in $G$. Furthermore, if the space is connected, we easily see that the group $G'$ is transitive. Then, the space admits a Kähler metric of constant holomorphic sectional curvature. Consequently, the space is homeomorphic to the complex projective space of complex dimension $m$. Thus, we obtain

THEOREM 7.3. *Under the condition as in Theorem 2.7 the space is homeomorphic to the complex projective space of complex dimension m.*

By virtue of Theorems 6.1 and 6.2, we have easily the following theorems.

THEOREM 7.4. *Let G be an effective group of almost analytic H-projective transformations of a half-symmetric F-connexion. If the F-connexion is not semi-symmetric, then dim* $G\leqq 2m^2$, *where the space is of dimension* $n=2m>2$.

THEOREM 7.5. *Let G be an effective group of almost analytic H-projective transformations of a half-symmetric F-connexion. If the Nijenhuis tensor* $N_{ji}{}^h$ *of the space does not vanish, then dim* $G\leqq 2m^2$, *where the space is of dimension* $n=2m>2$.

## § 8. *H*-projective vectors in Kähler spaces†

In a Kähler space, let $v^h$ be an $H$-projective vector of $\left\{ {h \atop j\ i} \right\}$. Then, we say for brevity that $v^h$ is *H-projective* in the Kähler space. Now we have

$$\underset{v}{\mathfrak{L}} \left\{ {h \atop j\ i} \right\} = \varrho_j A_i^h + \varrho_i A_j^h - \varrho_s F_j{}^s F_i{}^h - \varrho_s F_i{}^s F_j{}^h . \tag{8.1}$$

Transvecting (8.1) with $g_{ji}$, we have

$$\nabla^i \nabla_i v^h + K_i{}^h v^i = 0. \tag{8.2}$$

Thus, by virtue of the well known theorem on analytic vectors, we obtain

THEOREM 8.1. *In a compact Kähler space, any H-projective vector is necessarily analytic.*

Next, from Theorems 4.6 and 4.7 we have respectively

THEOREM 8.2. *In an irreducible Kähler space admitting no quaternion structure, any H-projective vector is analytic.*

THEOREM 8.3. *In an irreducible Kähler space having non-vanishing Ricci tensor, any H-projective vector is analytic.*

This Theorem implies immediately

THEOREM 8.4. *In an irreducible Kähler–Einstein space, if its scalar curvature is non-vanishing, any H-projective vector is analytic.*

Now we assume that $v^h$ is analytic. Then, by means of (8.1), we find

$$\nabla_k \underset{v}{\mathfrak{L}} g_{ji} = \varrho_j g_{ki} + \varrho_i g_{kj} - \varrho_t F_j{}^t F_{ki} - \varrho_s F_i{}^s F_{kj} + 2\varrho_k g_{ji} . \tag{8.3}$$

If we substitute (8.1) in the well-known formula

$$\underset{v}{\mathfrak{L}} K_{kji}{}^h = \nabla_k \underset{v}{\mathfrak{L}} \left\{ {h \atop j\ i} \right\} - \nabla_j \underset{v}{\mathfrak{L}} \left\{ {h \atop k\ i} \right\}, \tag{8.4}$$

† See S. Tachibana and S. Ishihara [1].

we obtain

$$\underset{v}{\mathfrak{L}} K_{kji}{}^h = A_j^h \nabla_k \varrho_i - A_k^h \nabla_j \varrho_i - (F_j{}^h \nabla_k \varrho_s - F_k{}^h \nabla_j \varrho_s) F_i{}^s - (\nabla_k \varrho_t F_j{}^t - \nabla_j \varrho_t F_k{}^t) F_i{}^h. \quad (8.5)$$

Contracting with respect to $h$ and $k$, we find

$$\underset{v}{\mathfrak{L}} K_{ji} = -n \nabla_j \varrho_i - 2 F_j{}^t F_i{}^s \nabla_t \varrho_s. \quad (8.6)$$

Since $K_{ji}$ is hybrid if $n > 2$, (8.6) implies that $\nabla_j \varrho_i$ is also hybrid. This means that the associated vector $\varrho^h = g^{hr} \varrho_r$ is contravariant analytic. Moreover, since $\nabla_j \varrho_i$ is hybrid, we have

$$\nabla_j(\varrho_t F_i{}^t) + \nabla_i(\varrho_t F_j{}^t) = 0,$$

which implies that $\varrho^r F_r{}^h$ is a Killing vector. Thus we obtain

THEOREM 8.5. *In a Kähler space of dimension $n > 2$, if $\varrho_i$ is the associated vector of an analytic H-projective vector, then $\varrho^h$ is contravariant analytic and $\varrho^r F_r{}^h$ is a Killing vector.*

Here we shall give some formulae for later use. From (8.5) and $O_{ji}^{ts} \nabla_t \varrho_s = 0$ it follows

$$\underset{v}{\mathfrak{L}} K_{ji} = -(n+2) \nabla_j \varrho_i. \quad (8.7)$$

If we substitute (8.1) and (8.7) in the identity

$$\underset{v}{\mathfrak{L}} \nabla_k K_{ji} - \nabla_k \underset{v}{\mathfrak{L}} K_{ji} = -K_{si} \underset{v}{\mathfrak{L}} \left\{ {s \atop k\ j} \right\} - K_{js} \underset{v}{\mathfrak{L}} \left\{ {s \atop k\ i} \right\}, \quad (8.8)$$

we obtain

$$\underset{v}{\mathfrak{L}} \nabla_k K_{ji} = -(n+2) \nabla_k \nabla_j \varrho_i - K_{ki} \varrho_j - K_{kj} \varrho_i - H_{ki} \varrho_t F_j{}^t - H_{kj} \varrho_s F_i{}^s - 2 K_{ji} \varrho_k. \quad (8.9)$$

Hence, if we put

$$P_{kji} = \frac{1}{n+2} (\nabla_k K_{ji} - \nabla_j K_{ki}),$$

we find

$$\underset{v}{\mathfrak{L}} P_{kji} = P_{kji}{}^h \varrho_h. \quad (8.10)$$

Now, we consider an analytic $H$-projective vector $v^h$ which leaves the tensor $\nabla_l P_{kji}{}^h$ invariant, where $P_{kji}{}^h$ is the $H$-projective curvature tensor of $\left\{ {h \atop j\ i} \right\}$. By means of the well-known formula

$$\underset{v}{\mathfrak{L}} \nabla_l P_{kji}{}^h - \nabla_l \underset{v}{\mathfrak{L}} P_{kji}{}^h \tag{8.11}$$

$$= P_{kji}{}^r \underset{v}{\mathfrak{L}} \left\{ {h \atop l\ r} \right\} - P_{kjs}{}^h \underset{v}{\mathfrak{L}} \left\{ {s \atop l\ i} \right\} - P_{kti}{}^h \underset{v}{\mathfrak{L}} \left\{ {t \atop l\ j} \right\} - P_{rji}{}^h \underset{v}{\mathfrak{L}} \left\{ {r \atop l\ k} \right\},$$

taking account of $\underset{v}{\mathfrak{L}} P_{kji}{}^h = 0$ and $\underset{v}{\mathfrak{L}} \nabla_l P_{kji}{}^h = 0$, we find

$$\begin{aligned} A_l^h P_{kji}{}^r \varrho_r - 2\varrho_l P_{kji}{}^h - \varrho_k P_{lji}{}^h - \varrho_j P_{kli}{}^h - \varrho_i P_{kjl}{}^h \\ - F_l^{\ h} P_{kji}{}^r F_r^{\ s} \varrho_s + F_l^{\ s} \varrho_r (F_k^{\ r} P_{sji}{}^h + F_j^{\ r} P_{ksi}{}^h + F_i^{\ r} P_{kjs}{}^h) = 0. \end{aligned} \tag{8.12}$$

Contracting with respect to $h$ and $l$, we can verify

$$P_{kji}{}^r \varrho_r = 0. \tag{8.13}$$

Transvecting (8.12) with $F_h^{\ l}$, we have

$$P_{kji}{}^r F_r^{\ s} \varrho_s = 0. \tag{8.14}$$

If we substitute (8.13) and (8.14) in (8.12), we find

$$\begin{aligned} 2\varrho_l P_{kji}{}^h + \varrho_k P_{lji}{}^h + \varrho_j P_{kli}{}^h + \varrho_i P_{kjl}{}^h \\ = F_l^{\ s} \varrho_r (F_k^{\ r} P_{sji}{}^h + F_j^{\ r} P_{ksi}{}^h + F_i^{\ r} P_{kjs}{}^h). \end{aligned} \tag{8.15}$$

Now, transvecting both sides with $\varrho^l P^{kji}{}_h = \varrho^l P_{uts}{}^r g^{ku} g^{jt} g^{is} g_{rh}$, we obtain

$$(\varrho_l P_{kji}{}^h)(\varrho^l P^{kjih}) + 2(\varrho^r P_{rjih})(\varrho_s P^{sjih}) + (\varrho^r P_{kjrs})(\varrho_t P^{kjts}) = 0, \tag{8.16}$$

where we have used the following properties of $P_{kji}{}^h$:

$$\left.\begin{aligned} P_{(kj)i}{}^h &= 0, & P_{[kji]}{}^h &= 0, \\ P_{rji}{}^r &= 0, & P_{kjr}{}^r &= 0, \\ F_s^{\ r} P_{rji}{}^s &= 0, & P_{kjs}{}^r F_r^{\ s} &= 0, \\ {}^*O_{ir}^{sh} P_{kjs}{}^r &= 0, & O_{kj}^{ts} P_{tsi}{}^h &= 0. \end{aligned}\right\} \tag{8.17}$$

The equation (8.16) implies $P_{kji}{}^h \varrho_h = 0$, from which, by making use of Theorem 3.4, we get

THEOREM 8.6. *If in a Kähler space an analytic H-projective vector, which is not affine, leaves the tensor* $\nabla_l P_{kji}{}^h$ *invariant, then the space is of constant holomorphic sectional curvature.*

In a symmetric Kähler space, $\nabla_l P_{kji}{}^h = 0$ holds good. Then, we obtain from Theorem 8.6,

THEOREM 8.7. *If a symmetric Kähler space admits an analytic H-projective vector, which is not affine, then the space is of constant holomorphic sectional curvature.*

Now, we shall give a theorem in Riemannian geometry.

THEOREM 8.8. *A necessary and sufficient condition for a Riemannian space to be an Einstein one is that* $K_{ji}K^{ji} = K^2/n$ *holds.*

*Proof.* We consider a tensor $Z_{ji} = K_{ji} - (K/n)g_{ji}$. Then it is easily seen that

$$Z_{ji}Z^{ji} = K_{ji}K^{ji} - \frac{K^2}{n}, \tag{8.18}$$

which implies the theorem.

In a Kähler space such that $\nabla_k K_{ji} = 0$, let $v^h$ be an analytic $H$-projective vector. Then, from (8.9) it follows

$$(n+2)\nabla_k\nabla_j\varrho_i = -K_{ki}\varrho_j - K_{kj}\varrho_i - H_{ki}\varrho_t F_j{}^t - H_{kj}\varrho_s F_i{}^s - 2K_{ji}\varrho_k. \tag{8.19}$$

Transvecting (8.19) with $g^{kj}$, we get

$$\nabla^r\nabla_r\varrho_i = -\frac{1}{n+2}(2K_i{}^r\varrho_r + K\,\varrho_i). \tag{8.20}$$

On the other hand, $\varrho^h$ being contravariant analytic, if we compare (8.2) and (8.20), we find

$$K_i{}^r\varrho_r = \frac{K}{n}\varrho_i. \tag{8.21}$$

We can now prove

THEOREM 8.9. *If a Kähler space satisfying* $\nabla_k K_{ji} = 0$ *is not an Einstein space, then the associated vector* $\varrho_i$ *of an analytic H-projective vector satisfies* $\nabla_i\varrho^i = 0$.

*Proof*. On applying the Ricci's identity to $K_{ji}$, we find

$$K_{lkj}{}^{h}K_{hi}+K_{lki}{}^{h}K_{jh}=0. \tag{8.22}$$

Transvecting (8.22) with $g^{ki}$, we get

$$K_{lsrj}K^{sr}=K_{l}{}^{r}K_{jr}. \tag{8.23}$$

If we apply $\underset{v}{\mathfrak{L}}$ to (5.5) and transvect the equation obtained with $K^{jk}g^{il}$, on using (8.23), we obtain

$$(K_{h}{}^{k}K^{ij}+K_{rh}K^{rj}g^{ik})\underset{v}{\mathfrak{L}}K_{kjr}{}^{h}=0. \tag{8.24}$$

If $v^h$ is an analytic $H$-projective vector, substituting (8.5) in (8.24), we obtain

$$(\nabla_r\varrho^r)K_{ji}K^{ji}-K\,K_{ji}\nabla^j\varrho^i=0. \tag{8.25}$$

Here from (8.21) and (8.25) it follows

$$\left(K_{ji}K^{ji}-\frac{K^2}{n}\right)\nabla_r\varrho^r=0, \tag{8.26}$$

which implies the theorem.

THEOREM 8.10. *If a Kähler space satisfying $\nabla_k K_{ji}=0$ admits an analytic H-projective vector which is not affine, it is a Kähler–Einstein space.*

*Proof*. Since $\nabla_k K_{ji}=0$, $K_{ji}K^{ji}$ is a constant so that for a vector field $v^h$ we have

$$0=\underset{v}{\mathfrak{L}}(K_{ji}K^{ji})=2(\underset{v}{\mathfrak{L}}K_{ji})K^{ji}+K_{j}{}^{s}K_{sr}\underset{v}{\mathfrak{L}}g^{jr}. \tag{8.27}$$

Then, if $v^h$ is analytic and $H$-projective, taking account of (8.7) we find

$$-\frac{n+2}{n}K\,\nabla_i\varrho^i+K_j{}^{s}K_{sr}\,\mathfrak{L}\,g^{jr}=0. \tag{8.28}$$

If we now assume that our space is not an Einstein space, then from Theorem 8.8 it follows

$$K_j{}^{s}K_{sr}\underset{v}{\mathfrak{L}}g^{jr}=0, \tag{8.29}$$

which can be written as

$$K^{js}K_s{}^r \underset{v}{\mathfrak{L}} g_{jr}=0. \tag{8.30}$$

Applying $\nabla_i$ to the two sides and then substituting (8.3), we obtain

$$\frac{1}{2}(\varrho_i\varrho^i)(K_{jh}K^{jh})+(K_{jh}\varrho^j)(K_i{}^h\varrho^i)=0, \tag{8.31}$$

which implies $\varrho_iK_{jh}=0$. This is a contradiction.

## § 9. Kähler–Einstein spaces with non-vanishing scalar curvature†

By making use of the irreducible decomposition of a Riemannian space and taking account of Obata's theorem, we can prove

THEOREM 9.1. *In a Kähler space with positive (or negative) definite Ricci form, any infinitesimal affine transformation is contravariant analytic.*

This theorem implies immediately

THEOREM 9.2. *In a Kähler–Einstein space whose scalar curvature does not vanish, any Killing vector is contravariant analytic.*

By making use of these theorems, we have

THEOREM 9.3. *In a Kähler–Einstein space with non-vanishing scalar curvature, an H-projective vector is analytic if and only if its associated vector is contravariant analytic.*

*Proof.* The associated vector of an analytic $H$-projective vector is contravariant analytic by means of Theorem 8.3. Conversely, we suppose that the associated vector $\varrho^h$ of an $H$-projective vector $v^h$ is contravariant analytic. Then from (8.7) and $O_{ji}^{ts}\nabla_t\varrho_s=0$ it follows

$$\underset{v}{\mathfrak{L}} K_{ji}=-(n+2)\nabla_j\varrho_i\,. \tag{9.1}$$

† See S. Tachibana and S. Ishihara [1].

The space being an Einstein space we have,

$$\underset{v}{\mathfrak{L}} g_{ji} = \frac{1}{k} \nabla_j \varrho_i, \qquad k = -\frac{K}{n(n+2)}. \tag{9.2}$$

Taking account of (9.2), and putting

$$p_i = v_i - \frac{1}{2k} \varrho_i, \tag{9.3}$$

we see that $p^h$ is a Killing vector. Then, by virtue of Theorem 9.2, the vector $p^h$ is contravariant analytic and so is $v^h$.

Let $v^h$ be an analytic $H$-projective vector in a Kähler–Einstein space. Then from (8.7) we have (9.2). If we define a vector field $p^h$ by (9.3), we see that $p^h$ is a Killing vector. Next, if we put $q^h = (1/2k)\ \varrho^r F_r{}^h$, then we have

$$q^r F_r{}^h = -\frac{1}{2k} \varrho^h, \tag{9.4}$$

$$v^h = p^h + F_r{}^h q^r. \tag{9.5}$$

Thus, taking account of Theorem 8.3, we obtain

THEOREM 9.4. *In a Kähler–Einstein space with $K \neq 0$ any analytic $H$-projective vector $v^h$ is uniquely decomposable in the form*

$$v^h = p^h + F_r{}^h q^r,$$

*where $p^h$ and $q^h$ are Killing vectors.*

We note here that a contravariant analytic vector $u^h$ is a gradient vector, i. e. $g_{ri} u^r$ is a gradient vector, if and only if $u^r F_r{}^h$ is a Killing vector. Then, the vector $F_r{}^h q^r$ of Theorem 9.4 is contravariant analytic and $F_r{}^h q^r$ is a gradient vector. From (8.5) it follows that

$$\underset{v}{\mathfrak{L}} \left\{ {h \atop j\ i} \right\} = -\underset{Fq}{\mathfrak{L}} \left\{ {h \atop j\ i} \right\}. \tag{9.6}$$

If we substitute (8.1) and (9.4) in (9.6), we find

$$\nabla_j\nabla_i\varrho^h + K_{kji}{}^h\varrho^k = 2k(\varrho_j A_i^h + \varrho_i A_j^h - \varrho_s F_j{}^s F_i{}^h - \varrho_s F_i{}^s F_j{}^h). \quad (9.7)$$

Thus we obtain

THEOREM 9.5. *In a Kähler–Einstein space with* $K\neq 0$, *the associated vector* $\varrho^h$ *of an analytic H-projective vector is also contravariant analytic and H-projective. Further,* $\varrho_i$ *is a gradient vector.*

Let $L_H$, $L_I$ and $L'$ be respectively the Lie algebras consisting of all analytic $H$-projective vectors, of all Killing vectors and of all contravariant analytic vector fields $u^h$ such that $u_i = g_{ir}u^r$ is a gradient. Then we have

THEOREM 9.6. *In a Kähler–Einstein space with* $K\neq 0$, *the following relations hold:*

$$L_H = L_I + L' \ (\textit{direct sum}),$$
$$[L_I, L_I]\subset L_I, \ [L_I, L']\subset L', \ [L', L']\subset L_I. \quad (9.8)$$

From (9.7) follows

$$\nabla_j\nabla_i\varrho_h + K_{kjih}\varrho^k = 2k(g_{jh}\varrho_i + g_{ih}\varrho_j - F_{jh}F_i{}^s\varrho_s - F_{ih}F_j{}^t\varrho_t). \quad (9.9)$$

Taking the alternating part with respect to $i$ and $h$, we get

$$K_{kjih}\varrho^k = k(g_{ki}g_{jh} - g_{ji}g_{kh} + F_{hi}F_{jh} - F_{ji}F_{kh} + 2F_{kj}F_{ih})\varrho^k, \quad (9.10)$$

which implies

THEOREM 9.7. *If a Kähler–Einstein space with* $K\neq 0$ *admits an analytic H-projective vector which is not affine, then its local homogeneous holonomy group at any point is the full unitary group* $U(n/2)$.

An analytic $H$-projective vector $\varrho^h$, if $\varrho_i = g_{is}\varrho^s$ is a gradient, satisfies (9.10). We suppose that $L'$ is transitive at each point $p$ of the space, i.e. that for each point $p$ the vector space $L'_p = \{v(p)\,|\,v\in L'\}$ is of dimension $n$, $v(p)$ denoting the value of the vector field $v$ at $p$. Then, from (9.10) it follows that the space is of constant holomorphic sectional curvature. We thus obtain

THEOREM 9.8. *In a Kähler–Einstein space with $K \neq 0$, if the vector space $L'$ is transitive at each point, then the space is of constant holomorphic sectional curvature.*

This theorem can also be proved by making use of (8.10). In fact, by means of the condition of the theorem, (8.10) implies $P_{kji}{}^h = 0$.

Theorem 9.8 implies immediately

THEOREM 9.9. *If a homogeneous Kähler–Einstein space with $K \neq 0$ admits an analytic H-projective vector which is not affine, and if its linear isotropy group is irreducible, then it is a space of constant holomorphic sectional curvature.*

We consider a compact space of constant holomorphic sectional curvature. Since a compact Kähler–Einstein space with $K < 0$ admits no non-trivial analytic vector, we may assume that the scalar curvature $K$ of the space is positive. Then, the space being an Einstein space by virtue of Matsushima's theorem, an analytic vector $v^h$ is decomposed uniquely in the form

$$v^h = p^h + F_r{}^h q^r, \tag{9.11}$$

where $p^h$ and $q^h$ are Killing vectors. Hence we have

$$\underset{v}{\mathfrak{L}} \left\{ {h \atop j\ i} \right\} = \underset{Fq}{\mathfrak{L}} \left\{ {h \atop j\ i} \right\} = \nabla_j \nabla_i (F_r{}^h q^r) + K_{kji}{}^h F_s{}^k q^s. \tag{9.12}$$

Since $q^h$ is a Killing vector, we find

$$\nabla_j \nabla_i q^h + K_{kji}{}^h q^k = 0. \tag{9.13}$$

Substituting (9.13) in (9.12), we obtain

$$\underset{v}{\mathfrak{L}} \left\{ {h \atop j\ i} \right\} = (-F_r{}^h K_{kji}{}^r + F_k{}^r K_{rji}{}^h) q^k. \tag{9.14}$$

Next, if we substitute (8.5) in (9.14), we find

$$\underset{v}{\mathfrak{L}} \left\{ {h \atop j\ i} \right\} = \varrho_j A_i^h + \varrho_i A_j^h - \varrho_s F_j{}^s F_i{}^h - \varrho_s F_i{}^s F_j{}^h, \tag{9.15}$$

where we have put $\varrho_i = 2k\, F_i{}^s q_s$. Thus $v^h$ is $H$-projective. Therefore, taking account of Theorem 8.1, we have

THEOREM 9.10. *In a compact Kähler space of constant holomorphic sectional curvature with $K > 0$, a necessary and sufficient condition for a vector field to be analytic is that it is H-projective.*

We note here that Theorem 9.10 can be proved easily by making use of Matsushima's theorem.

## § 10. Almost Tachibana space†

In an almost Tachibana space, let $v$ be an almost analytic $H$-projective vector. Then we find

$$\begin{aligned}\underset{v}{\mathfrak{L}}\, K_{kji}{}^h = {} & A_j^h \nabla_k \varrho_i - A_k^h \nabla_j \varrho_i - F_j{}^h \nabla_k(\varrho_s F_i{}^s) + F_k{}^h \nabla_j(\varrho_s F_i{}^s) \\ & - F_i{}^h [\nabla_k(\varrho_t F_j{}^t) - \nabla_j(\varrho_t F_k{}^t)] \\ & - \varrho_t F_j{}^t \nabla_k F_i{}^h + \varrho_t F_k{}^t \nabla_j F_i{}^h - 2\varrho_s F_i{}^s \nabla_k F_j{}^h, \end{aligned} \tag{10.1}$$

where $\varrho_i$ is the associated vector of $v^h$. Contracting with respect to $h$ and $k$ and taking account of $\nabla_{(j} F_{i)}{}^h = 0$, $\nabla_r F_i{}^r = 0$, we find

$$\underset{v}{\mathfrak{L}}\, K_{ji} = -n \nabla_j \varrho_i - 2F_j{}^r F_i{}^t \nabla_r \varrho_t. \tag{10.2}$$

On the other hand, applying $\underset{v}{\mathfrak{L}}$ to $O_{ji}^{ts} K_{ts} = 0$, we have

$$\underset{v}{\mathfrak{L}}\, K_{ji} = -n F_j{}^r F_i{}^t \nabla_r \varrho_t - 2\nabla_j \varrho_i. \tag{10.3}$$

From (10.2) and (10.3) follows

$$O_{ji}^{ts} \nabla_t \varrho_s = 0, \tag{10.4}$$

which implies, together with $\nabla_{(j} F_{i)}{}^h = 0$,

$$\nabla_j(\varrho_s F_i{}^s) + \nabla_i(\varrho_s F_j{}^s) = 0. \tag{10.5}$$

† See S. Tachibana [2].

THEOREM 10.1. *In an almost Tachibana space, if* $\varrho_i$ *is the associated vector of an almost analytic H-projective vector, then* $\varrho_s F_i^{\ s}$ *is a Killing vector.*

From (10.2) and (10.4) follows

$$\underset{v}{\mathfrak{L}} K_{ji} = -(n+2)\nabla_j \varrho_i. \tag{10.6}$$

In an almost Tachibana–Einstein space with non-vanishing scalar curvature $K$, let $v^h$ be an almost analytic $H$-projective vector. Then, by means of (10.6) we find

$$\nabla_j\left(v_i - \frac{1}{2k}\varrho_i\right) + \nabla_i\left(v_j - \frac{1}{2k}\varrho_j\right) = 0, \tag{10.7}$$

where $k = -K/n(n+2)$. Therefore, if we define $p_i$ by $v_i = p_i + (1/2k)\varrho_i$ then $p^h$ is a Killing vector. Now, if we put $q_i = -(1/2k)\varrho_s F_i^{\ s}$, then $q^h$ is also a Killing vector because of Theorem 10.1. Hence we obtain

THEOREM 10.2. *In an almost Tachibana–Einstein space with* $K \neq 0$ *an almost analytic H-projective vector* $v^h$ *is uniquely decomposed into the form*

$$v^h = p^h + F_r^{\ h} q^r, \tag{10.8}$$

*where* $p^h$ *and* $q^h$ *are both Killing vectors and* $q_s F_i^{\ s}$ *is a gradient.*

By means of (10.8) we have

$$\underset{v}{\mathfrak{L}}\left\{ {h \atop j\ i} \right\} = \underset{p}{\mathfrak{L}}\left\{ {h \atop j\ i} \right\} + \underset{Fq}{\mathfrak{L}}\left\{ {h \atop j\ i} \right\} = \frac{1}{2k}\underset{v}{\mathfrak{L}}\left\{ {h \atop j\ i} \right\}, \tag{10.9}$$

which implies

$$\nabla_j \nabla_i \varrho^h + K_{kji}^{\ \ \ h}\varrho^k = 2k(\varrho_j A_i^h - \varrho_i A_j^h - \varrho_s F_j^{\ s} F_i^{\ h} - \varrho_s F_i^{\ s} F_j^{\ h}). \tag{10.10}$$

Using (10.10) we obtain

THEOREM 10.3. *In an almost Tachibana–Einstein space, the associated vector of an almost analytic H-projective vector is also H-projective.*

Now let $\xi^h = \xi^h(s)$ be a geodesic such that $\varrho_h(d\xi^h/ds) \neq 0$ at a point on it, where $s$ denotes its arc length. If we define a function $f(s) = \varrho_h(d\xi^h/ds)$ along the geodesic, then it follows that $f''(s)=4kf(s)$ by virtue of (10.10). If $K<0$, by solving this differential equation, we find $f(s)=A\exp(2\sqrt{k}\,s)+B\exp(-2\sqrt{k}\,s)$, where $A$ and $B$ are constant. Therefore, we get

THEOREM 10.4. *In an almost Tachibana–Einstein space with $K<0$, if it is complete, the length of the associated vector of an almost analytic H-projective vector cannot be bounded.*

From (10.10) follows

$$\nabla_j\nabla_i\varrho_k+K_{kjih}\varrho^k=2k(\varrho_j g_{ih}+\varrho_i g_{jh}-\varrho_t F_j{}^t F_{ih}-\varrho_t F_i{}^t F_{jh}). \quad (10.11)$$

Taking the symmetric and alternating parts of (10.11) with respect to $i$ and $h$, we obtain respectively

$$\nabla_j\nabla_i\varrho_h=k(\varrho_i g_{jh}+\varrho_h g_{ji}-\varrho_s F_i{}^s F_{jh}-\varrho_r F_h{}^r F_{ji}+2\varrho_j g_{ih}), \quad (10.12)$$

$$K_{kjih}\varrho^k=k(\varrho_i g_{jh}-\varrho_h g_{ji}-\varrho_s F_i{}^s F_{jh}+\varrho_r F_h{}^r F_{ji}-2\varrho_t F_j{}^t F_{ih}). \quad (10.13)$$

If we transvect (10.12) with $g^{ji}$, we find

$$\nabla^r\nabla_r\varrho_h-\frac{K}{n}\varrho_h=0. \quad (10.14)$$

On the other hand, if we transvect (10.13) with $F^{hi}F_r{}^j$, we get

$$K^*_{ri}\varrho^i=-\frac{K}{n}\varrho_r. \quad (10.15)$$

By virtue of (10.14) and (10.15) we have

THEOREM 10.5. *In a compact almost Tachibana–Einstein space the associated vector of an almost analytic H-projective vector is also an almost analytic H-projective vector.*

If we take account of (10.13), we obtain

THEOREM 10.6. *If an almost Tachibana–Einstein space with $K\neq 0$ admits an almost analytic non-affine H-projective vector, then the restricted homogeneous holonomy group contains the full unitary group $U(n/2)$.*

We consider an almost Tachibana space of constant sectional curvature with $K\neq 0$. Then we find

$$K^*_{ji}=-ag_{ji}, \qquad a=-K/n(n-1). \quad (10.16)$$

Let $v^h$ be an almost analytic $H$-projective vector and $\varrho_i$ its associated vector, then (10.15) is valid. If we substitute (10.16) into (10.15) then we get $(n-2)K\varrho_i/n(n-1) = 0$, which implies $\varrho_i = 0$. On the other hand, it is known that there exists no almost Tachibana space of negative constant curvature.

Thus we obtain

THEOREM 10.7. *In an almost Tachibana space of positive constant curvature, any almost analytic $H$-projective vector is necessarily affine and hence a Killing vector.*

## § 11. Kähler spaces with parallel Ricci tensor†

In a complex space of dimension $n > 2$, we suppose that there are given two Kähler metrics $g_{ji}$ and $\bar{g}_{ji}$. We assume further that these two metrics are $H$-projectively related to each other. Then by virtue of (3.4) we obtain

$$\overline{\left\{ {h \atop j\ i} \right\}} = \left\{ {h \atop j\ i} \right\} + \varrho_j A_i^h + \varrho_i A_j^h - \varrho_s F_j{}^s F_i{}^h - \varrho_s F_i{}^s F_j{}^h \quad (11.1)$$

for a certain vector field $\varrho_i$, where $\left\{ {h \atop j\ i} \right\}$ and $\overline{\left\{ {h \atop j\ i} \right\}}$ denote the Christoffel symbols of $g_{ji}$ and $\bar{g}_{ji}$ respectively.

If we denote by $K_{ji}$ and $\bar{K}_{ji}$ the Ricci tensors of $g_{ji}$ and $\bar{g}_{ji}$ respectively, taking account of (3.7), we obtain

$$\bar{K}_{ji} = K_{ji} - n\varrho_{ji} + 2\varrho_{ts} F_j{}^t F_i{}^s, \quad (11.2)$$

where $$\varrho_{ji} = \nabla_j \varrho_i - 2O_{ji}^{ts} \varrho_t \varrho_s .$$

On the other hand, we know that $K_{ji}$ and $\bar{K}_{ji}$ are both hybrid. Then we find from (11.2)

$$O_{ji}^{ts} \varrho_{ts} = 0, \quad (11.3)$$

† See S. Ishihara and S. Tachibana [1].

which implies together with (11.2)

$$\bar{K}_{ji} = K_{ji} - (n+2)\varrho_{ji} \tag{11.4}$$

and

$$\nabla_j(\varrho_s F_i{}^s) + \nabla_i(\varrho_s F_j{}^s) = 2(\varrho_j \varrho_s F_i{}^s + \varrho_i \varrho_s F_j{}^s). \tag{11.5}$$

We consider a geodesic $g: \xi^h = \xi^h(s)$ of $g_{ji}$, $s$ being its affine parameter. If we define a function $f(s)$ along $g$ by

$$f(s) = \varrho_s F_r{}^s \frac{d\xi^r}{ds}, \tag{11.6}$$

then we obtain by making use of (11.5)

$$\frac{df}{ds} = 2\left(\varrho_h \frac{d\xi^h}{ds}\right) f. \tag{11.7}$$

The equations (11.7) shows that the function $f$ vanishes identically along $g$ if there exists a zero point of $f$ in $g$. Consequently we know that there exists a geodesic along which the function $f(s)$ vanishes identically. We call such a geodesic a *$\varrho$-geodesic* of $g_{ji}$. We can define a $\varrho$-geodesic of $\bar{g}_{ji}$ in the same way. We then have

THEOREM 11.1. *A $\varrho$-geodesic of $g_{ji}$ is at the same time a $\varrho$-geodesic of $\bar{g}_{ji}$.*

Taking a $\varrho$-geodesic $g$, we denote by $s$ and $\bar{s}$ its affine parameters with respect to $g_{ji}$ and $\bar{g}_{ji}$ respectively. Then we find

$$\bar{s} = C \int_0^s A(u)du + B, \tag{11.8}$$

$$A(u) = \exp\left(\int_0^u \varrho_i d\xi^i\right), \tag{11.9}$$

$B$ and $C$ being constant, where the line integral in (11.9) is taken along $g$.

If we put $C = 1$ and $B = 0$, then we have

$$\bar{s} = \int_0^s A(u)du. \tag{11.10}$$

THEOREM 11.2. *If the affine parameter $\bar{s}$ defined by* (11.10) *has the form $\bar{s}=as$, $a$ being a constant, along any $\varrho$-geodesic, then the two metrics $g_{ji}$ and $\bar{g}_{ji}$ are affinely related.*

*Proof.* Take a $\varrho$-geodesic $g$ arbitrarily and assume that the parameter $\bar{s}$ defined by (11.10) has the form $\bar{s}=as$, where $a$ is constant. Then the function $A(u)$ must be constant along $g$. This implies $\varrho_i(d\xi^i/ds)=0$. Take an arbitrary point $p$ in the space. Then, since the last equation holds for any $\varrho$-geodesic passing through $p$, we find $\varrho_i v^i=0$ at $p$ for any vector $v^h$ such that $\varrho_s F_r{}^s v^r=0$. Hence we can conclude that $\varrho_i$ vanishes at $p$ since $g^{ji}\varrho_j\varrho_r F_i{}^r=0$. The point $p$ being arbitrary, $\varrho_i$ vanishes identically in the space. Thus the two metrics are affinely related.

Let us consider a geodesic $g$ of $g_{ji}$. We define along $g$ a parameter $t$ as a solution of the equation

$$\{t\}_s=\frac{2}{n+2}K_{ji}\frac{d\xi^j}{ds}\frac{d\xi^i}{ds}, \tag{11.11}$$

where $s$ is an affine parameter of $g$ and

$$\{t\}_s=\frac{\dfrac{d^3t}{ds^3}}{\dfrac{dt}{ds}}-\frac{3}{2}\left(\frac{\dfrac{d^2t}{ds^2}}{\dfrac{dt}{ds}}\right)^2. \tag{11.12}$$

Now we shall call such a parameter $t$ an $H$-projective parameter along $g$ with respect to $g_{ji}$. It is well known that the general solution of (11.11) is given by

$$t=\frac{a\tau(s)+b}{c\tau(s)+d}, \qquad ad-cb\neq 0, \tag{11.13}$$

$\tau(s)$ being a solution of (11.11), where $a$, $b$, $c$ and $d$ are constant.

THEOREM 11.3. *If two Kähler metrics are $H$-projectively related to each other, then they have all $H$-projective parameters in common along any $\varrho$-geodesics.*

*Proof*. Consider a $\varrho$-geodesic $\xi^h=\xi^h(s)$, $s$ being an affine parameter with respect to $g_{ji}$. If we take $\bar{s}$ defined by (11.10), then it follows from (11.9) and (11.10),

$$\{\bar{s}\}_s=2\left[\nabla_j\varrho_i\frac{d\xi^j}{ds}\frac{d\xi^i}{ds}-\left(\varrho_i\frac{d\xi^i}{ds}\right)^2\right]. \tag{11.14}$$

If we substitute (11.14) in the well-known identity

$$\{t\}_{\bar{s}}\left(\frac{d\bar{s}}{ds}\right)^2=\{t\}_s-\{\bar{s}\}_s, \tag{11.15}$$

then we get

$$\{t\}_{\bar{s}}=\frac{2}{n+2}K_{ji}\frac{d\xi^j}{ds}\frac{d\xi^i}{ds}, \tag{11.16}$$

which proves the theorem.

Now we will prove

THEOREM 11.4. *Let $g_{ji}$ and $\bar{g}_{ji}$ be two complete Kählerian metrics on a complex space of dimension $n>2$ such that $\nabla_k K_{ji}=0$ and $\bar{\nabla}_k K_{ji}=0$, where $\bar{\nabla}$ denotes covariant differentiation with respect to $\overline{\left\{ {h \atop j\ i} \right\}}$. If $g_{ji}$ and $\bar{g}_{ji}$ are H-projectively related, then*

(i) *in the case where the Ricci form $K(v)=K_{ji}v^jv^i$ of $g_{ji}$ is negative semi-definite, these two metrics are affinely related and hence their Ricci tensors coincide;* or

(ii) *in the case when $K(v)$ is positive semi-definite, so is the Ricci form $\bar{K}(v)=\bar{K}_{ji}v^jv^i$ of $\bar{g}^{ji}$ also.*

*Proof*. Consider a $\varrho$-geodesic $g:\xi^h=\xi^h(s)$, $s$ being an affine parameter of $g$ with respect to $g_{ji}$. Then by means of (11.10)

$$\bar{s}=\int_0^s A(u)du, \qquad A(u)=\exp\left(\int_0^u \varrho_i d\xi^i\right) \tag{11.17}$$

is an affine parameter of $g$ with respect to $\bar{g}_{ji}$. Evidently, $\bar{s}$ is an increasing function of $s$ and $\bar{s}=0$ if and only if $s=0$. Thus, taking account of the completeness of $g_{ji}$

and $\bar{g}_{ji}$, we find

$$\begin{aligned} \bar{s}=0 &\quad \text{if and only if} \quad s=0, \\ \bar{s}\to+\infty &\quad \text{if and only if} \quad s\to+\infty, \\ \bar{s}\to-\infty &\quad \text{if and only if} \quad s\to-\infty. \end{aligned} \tag{11.18}$$

Now, if we put

$$K(g)=\frac{2}{n+2}K_{ji}\frac{d\xi^j}{ds}\frac{d\xi^i}{ds},\qquad \bar{K}(g)=\frac{2}{n+2}\bar{K}_{ji}\frac{d\xi^j}{ds}\frac{d\xi^i}{ds},$$

then they are constant along $g$, because of $\nabla_k K_{ji}=0$ and $\bar{\nabla}_k \bar{K}_{ji}=0$.

An $H$-projective parameter $t$ of $g$ satisfies the equations

$$\{t\}_s=K(g), \tag{11.19}$$

$$\{t\}_{\bar{s}}=\bar{K}(g). \tag{11.20}$$

In the first place we consider the case $K(g)=0$. In this case, the affine parameter $s$ itself satisfies (11.19), i.e. $t=s$ is a solution of (11.19). Hence $t=s$ is a solution of (11.20) by virtue of Theorem 11.3.

Now we shall consider the following three cases.

(i) When $\bar{K}(g)=0$; since $\bar{t}=\bar{s}$ is a solution of (11.20), taking account of (11.13), we have

$$\bar{s}=\frac{as+b}{cs+d},\quad ad-bc\neq 0. \tag{11.21}$$

By means of (11.18), we obtain $\bar{s}=as$.

(ii) When $\bar{K}(g)>0$; if we put $\sqrt{\bar{K}(g)/2}=\lambda$, then $\bar{t}==(1/\lambda)\tan\lambda s$ is a solution of (11.20). Hence we find

$$\frac{1}{\lambda}\tan\lambda s=\frac{as+b}{cs+d},\quad ad-bc\neq 0, \tag{11.22}$$

which contradicts (11.18).

(iii) When $\bar{K}(g)<0$; if we put $\sqrt{-\bar{K}(g)/2}=\lambda$, then $\bar{t}=(1/\lambda)\tanh\lambda s$ is a solution of (11.20). The same arguments as in (ii) lead to a contradiction.

Next, we consider the case where $K(g)<0$. In this case if we put $\sqrt{-K(g)/2}=\lambda$, then $t=(1/\lambda)\tan\lambda s$ is a solution of (11.19). Let us consider the following two cases.

(iv) When $\bar{K}(g)>0$; since $\bar{t}$ defined in (ii) is a solution of (11.20) we obtain

$$\bar{t}=\frac{at+b}{ct+d},\quad ad-bc\neq 0, \tag{11.23}$$

which contradicts (11.18).

(v) When $\bar{K}(g)<0$; since $\bar{t}$ defined in (ii) is a solution of (11.20), we have (11.23). As we have $\bar{t}=0$ at $t=0$, it follows that $b=0$. If we consider the limiting case $s\to\pm\infty$, then we know that $a=d$ and $c=0$, from which we get $\bar{t}=t$. Hence we find

$$\frac{1}{\bar{\lambda}}\tanh\bar{\lambda}s=\frac{1}{\lambda}\tanh\lambda s.$$

Again taking account of the limiting cases $s\to\pm\infty$, we obtain $\bar{\lambda}=\lambda$ and $\bar{s}=s$.

In the same way we have a contradiction in the following cases: $K(g)<0$, $\bar{K}(g)=0$; $K(g)>0$, $\bar{K}(g)=0$; $K(g)>0$, $\bar{K}(g)>0$. Therefore the theorem is proved completely.

Theorem 11.4 implies

THEOREM 11.5. *Let $g_{ji}$ and $\bar{g}_{ji}$ be two complete Kähler–Einstein metrics on a complex space of dimension $n>2$. If $g_{ji}$ and $\bar{g}_{ji}$ are H-projectively related, then the scalar curvatures $K$ and $\bar{K}$ satisfy*

(1) $K=0$, $\bar{K}=0$, *or*

(2) $K<0$, $\bar{K}<0$, *or*

(3) $K>0$, $\bar{K}>0$.

In cases (1) and (2), the two metrics are affinely related.

## § 12. Holomorphic planes and the axiom of holomorphic planes†

In an almost complex space the structure $F_j^{\ h}$ induces naturally an automorphism $F_p$ in the tangent space $T_p$ at each point $p$. A subspace is said to be *invariant* if its tangent space is invariant under the automorphism $F_p$ at each point $p$. Assuming that there is given an $F$-connexion in the space, we consider an invariant subspace $\xi^h = \xi^h(u^1, u^2)$ of two dimensions, along which its tangent spaces are parallel to themselves. Such a subspace $\xi^h = \xi^h(u^1, u^2)$ is a solution of the differential equations

$$\frac{\partial^2 \xi^h}{\partial u^c \partial u^b} + \Gamma_{j\ i}^{\ h} \frac{\partial \xi^j}{\partial u^c} \frac{\partial \xi^i}{\partial u^b} = {}'\Gamma_{cb}^a \frac{\partial \xi^h}{\partial u^a}, \quad (a, b, c, \ldots = 1, 2), \tag{12.1}$$

${}'\Gamma_{cb}^a$ being certain functions of the parameters $u^a$, with the additional algebraic condition

$$F_i^{\ h} B_b^{\ i} = {}'F_b^{\ a} B_a^{\ h}, \qquad B_a^{\ h} = \partial_a \xi^h \tag{12.2}$$

with some functions ${}'F_b^{\ a}$ of $u^a$. (12.2) means that the subspace is invariant. Such an invariant subspace is by definition a *holomorphic plane.*

Given an $F$-connexion $\Gamma_{j\ i}^{\ h}$, then the connexion $\bar{\Gamma}_{j\ i}^{\ h}$ defined by

$$\bar{\Gamma}_{j\ i}^{\ h} = \Gamma_{j\ i}^{\ h} - O_{ji}^{tk} O_{kr}^{sh} S_{ts}^{\ \ r} \tag{12.3}$$

is a half-symmetric $F$-connexion because of (11.1), where $S_{ji}^{\ \ h}$ is the torsion tensor of $\Gamma_{j\ i}^{\ h}$. If an invariant subspace $\xi^h = \xi^h(u^1, u^2)$ of two dimensions is a solution of (12.1), then the subspace satisfies also the differential equation obtained by substituting $\bar{\Gamma}_{j\ i}^{\ h}$ for $\Gamma_{j\ i}^{\ h}$ in (12.1), since we find easily $(O_{jm}^{th} O_{kr}^{sh} S_{ts}^{\ \ r})\ F_i^{\ m} v^j v^i = 0$ for any vector $v^h$. Thus, $\bar{\Gamma}_{j\ i}^{\ h}$ has all holomorphic planes in common with $\Gamma_{j\ i}^{\ h}$. Then we have

THEOREM 12.1. *In an almost complex space, given an arbitrary $F$-connexion, then there always exists a half-sym-*

† See S. Ishihara [5]; J. A. Schouten and K. Yano [3].

*metric F-connexion which has all holomorphic planes and all holomorphically planar curves in common with the given F-connexion.*

Next, from the definition we obtain

THEOREM 12.2. *Any two half-symmetric F-connexions have all holomorphic planes in common, if they are H-projectively related to each other.*

The functions $'F_b{}^a$ defined on a holomorphic plane by (12.2) determines an almost complex structure in the holomorphic plane, that is, $'F_b{}^{c'}F_c{}^a = -A_b^a$. Further we have from (12.1), $S_{ji}{}^h$ and $R_{kji}{}^h$ being respectively the torsion tensor and the curvature tensor of the $F$-connexion $\Gamma_{j\,i}^{\,h}$,

$$S_{ji}{}^h B_c{}^j B_b{}^i = \sum{}_{cb}{}^a B_a{}^h,$$

$$R_{kji}{}^h B_d{}^k B_c{}^j B_b{}^i = Q_{dcb}{}^a B_a{}^h, \tag{12.4}$$

where $\sum_{cb}{}^a$ and $Q_{dcb}{}^a$ are certain functions of $u^a$.

In an almost complex space with an $F$-connexion $\Gamma_{j\,i}^{\,h}$, we assume that we can always draw a holomorphic plane passing through an arbitrary point, and being tangent to an arbitrarily given holomorphic section at this point. If this is the case, we say that the $F$-connexion satisfies the axiom of holomorphic planes. By means of Theorem 12.1 we obtain

THEOREM 12.3. *In an almost complex space, if an F-connexion satisfies the axiom of holomorphic planes, then there always exists a half-symmetric F-connexion satisfying the axiom of holomorphic planes and having all holomorphic planes in common with the original one.*

Further, making use of Theorem 12.2, we have

THEOREM 12.4. *In an almost complex space, if a half-symmetric F-connexion satisfies the axiom of holomorphic planes, so does any half-symmetric F-connexion which is H-projectively related to the given one.*

We shall give the following theorem without proof.

THEOREM 12.5. *In an almost complex space, an F-connexion satisfies the axiom of holomorphic planes, if and only if its torsion tensor* $S_{ji}{}^h$ *and its curvature tensor* $R_{kji}{}^h$ *satisfy respectively*

$$F_j{}^t S_{ti}{}^h v^j v^i = a v^h + b F_r{}^h v^r, \tag{12.5}$$

$$F_k{}^t R_{tji}{}^h v^k v^j v^i = r v^h + s F_r{}^h v^r \tag{12.6}$$

*for any vector* $v^h$ *at all points, where a, b, r and s are certain functions depending not only on the point but also on* $v^h$.

By virtue of Theorem 12.5 we obtain

THEOREM 12.6. *A half-symmetric F-connexion satisfies* (12.5) *for any vector* $v^h$ *at all points, if and only if*

$$S_{ji}{}^h = \frac{1}{8} N_{ji}{}^h + {}^*O_{ji}^{ts} T_{[t} A_{s]}^h, \tag{12.7}$$

$T_i$ *being a certain vector field.*

*Proof.* We suppose first that $S_{ji}{}^h$ satisfies (12.5) for any vector $v^h$ at all points. Then we find

$$F_{(j}{}^t S_{|t|i)}{}^h = R_{(j} A_{i)}^h + S_{(j} F_{i)}{}^h, \tag{12.8}$$

$R_i$ and $S_i$ being certain vector fields. This implies

$$2{}^*O_{ji}^{ts} S_{ts}^h = -F_j{}^s R_s A_i^h - R_i A_j^h - F_j{}^s S_s F_i{}^h + S_i F_j{}^h. \tag{12.9}$$

Making use of $S_{(ji)}{}^h = 0$, we have $S_i = F_i{}^s R_s$. Then we get

$${}^*O_{ji}^{ts} S_{ts}{}^h = {}^*O_{ji}^{ts} T_{[t} A_{s]}^h, \tag{12.10}$$

where

$$T_i = -2 F_i{}^s A_s.$$

On the other hand, by virtue of Theorem 1.1 we find $8 O_{ji}^{ts} S_{ts}{}^h = N_{ji}{}^h$, which implies together with (12.10) the required (12.7).

Conversely, if $S_{ji}{}^h$ is given by (12.7), $S_{ji}{}^h$ satisfies (12.5) for any $v^h$ at all points. In fact, ${}^*O_{ji}^{ts} N_{ts}{}^h = 0$ and $N_{(ji)}{}^h = 0$ imply $F_{(j}{}^t N_{|t|i)}{}^h = 0$.

As a corollary of Theorem 12.6, we obtain

THEOREM 12.7. *When the Nijenhuis tensor* $N_{ji}{}^h$ *vanishes identically, the torsion tensor* $S_{ji}{}^h$ *of a half-symmetric* $F$*-connexion satisfies* (12.5) *identically if and only if the given half-symmetric* $F$*-connexion is semi-symmetric.*

THEOREM 12.8. *In an almost complex space, if a half-symmetric* $F$*-connexion satisfies the axiom of holomorphic planes, then there exists in the space an* $H$*-projectively related half-symmetric* $F$*-connexion satisfying also the axiom of holomorphic planes and having* $\frac{1}{8}N_{ji}{}^h$ *as its torsion tensor.*

Next, we shall prove

THEOREM 12.9. *In order that the curvature tensor* $R_{kji}{}^h$ *of a symmetric* $F$*-connexion satisfies* (12.6) *for any* $v^h$ *at all points, it is necessary and sufficient that*

$$*O_{kj}^{ts}P_{tsi}{}^h=0, \tag{12.11}$$

*where* $P_{kji}{}^h$ *is the* $H$*-projective curvature tensor of the given symmetric* $F$*-connexion.*

*Proof*. We suppose that $R_{kji}{}^h$ satisfies (12.6) for any $v^h$ at all points. Then we find

$$F_{(k}{}^tR_{|t|ji)}{}^h=T_{(kj}A_{i)}^h+S_{(kj}F_{i)}{}^h, \tag{12.12}$$

$T_{ji}$ and $S_{ji}$ being certain tensors such that

$$T_{[ji]}=0, \qquad S_{[ji]}=0;$$
$$O_{ji}^{ts}T_{ts}=0, \qquad O_{ji}^{ts}S_{ts}=0. \tag{12.13}$$

Making use of (12.13), if we contract $h$ and $i$ in (12.12), we have

$$T_{kj}=\frac{1}{m+1}F_k{}^r*O_{jr}^{ts}R_{[ts]} \tag{12.14}$$

where $R_{ji}=R_{rji}{}^r$ and the space is of dimension $n=2m$. If we transvect (12.12) with $F_h{}^i$, we get

$$S_{kj}=\frac{1}{m+1}*O_{kj}^{ts}R_{[ts]}. \tag{12.15}$$

If we substitute (12.14) and (12.15) in (12.12), after some calculations we get (12.11).

Conversely, if (12.11) is established, we can see easily that $R_{kji}{}^h$ satisfies (12.6) for any $v^h$ at all points.

Theorems 12.5, and 12.9 imply immediately

THEOREM 12.10. *In an almost complex space in order that a symmetric F-connexion satisfies the axiom of holomorphic planes it is necessary and sufficient that its H-projective curvature tensor* $P_{kji}{}^h$ *satisfies*

$$O_{kj}^{ts} P_{tsi}{}^h = 0.$$

THEOREM 12.11. *In order that a Kähler space satisfies the axiom of holomorphic planes, it is necessary and sufficient tha it is of constant holomorphic sectional curvature.*

# BIBLIOGRAPHY


APTE, M.

[1] Sur certaines variétés hermitiques. *C. R. Acad. Sci. Paris*, **240** (1954), 101–103.

[2] Sur les isométries des variétés presque kählériennes. *C. R. Acad. Sci. Paris*, **242** (1956), 63–65.

APTE, M. and A. LICHNEROWICZ

[1] Sur les transformations affines d'une variété hermitienne compacte. *C. R. Acad. Sci. Paris*, **242** (1956), 337–339.

ARAGNOL, A.

[1] Connexions euclidiennes canoniquement associées à certaines structures presque-produit. *C. R. Acad. Sci. Paris*, **242** (1956), 339–341.

BERGMANN, S.

[1] Über eine in der Theorie der Funktionen von zwei komplexen Veränderlichen auftretende unitäre Geometrie. *Proc. Kon. Ned. Akad. Amsterdam*, **36** (1933), 307–313.

BOCHNER, S.

[1] Remarks on the theorem of Green. *Duke Math. J.* **3** (1937) 334–338.

[2] Vector fields and Ricci curvature. *Bull. Amer. Math. Soc.*, **52** (1946), 776–797.

[3] Curvature in Hermitian metric. *Bull. Amer. Math. Soc.*, **53** (1947), 179–195.

[4] On compact complex manifolds. *J. Ind. Math. Soc.*, **11** (1947), 1–21.

[5] Curvature and Betti numbers. *Ann. of Math.*, **49** (1948), 379–390.

[6] Curvature and Betti numbers, II. *Ann of Math.*, **50** (1949), 77–93.

[7] Euler–Poincaré characteristic for locally homogeneous and complex spaces. *Ann. of Math.*, **51** (1950), 241–261.

[8] Vector fields on complex and real manifolds. *Ann. of Math.*, **52** (1950), 642–649.

[9] Complex spaces with transitive commutative groups of transformations. *Proc. Nat. Acad. Sci., U. S. A.*, **37** (1951), 356–357.

[10] Tensorfields and Ricci curvature in Hermitian metric. *Proc. Nat. Acad. Sci. U.S.A.*, **37** (1951), 704–706

BOCHNER, S. and K. YANO.

[1] Tensor-fields in non-symmetric connections. *Ann. of Math.*, **56** (1952), 504–519.

BOOTHBY, W. M.

[1] Some fundamental formulae for Hermitian manifolds with non-vanishing torsion. *Amer. J. Math.*, **76** (1954), 509–534.

[2] Hermitian manifolds with zero curvature. *Mich. Math. J.*, **56** (1958), 229–233.

BOOTHBY, W. M. and H. C. WANG.

[1] On contact manifolds. *Ann. of Math.*, **68** (1958), 721–734.

BOREL, A.

[1] Kählerian coset spaces of semi-simple Lie groups. *Proc. Nat. Acad. Sci., U. S. A.*, **40** (1954), 1147–1151.

[2] On the curvature tensor of the Hermitian symmetric manifolds. *Ann. of Math.*, **71** (1960), 508–521.

BOREL, A. and A. LICHNEROWICZ.

[1] Espaces riemanniens et hermitiens symmétriques. *C. R. Acad. Sci. Paris*, **234** (1952), 2332–2334.

CALABI, E.

[1] The space of Kähler metrics. *Proc. Int. Congress. Amsterdam*, **2** (1954), 206.

[2] Construction and properties of some 6-dimensional almost complex manifolds. *Trans. Amer. Math. Soc.*, **67** (1958), 401–438.

[3] On Kähler manifolds with vanishing canonical class. *Alg. Geom. and Top.* Princeton (1957), 78–89.

CALABI, E. and B. ECKMANN.

[1] A class of compact complex manifolds which are not algebraic. *Ann. of Math.*, **58** (1953), 494–500.

CALABI, E. and D. C. SPENCER,

[1] Completely integrable almost complex manifolds. *Bull. Amer. Math. Soc.*, **57** (1951), 254–255.

CALABI, E. and E. VESENTINI.

[1] On compact locally symmetric Kähler manifolds. *Ann. of Math.*, **71** (1960), 472–501.

CARTAN, E.

[1] *Leçons sur la géométrie des espaces de Riemann. Paris*, (1928).

CARTAN, H.

[1] *Variétés riemanniennes, variétés analytiques complexes.* Sem. Ecole Norm. Sup., Paris (1951–1952).

CHERN, S.S.

[1] Characteristic classes of Hermitian manifolds. *Ann. of Math.*, **47** (1946), 85–121.
[2] *Topics in differential geometry.* Institute for Advanced Study, (1951).
[3] *Differentiable manifolds.* Chicago lecture note (1952).
[4] Several complex variables. Scientific Report on the Second Summer Institute, Part II, *Bull. Amer. Math. Soc.*, **62** (1956) 101–117.
[5] *Complex manifolds.* Lecture Notes (1956).
[6] On a generalization of Kähler geometry. *Alg. Geom. Top., Princeton* (1957).

COBURN, N.

[1] Unitary spaces with corresponding geodesics. *Bull. Amer. Math. Soc.*, **47** (1941), 901–910.
[2] Conformally unitary spaces. *Trans. Amer. Math. Soc.*, **50** (1941), 26–39.

COUTY, R.

[1] Vecteurs et tenseurs invariants sur un espace homogène. *C. R. Acad. Sci. Paris*, **246** (1958), 2569–2571.
[2] Transformations infinitésimales projectives. *C. R. Acad. Sci. Paris*, **247** (1958), 804–806.
[3] Sur les transformations des variétés riemanniennes et kähleriennes. *Ann. Inst. Fourier, Grenoble*, **9** (1959), 147–248.
[4] Transformations projectives des variétés presque kähleriennes. *C. R. Acad. Sci. Paris*, **254** (1962), 4132–4135.

ECKMANN, B.

[1] Sur les structures complexes et presque complexes. Géom. *Diff. Coll. Intern. de C. N. R. S. Strasburg*, (1953), 151–159.

ECKMANN, B. and A. FRÖLICHER

[1] Sur l'intégrabilité de structures presque complexes. *C. R Acad. Sci. Paris*, **232** (1951), 2284–2286.

Eckmann, B. and H. Guggenheimer

[1] Formes différentielles et métrique sans torsion, I. Structure complexe, formes pures. *C. R. Acad. Sci. Paris,* **229** (1949) 464–466.

[2] Formes differentielles et métrique sans torsion, II. Formes de classe K, formes analytiques. *C. R. Acad. Sci. Paris,* **229** (1949), 489–491.

[3] Sur les variétés closes à métrique hermitienne sans torsion. *C. R. Acad. Sci. Paris,* **229** (1949), 503–505.

[4] Quelques propriétés globales des variétés hermitiennes. *C. R. Acad. Sci. Paris,* **229** (1949), 577–579.

Ehresmann, C.

[1] Sur la théorie des espaces fibrés. *Coll. Int. C. N. R. S. Topologie algébrique, Paris* (1947), 3–35.

[2] Sur les variétés presque complexes *Proc. Int. Congr. Math.,* II (1950), 412–419.

Ehresmann, C. and P. Libermann

[1] Sur les structures presque hermitiennes. *C. R. Acad. Sci. Paris,* **232** (1961), 1281–1283.

Eisenhart, L. P.

[1] *Riemannian geometry.* Princeton (1949).

Frankel, T.

[1] Fixed points and torsion on Kähler manifolds. *Ann. of Math.*. **70** (1961), 1–8.

[2] Manifolds with positive curvature. *Pac. J. Math.,* **11** (1961), 165–174.

Frölicher, A.

[1] Zur Differentialgeometrie der komplexen Strukturen *Math. Ann.* **129** (1955), 50–95.

Fukami, T.

[1] Invariant tensors under the real representation of the unitary group and their applications. *J. Math. Soc. Japan,* **10** (1958), 135–144.

[2] Invariant tensors under the real representation of symplectic group and their applications. *Tôhoku Math. J.,* **10** (1958), 81–90

[3] Affine connections in almost product manifolds with some structures. *Tôhoku Math. J.,* **11** (1959), 430–446.

Fukami, T. and S. Ishihara.

[1] Almost Hermitian structure on $S^6$. *Tôhoku Math. J.,* **7** (1955), 151–156.

GARABEDIAN, P. R. and D. C. SPENCER

[1] *A complex tensor calculus for Kähler manifold.* Technical Report 17, Stanford University, (1951).

[2] A complex tensor calculus for Kähler manifolds. *Acta Math.*, 81 (1953), 279–331.

GOLDBERG, S. I.

[1] Tensor fields and curvature in Hermitian manifolds with torsion. *Ann. of Math.* **63** (1956), 64–76.

[2] On pseudo-harmonic and pseudo-Killing vectors in metric manifolds with torsion. *Ann. of Math.* **64** (1956), 364–373.

[3] Note on projectively Euclidean Hermitian manifolds. *Proc. Nat. Acad. Sci. U. S. A.*, (1956), 128–130.

[4] Conformal transformations of Kähler manifolds. *Bull. Amer. Math. Soc.*, **66** (1960), 54–58.

[5] Groups of automorphisms of almost Kähler manifolds. *Bull. Amer. Math. Soc.*, **66** (1960), 180–183.

[6] Groups of transformations of Kähler and almost Kähler manifolds. *Comm. Math. Helv.* **35** (1961), 35–46.

GOLDBERG, S. I. and S. KOBAYASKI

[1] On compact Kähler manifolds with positive definite Ricci tensor. *Ann. Math.*, **74** (1961), 570–574.

[2] The conformal transformation group of a compact Riemannian manifolds. *Amer. J. Math.*, **84** (1962), 170–174.

GRAY, J. W.

[1] Some global properties of contact structures. *Ann. of Math.*, **69** (1959), 421–450.

GUGGENHEIMER, H.

[1] Sur les variétés qui possèdent une forme extérieure quadratique fermée. *C. R. Acad. Sci. Paris*, **232** (1951), 470–472.

[2] A note on curvature and Betti numbers. *Proc. Amer. Math. Soc.*, **2** (1951), 867–890.

[3] Über complex-analytische Mannigfaltigkeiten mit Kählerscher Metrik. *Comm. Math. Helv.* **25** (1951), 257–297.

[4] Formes et vecteurs analytiques. *Annali di Mat. Pura ed Appl.*, (4) **36** (1954), 223–246.

HANO, J.

[1] On affine transformations of a Riemannian manifold. *Nagoya Math. J.*, **9** (1955), 99–109.

[2] On Kählerian homogeneous spaces of unimodular Lie groups. *Amer. J. Math.*, **79** (1957), 885–900.

HATAKEYAMA, Y.

[1] On the existence of Riemann metric associated with a 2-form of rank 2r. *Tôhoku Math. J.*, **14** (1962), 162–166.

HAWLEY, N. S.

[1] Constant holomorphic curvature. *Canadian J. Math.*, **5** (1953), 53–56.

HAYDEN, H. A.

[1] Subspaces of a space with torsion. *Proc. London Math. Soc.* (2) **34** (1934), 27–50.

HIRAMATU, H.

[1] On some properties of groups of homothetic transformations in Riemannian and Finslerian spaces. *Tensor*, **4** (1954), 28–39.

HIRZEBRUCH, F.

[1] Problems on differentiable and complex manifolds. *Ann. of Math.*, **60** (1954), 213–236.

HODGE, W. V. D.

[1] *Theory and appliciations of harmonic integrals.* Cambridge (1941). 2nd edition (1951).

[2] A special type of Kähler manifold. *Proc. London Math. Soc.*, (3) **1** (1951), 104–119.

[3] Structure problems for complex manifolds. *Rend. Mat. Ser. V*, **11** (1952), 101–110.

HOMBU, H.

[1] Zur Theorie der unitäre Geometrie. *J. Fac. Sci. Hokkaido Univ.*, (1) **3** (1935), 27–42.

HOPF, E.

[1] Elementäre Bemerkungen über die Lösungen partieller Differentialgleichungen zweiter Ordnungen vom elliptischen Typus. *Sitzungsb. Preuss. Akad. Wiss.*, **19** (1927), 141–152.

HOPF, H.

[1] *Zur Topologie der komplexen Mannigfaltigkeiten.* Courant Anniversary Volume, (1948), 167–185.

[2] Über komplex-analytische Mannigfaltigkeiten. *Rend. Mat. et Appl.* **10** (1951), 169–182.

HSU, C. J.

[1] On some structures which are similar to the quaternion structure. *Tôhoku Math. J.*, **12** (1960), 403–428.

[2] On some properties of $\pi$- structures on differentiable manifold. *Tôhoku Math. J.*, **12** (1960), 429–454.

[3] Note on the integrability of a certain structure on differentiable manifold. *Tôhoku Math. J.* **12** (1960), 349–360.

[4] Remarks on automorphisms in certain compact almost Hermitian spaces. *Tôhoku Math. J.*, **13** (1961), 186–192.

ISHIHARA, S.

[1] Groups of isometries of pseudo-Hermitian spaces, I, II. *Proc. Japan Acad.*, **30** (1954), 940–945; **31** (1955), 418–420.

[2] Holomorphically projective changes and their groups in an almost complex manifold. *Tôhoku Math. J.* **9** (1959), 273–297.

[3] On holomorphic planes. *Ann. di Mat. Pure ed Appl.*, XLVII (1959), 197–241.

ISHIHARA, S. and M. OBATA.

[1] Affine transformations in a Riemannian manifold. *Tôhoku Math. J.*, **16** (1955), 146–150.

ISHIHARA, S. and S. TACHIBANA

[1] A note on holomorphically projective transformations of a Kählerian space with parallel Ricci tensor. Tôhoku M. J., **13** (1961), 193–200.

KÄHLER, E.

[1] Über eine bemerkenswerte Hermitische Metrik. *Abh. Math. Sem. Hamburg*, **9** (1933), 173–186.

KILLING, W.

[1] Über die Grundlagen der Geometrie. *J. für den reine und angew. Math.*, **109** (1892), 121–186.

KLINGENBERG, W.

[1] On compact Kählerian manifolds with positive holomorphic curvature. *Proc. Amer. Math. Soc.*, **12** (1961), 350–356.

KNEBELMAN, M. S.

[1] On the equations of motions in a Riemannian space. *Bull. Amer. Math. Soc.*, **51** (1945), 682–685.

KNEBELMAN, M. S. and K. YANO.

[1] On homothetic mappings of Riemannian spaces. To appear in *Proc. Amer. Math. Soc.*

KOBAYASHI, S.

[1] A theorem on the affine transformation group of a Riemannian manifold. *Nagoya Math. J.*, **9** (1955), 39–41.

[2] On compact Kähler manifolds with positive definite Ricci tensor. *Ann. of Math.*, **74** (1961), 570–574.

KOBAYASHI, S. and K. NOMIZU,

[1] On automorphisms of a Kählerian structure. *Nagoya Math. J.*, **11** (1957), 115–124.

KODAIRA, K. and D. C. SPENCER.

[1] On the variation of almost complex structure. *Alg. Geom. and Top.* Princeton (1957). 139–150.

KOJYO, H.

[1] On infinitesimal holomorphically projective transformation. *J. Fac. Sci. Hokkaidô Univ. Series I*, xvi (1962), 1–4.

KOSTANT, B.

[1] Holonomy and the Lie algebra of infinitesimal motions of a Riemannian manifold. *Trans. Amer. Math. Soc.*, **80** (1955), 528–542.

KOSZUL, I.

[1] Sur la forme hermitienne canonique des espaces homogènes complexes. *Can. J. of Math.*, **7** (1955), 562–576

KOTŌ, S.

[1] Somo theorems on almost Kählerian spaces. *J. Math. Soc. Japan*, **12** (1960), 422–433.

[2] Curvatures in Hermitian spaces. *Mem. of the Fac. Ed. Niigata Univ.*, **2** (1960), 15–25.

[3] On harmonic tensors in an almost Tachibana space. *Tôhoku Math. J.*, **13** (1961), 423–426.

[4] On invariant subspaces in almost Hermitian spaces. *Mem. Fac. Ed. Niigata Univ.*, **3** (1961), 1–5.

[5] Almost complex spaces with almost analytic tensors. *Mem. of the Fac. Ed. Niigata Univ.*, **4** (1962), 1–3.

[6] On almost analytic tensors in almost complex spaces. *Tensor, New Series*, **12** (1962), 110–132.

LEE, E. H.

[1] On even dimensional skew symmetric spaces and their groups of transformations. *Amer. J. Math.*, **67** (1945), 321–328.

LEGRAND, G.

[1] Etude d'une généralisation des structures presque complexes sur les variétés diffèrentiables. *Rend. Circ. Mat. Palermo*, **7** (1956), 323–354; **8** (1957), 5–48.

LIBERMANN, P.

[1] Sur le problème d'équivalence de certaines structures infinitésimales. Thèse (1954).

[2] Sur les structures presgue complexes et les autres structures infinitésimales régulières. *Bull. Soc. Math. France*, **83** (1955), 195–224.

LICHNEROWICZ, A.

[1] Courbure et nombres de Betti d'une variété riemannienne compacte. *C. R. Acad. Sci. Paris*, **226** (1948), 1678–1680.

[2] Dérivation covariante et nombres de Betti. *C. R. Acad. Sci. Paris*, **230** (1950), 1248–1250.

[3] Théorèmes de reductibilité des variétés kählériennes et applications. *C. R. Acad. Sci. Paris*, **231** (1950), 1280–1282.

[4] Sur les variétés riemanniennes admettant une forme quadratique à derivée covariante nulle. *C. R. Acad. Sci. Paris,* **231** (1950), 1413–1415.
[5] Formes à dérivée covariante nulle sur une variété riemannienne. *C. R. Acad. Sci. Paris,* **232** (1951), 146–147.
[6] Sur les variétés riemanniennes admettant une forme à dérivée covariante nulle. *C. R. Acad. Sci. Paris,* **232** (1951), 611–619.
[7] Sur les formes harmoniques des variétés riemanniennes localement réductibles *C. R. Acad. Sci. Paris,* **232** (1951) 1634–1636.
[8] Généralisation de la géométrie kaehlérienne globale. *Colloque de Géom. Diff.* (1951), 99–122.
[9] Courbure, nombres de Betti et espaces symmétriques. *Proc. Int. Congress. Math.,* II (1952), 216–223.
[10] Espaces homogènes kaehlériennes. *Coll. Int. Geom. Diff. Strasbourg.* (1953), 172–184.
[11] Sur les groups d'automorphismes de certaines variétés kählériennes. *C. R. Acad. Sci. Paris,* **239** (1954), 1344–1347.
[12] Un théorème sur les espaces homogènes complexes. *Archiv der Math.,* **5** (1954), 207–215.
[13] Transformations infinitésimales conformes de certaines variétés riemanniennes compactes. *C. R. Acad. Sci. Paris,* **241** (1955), 726–729.
[14] *Théorie globale des connexions et des groupes d'holonomie.* Ed. Cremonese. Rome (1955).
[15] *Some problems on transformations of Riemannian and Káhlerian manifold.* Mimeographed notes, Princeton (1955).
[16] Sur les transformations affines des variétés riemanniennes. *C. R. Acad. Sci. Paris,* **242** (1956), 1568–1570.
[17] Sur la reductibilité des espaces homogénes riemanniens. *C. R. Acad. Sci. Paris,* **243** (1956), 646–672.
[18] Sur les transformations analytiques des variétés kählèriennes compactes. *C. R. Acad. Sci. Paris,* **244** (1957), 3011–3019.
[19] Transformations analytiques d'une variété kählerienne. C. *R. Acad. Sci. Paris,* **245** (1957), 953–956.
[20] Transformations analytiques et isométries d'une variété kählérienne compacte. *C. R. Acad. Sci. Paris,* **241** (1958), 855–857.
[21] *Géométrie des groupes de transformations.* Dunod Paris (1958).

MATSUSHIMA, Y.
[1] Sur la structure du groupe d'homéomorphismes analytiques d'une certaine variété kählérienne. *Nagoya Math. J.,* **11** (1957), 145–150.
[2] Sur les espaces homogènes kählériennes d'un groupe de Lie reductif. *Nagoya. Math. J.,* **11** (1957), 53–60.
[3] Sur certaines variétés homogènes complexes. *Nagoya Math. J.,* **18** (1961), 1–12.

MATSUSHIMA, Y. and J. HANO.

[1] Some studies on Kählerian homogeneous spaces. *Nagoya Math. J.*, **116** (1957), 1–16.

MIZUSAWA, H.

[1] Notes on certain Hermitian spaces. *J. Fac. Sci., Niigata Univ.*, **3** (1960), 45–55.

[2] On infinitesimal holomorphically projective transformations in *$O$-spaces. *Tôhoku Math. J.*, **13** (1961), 466–480.

MIZUSAWA, H. and S. KOTŌ.

[1] Holomorphically projective curvature tensors in certain almost Kählerian spaces. *J. Fac. Sci. Niigata Univ.*, **2** (1960), 33–43.

MOGI, I.

[1] On harmonic field in Riemannian manifold. *Kōdai Math. Sem. Rep.*, **2** (1950), 61–66.

MUTŌ, Y.

[1] On some almost Kählerian spaces. *Tôhoku Math. J.*, **14** (1962), 344–364.

MYERS, S. B.

[1] Riemannian manifolds with positive mean curvature. *Duke Math. J.*, 8 (1941), 401–404.

MAYERS, S. B. and N. E. STEENROD.

[1] The group of isometrics of a Riemannian manifold. *Ann. of Math.*, **40** (1939), 400–416.

NAGANO, T.

[1] On conformal transformations of Riemannian spaces. *J. Math. Soc. Japan*, **10** (1958), 79–93.

[2] Isometries on complex-product space. *Tensor*, **9** (1959), 47–61.

NAKAYAMA, S.

[1] Conformal relations in almost Hermitian spaces. *Tensor, New Series*, **12** (1962), 278–289.

NEWLANDER A. and L. NIRENBERG

[1] Complex analytic coordinates in almost complex manifolds. *Ann. of Math.*, **65** (1957), 391–404.

NIJENHUIS, A.

[1] $X_{n-1}$-forming sets of eigenvectors. *Proc. Kon. Ned. Akad. Amsterdam*, 54 = Indagationes Math., **13** (1951), 200—212.

NIJENHUIS, A. and W. B. WOOLF

[1] Three integration problems in almost complex manifolds. *Technical Reports*, **13** (1962), 1–114.

NOMIZU, K.

[1] On the group of affine transformations of an affinely connected manifold. *Proc. Amer. Math. Soc.*, **4** (1953), 816–828.

[2] Sur les transformations affines d'une variété riemannienne. *C. R. Acad. Sci. Paris*, **237** (1953), 1308–1310.

[3] *Remarques sur les groupes d'holonomie et d'isométries.* Colloque de Topologie de Strasbourg (1956).

[4] Invariant affine connections on homogeneous spaces. *Amer. J. Math.*, **76** (1954), 33–65.

[5] Studies on Riemannian homogeneous spaces. *Nagoya Math. J.*, **96** (1955), 42–50.

[6] *Lie groups and differential geometry.* Publications of the *Math. Soc. Japan.* (1956)

[7] On automorphisms of Kähler manifold. *Nagoya Math. J.* **116** (1957), 115–124.

OBATA, M.

[1] Affine connections on manifolds with almost complex, quaternion or Hermitian structure. *Jap. J. Math.*, **26** (1956), 43–77.

[2] Affine transformations in an almost complex manifold with a natural affine connexion. *J. Math. Soc. Japan*, **8** (1956), 345–362.

[3] Affine connections in a quaternion manifold and transformations preserving the structure. *J. Math. Soc. Japan*, **9** (1957), 406–416.

ŌTSUKI, T. and Y. TASHIRO.

[1] On curves in Kählerian spaces. *Math. J. Okayama Univ.* **4** (1954), 57–78.

PATTERSON, E. M.

[1] A characterization of Kähler manifolds in terms of parallel fields of planes. *J. London Math. Soc.*, **28** (1953), 260–269.

RAUCH, H. E.

[1] A contribution to differential geometry in the large. *Ann. of Math.*, **54** (1951), 38–55.

DE RHAM G.

[1] Sur la théorie des formes différentielles harmoniques. *Ann. de l'Univ. de Grenoble*, **22** (1946), 135–156.

[2] Remarques au sujet de la théorie des formes différentielles harmoniques. *Ann. de l'Univ. de Grenoble*, 23 (1941–48), 55–56.

[3] Sur la réductibilité d'un espace de Riemann. *Comm. Math. Helv.* **26** (1952), 328–344.
[4] *Variétés différentiables. Act. Sci. Hermann, Paris,* (1955).

DE RHAM G. and K. KODAIRA
[1] *Harmonic Integrals.* Institute for Advanced Study, Princeton, (1950).

SANG-SEUP EUM
[1] On infinitesimal holomorphically projective transformation in recurrent Kähler manifolds. *Kyungpook Math. J.*, **4** (1961), 1–23.

SASAKI, S.
[1] On the differential geometry of tangent bundles of Riemannian manifolds. I. *Tôhoku Math. J.*, **10** (1958) 338–353.
[2] On differentiable manifolds with certain structures which are closely related to almost contact structure, I. *Tôhoku Math. J.*, **12** (1960), 459–476.

SASAKI, S. and K. YANO.
[1] Pseudo-analytic vectors on pseudo-Kählerian manifolds. *Pacific J. Math.*, **5** (1955), 989–993.

SATO, I.
[1] On conformal Killing tensor fields. *Bull. Yamagata Univ.*, **3** (1956), 175–180.
[2] On a theorem of Tsukamoto. *Ibid.* **5** (1960), 133–135.
[3] On Kähler manifolds with positive constant holomorphic sectional curvature. Ibid. 137–142.
[4] On closed geodesic in a Kählerian manifold with positive constant holomorphic sectional curvature. Ibid. 259–261.

SAWAKI, S.
[1] On analytic tensors in certain Hermitian manifolds. *J. Fac. Sci. Niigata Univ.*, **3** (1960) 7–22.
[2] On almost analytic vectors in almost Hermitian spaces. *J. Fac. Sci. Niigata Univ.*, **3** (1960), 23–32.
[3] On almost analytic tensors in *$O$-spaces. *Tôhoku Math. J.*, **13** (1961), 154–178.
[4] On almost analytic tensors of mixed type in $K$-space. *J. Math. Soc. Japan.*, **13** (1961), 165–182.
[5] On Matsushima's theorem in a compact Einstein $K$-space. *Tôhoku Math. J.*, **13** (1961), 455–465.
[6] A generalisation of Matsushima's theorem. *Math. Ann.*, **146** (1962), 219–286.
[7] On infinitesimal transformations of almost Kählerian spaces and $K$-spaces. To appear in Kōdai Math. Sem. Rep.

SAWAKI, S. and S. KOTŌ.

[1] On the analytic tensor in a compact Kähler space. *J. Fac. Sci. Niigata Univ.*, **1** (1958), 77–84.

[2] On some *F*-connexion in almost Hermitian manifolds. *J. Fac. Sci. Niigata Univ.*, **1** (1958), 85–96.

SCHOUTEN, J. A.

[1] Über unitäre Geometrie *Proc. Kon. Akad. Amsterdam*, **32** (1929), 457–465.

[2] Sur les tenseurs de $V_n$ aux directions principales $V_{n-1}$-normales *Coll. de Géom. Diff. Louvain*, (1951), 67–70.

[3] *Ricci-Calculus*. Springer (1954).

SCHOUTEN, J. A. and D. VAN DANTZIG

[1] Über unitäre Geometrie. *Math. Ann.*, 103 (1930), 319–346.

[2] Über unitäre Geometrie konstanter Krümmung. *Proc. Kon. Ned. Akad., Amsterdam,* **34** (1931) 1293–1314.

SCHOUTEN, J. A. and K. YANO.

[1] On an intrinsic connexion in an $X_{2n}$ with an Hermitian structure. *Proc. Kon. Ned. Akad., Amsterdam,* **58** (1955), 1–9.

[2] On the geometric meaning of the vanishing of the Nijenhuis tensor in an $X_{2n}$ with an almost complex structure. *Ind. Math.*, **17** (1955), 132–138.

[3] On invariant subspaces in the almost complex $X_{2n}$. *Ind. Math.*, **17** (1955), 261–269.

[4] On pseudo-Kählerian spaces admitting a continuous group of motions. *Ind. Math.* **17** (1955), 565–570.

SHANKS, E. B.

[1] Homothetic correspondence between Riemannian spaces. *Duke Math. J.*, **17** (1950), 299–311.

SPENCER, D. C.

[1] *Differentiable manifolds*. Notes. Princeton University. (1954).

SUMITOMO, T.

[1] Projective and conformal transformations in compact Riemannian manifold. *Tensor,* **9** (1959), 113–135.

TACHIBANA, S.

[1] Note on conformally flat almost Kählerian spaces. *Natural Sci. Rep. of the Ochanomizu Univ.*, **10** (1958), 41–43.

[2] On infinitesimal holomorphically projective transformations in certain almost Hermitian spaces. *Natural Sci. Rep. of the Ochanomizu Univ.*, **10** (1958), 43–51.

[3] On almost analytic vectors in almost Kählerian manifolds. *Tôhoku Math. J.*, **11** (1959), 247–265.

[4] On almost analytic vectors in certain almost Hermitian manifolds. *Tôhoku Math. J.*, **11** (1959), 351–363.

[5] A remark on linear connections with some properties. *Natural Sci. Rep. of the Ochanomizu* Univ., **10** (1959), 1–6.

[6] Anylytic tensor and its generalisation. *Tôhoku Math. J.*, **12** (1960), 208–221.

[7] Some theorems on locally product Riemannian manifold. *Tôhoku Math. J.*, **12** (1960), 281–292.

[8] On infinitesimal conformal and projective transformations of almost *K*-spaces. *Tôhoku Math. J.*, **13** (1961), 386–392.

[9] On automorphisms of certain compact almost Hermitian spaces. *Tôhoku Math. J.*, **13** (1961), 179–185.

[10] On automorphisms of conformally flat *K*-spaces. *J. Math. Soc. Japan.*, **13** (1961), 183–188.

[11] On certain compact almost Hermitian space of constant curvature. To appear.

TACHIBANA, S. and S. ISHIHARA.

[1] On infinitesimal holomorphically projective transformations in Kählerian manifolds. *Tôhoku Math. J.*, **12** (1960), 17–101.

TACHIBAMA, S. and S. KOTÔ

[1] On almost analytic functions, tensors and invariant subspace. *Tôhoku Math. J.*, **14** (1962), 177–186.

TACHIBAMA, S. and M. OKUMURA

[1] On the almost-complex structure of tangent bundles of Riemannian spaces. *Tôhoku Math. J.*, **14** (1962), 156–161.

TASHIRO, Y.

[1] On holomorphically projective correspondences in an almost complex space. *Math. J. Okayama Univ.* **6** (1957), 167–152.

[2] On contact structure of hypersurfaces in complex manifold. *Tôhoku Math. J.*, **15** (1963), 50–62.

THOMAS, T. Y.

[1] Some applications of Green's theorem for compact Riemannian spaces. *Tôhoku Math. J.*, **46** (1940), 261–266.

TOMONAGA, Y.

[1] On Betti numbers of Riemannian spaces. *J. Math. Soc. Japan*, **26** (1950), 93–104.

TSUKAMOTO, Y.

[1] On Kählerian manifolds with positive holomorphic curvature. *Proc. Japan Acad.*, **33** (1951), 333–335.

WAKAKUWA, H.

[1] On almost complex symplectic manifolds and affine connections with restricted homogeneous holonomy group Sp (n, C). *Tôhoku Math. J.*, **12** (1960), 175–202.

[2] On linearly independent almost complex structures in a differentiable manifold. *Tôhoku Math. J.*, **13** (1961), 393–422.

WALKER, A. G.

[1] Connexions for parallel distributions in the large I. II. *Quart. J. Math.*, Oxford (2), **6** (1955), 301–308; **9** (1958), 221–231.

[2] Dérivation tensorielle et seconde torsion pour une structure presque complexe. *C. R. Acad. Sci. Paris*, **245** (1957), 1213–1215.

WATANABE, S.

[1] On special Kawaguchi spaces, VI. Some transformations in certain special Kawaguchi spaces. *Tensor, New Series*, **12** (1962), 244–253.

WEIL, A.

[1] Sur la théorie des formes différentielles attachées à une variété analytique complexe. *Conn. Math. Helv.*, **20** (1947), 110–116.

[2] *Variété kählérienne.* (1958)

WESTLAKE, W. J.

[1] Hermitian spaces in geodesic correspondence. *Proc. Amer. Math. Soc.*, **5** (1954), 301–306.

[2] Conformally Kähler manifolds. *Proc. Cambridge Philos. Soc.*, **50** (1954), 16–19.

WILLMORE, T. J.

[1] Connexion for systems of parallel distributions. *Quart. J. Math.*, Oxford (2), **7** (1956), 269–276

[2] Generalized torsional derivation. Sem. de Topologie et de Géom. Diff. Paris, (1959).

[3] Global theorems on manifolds which admit distributions. Sem. de Top. et de Géom. Diff. Paris, (1959).

[4] *Introduction to differential geometry.* Oxford Univ. Press (1960).

YANO, K.

[1] *Groups of transformations in generalized spaces.* Tokyo, (1949).

[2] On group of homothetic transformations in Riemannian spaces. *J. Ind. Math. Soc.*, **15** (1951), 105–117.

[3] On harmonic and Killing vector fields. *Ann. of Math.* **55** (1952). 38–45.

[4] Some remarks on tensor fields and curvature. *Ann. of Math.* **55** (1952), 328–341.

[5] On Killing vector fields in a Kählerian space. *J. Math. Soc. Japan*, **5** (1953), 6–12.

[6] Sur la correspondence projective entre deux espaces pseudo-hermitiens. *C. R. Acad. Sci Paris*, **239** (1956), 1346–1348.

[7] Geometria conforme in varietà quasi hermitiane. *Rend. Acad.-Lincei*, **16** (1954), 449–454.

[8] Some remarks on almost complex structures. *Proc. Inst. Congress of Math. Amsterdam,* II (1954), P. 268.

[9] On three remarkable affine connexions in almost Hermitian spaces. *Proc. Kon. Ned. Akad. Amsterdam,* **58** (1955), 24–32.

[10] Quelques remarques sur les variétés à structure presque complexe. *Bull. Soc. Math. France,* **83** (1955), 57–80.

[11] *The theory of Lie derivatives and its applications.* Amsterdam (1957).

[12] Sur un théorème de M. Matsushima. *Nagoya Math. J.* **12** (1957), 167–170.

[13] Some integral formulas and their applications. *Michigan J. Math.,* **5** (1958), 63–73.

[14] On Walker differentiation in almost product and almost complex spaces. *Ind. Math.,* **20** (1958), 573–580.

[15] Affine connections in an almost product space. *Kōdai Math. Sem. Rep.,* **11** (1959), 1–24.

[16] Champs de vecteurs dans un espace riemannien ou hermitien. *C. R. Acad. Sci. Paris,* **25** (1960), 194–195.

[17] Conformal transformations in Riemannian and Hermitian spaces. *Bull. Amer. Math. Soc.,* **66** (1960), 369–372.

[18] Eckmann–Frölicher connexions on almost analytic submanifolds. *Kōdai Math. Sem. Rep.,* **14** (1962), 53–58.

YANO, K. and M. AKO.

[1] Almost analytic vectors in almost complex spaces. *Tôhoku Math. J.,* **13** (1961), 24–45.

YANO, K. and S. BOCHNER.

[1] *Curvature and Betti numbers. Ann. of Math. Stud.* **32,** (1953).

YANO, K. and I. MOGI.

[1] Sur les variétés pseudo-Kählériennes à courbure holomorphique. constante. *C. R. Acad. Sci. Paris,* **239** (1953), 962–964.

[2] On real representations of Kählerien manifolds. *Ann. of Math.* **61** (1955), 170–189.

YANO, K. and T. NAGANO.

[1] Some theorems on projective and conformal transformations. *Ind. Math.,* 14 (1957), 451–458

[2] Einstein spaces admitting a one-parameter group of conformal transformations *Ann. of Math.,* **69** (1959), 451–461.

[3] The de Rham decomposition, isometries and affine transformations in Riemannian space. *Japanese J. Math.* **29** (1959), 173–184.

[4] On geodesic vector fields in a compact orientable Riemannian space. *Comm. Math. Helv.,* **35** (1961), 55–64.

# AUTHOR INDEX

# SUBJECT INDEX

## *OTHER TITLES IN THE SERIES ON PURE AND APPLIED MATHEMATICS*

Vol. 1. WALLACE – *An Introduction to Algebraic Topology*
Vol. 2. PEDOE – *Circles*
Vol. 3. SPAIN – *Analytical Conics*
Vol. 4. MIKHLIN – *Integral Equations*
Vol. 5. EGGLESTON – *Problems in Euclidean Space: Application of Convexity*
Vol. 6. WALLACE – *Homology Theory on Algebraic Varieties*
Vol. 7. NOBLE – *Methods Based on the Wiener–Hopf Technique for the Solution of Partial Differential Equations*
Vol. 8. MIKUSINSKI – *Operational Calculus*
Vol. 9. HEINE – *Group Theory in Quantum Mechanics*
Vol. 10. BLAND – *The Theory of Linear Viscoelasticity*
Vol. 11. KURTH – *Axiomatics of Classical Statistical Mechanics*
Vol. 12. FUCHS – *Abelian Groups*
Vol. 13. KURATOWSKI – *Introduction to Set Theory and Topology*
Vol. 14. SPAIN – *Analytical Quadrics*
Vol. 15. HARTMAN and MIKUSINSKI – *Theory of Lebesgue Measure and Integration*
Vol. 16. KULCZYCKI – *Non-Euclidean Geometry*
Vol. 17. KURATOWSKI – *Introduction to Calculus*
Vol. 18. GERONIMUS – *Polynomials Orthogonal on a Circle and Interval*
Vol. 19. ELSGOLC – *Calculus of Variations*
Vol. 20. ALEXITS – *Convergence Problems of Orthogonal Series*
Vol. 21. FUCHS and LEVIN – *Functions of a Complex Variable, Volume II*
Vol. 22. GOODSTEIN – *Fundamental Concepts of Mathematics*
Vol. 23. KEENE – *Abstract Sets and Finite Ordinals*

Vol. 24. DITKIN and PRUDNIKOV—*Operational Calculus in Two Variables and its Applications*
Vol. 25. VEKUA—*Generalized Analytic Functions*
Vol. 26. AMIR-MOÉZ and FASS—*Elements of Linear Spaces*
Vol. 27. GRADSTEIN—*Direct and Converse Theorems*
Vol. 28. FUCHS—*Partially Ordered Algebraic Systems*
Vol. 29. POSTNIKOV—*Foundations of Galois Theory*
Vol. 30. BERMANT—*A Course of Mathematical Analysis, Part II*
Vol. 31. LUKASIEWICZ—*Elements of Mathematical Logic*
Vol. 32. VULIKH—*Introduction to Functional Analysis for Technologists and Engineers*
Vol. 33. PEDOE—*An Introduction to Projective Geometry*
Vol. 34. TIMAN—*Theory of Approximation of Functions of a Real Variable*
Vol. 35. CSÁSZÁR—*Foundations of General Topology*
Vol. 36. BRONSHTEIN and SEMENDYAYEV—*A Guide-Book to Mathematics for Technologists and Engineers*
Vol. 37. MOSTOWSKI and STARK—*Introduction to Higher Algebra*
Vol. 38. GODDARD—*Mathematical Techniques of Operational Research*
Vol. 39. TIKHONOV and SAMARSKII—*Equations of Mathematical Physics*
Vol. 40. McLEOD—*Introduction to Fluid Dynamics*
Vol. 41. MOISIL—*The Algebraic Theory of Switching Circuits*
Vol. 42. OTTO—*Nomography*
Vol. 43. RANKIN—*An Introduction to Mathematical Analysis*
Vol. 44. BERMANT—*A Course of Mathematical Analysis, Part I*
Vol. 45. KRASNOSEL'SKII—*Topological Methods in the Theory of Nonlinear Integral Equations*
Vol. 46. KANTOROVICH and AKILOV—*Functional Analysis in Normed Spaces*
Vol. 47. JONES—*The Theory of Electromagnetism*
Vol. 48. FEJES TÓTH—*Regular Figures*

MADE IN GREAT BRITAIN